新编高等院校计算机科学与技术规划教材

Java 面向对象程序设计

（第 4 版）

张桂珠　主　编

北 京 邮 电 大 学 出 版 社
·北 京·

内容简介

本书是Java面向对象程序设计的一本经典教材，它将Java语言与面向对象程序设计的原理和方法相结合，使用Java最新类库，以大量实例详细介绍如何使用Java进行面向对象的程序设计、GUI的程序设计、网络通信应用的程序设计、数据库应用的程序设计和Web应用的程序设计。

本书可作为Java面向对象程序设计课程的教材，也可作为Java的GUI程序设计、网络应用程序设计、数据库应用开发、JSP的Web应用开发的入门和提高学习教材。本书的读者对象是各类编程人员、计算机相关专业的本科生和研究生，也可作为Java技术的自学者或短训班人员学习教材。为方便人员学习，本书还配有电子课件、实例代码、习题答案与实验。

图书在版编目(CIP)数据

Java面向对象程序设计 / 张桂珠主编. --4版. --北京：北京邮电大学出版社，2015.1(2023.8重印)

ISBN 978-7-5635-4250-5

Ⅰ. ①J… Ⅱ. ①张… Ⅲ. ①JAVA语言—程序设计—教材 Ⅳ. ①TP312

中国版本图书馆CIP数据核字(2014)第296184号

书　　名：Java面向对象程序设计(第4版)
著作责任者：张桂珠　主编
责 任 编 辑：张珊珊
出 版 发 行：北京邮电大学出版社
社　　址：北京市海淀区西土城路10号(邮编:100876)
发 行 部：电话:010-62282185　传真:010-62283578
E-mail：publish@bupt.edu.cn
经　　销：各地新华书店
印　　刷：唐山玺诚印务有限公司
开　　本：787 mm×1 092 mm　1/16
印　　张：23.25
字　　数：609千字
版　　次：2005年9月第1版　2007年8月第2版　2010年8月第3版　2015年1月第4版
2023年8月第9次印刷

ISBN 978-7-5635-4250-5　　定　价：46.00元

· 如有印装质量问题，请与北京邮电大学出版社发行部联系 ·

前　言

Java 语言自从 1995 年 Sun Microsystem 推出以来，因其面向对象，跨平台，稳定，多线程，支持用户图形界面、数据库应用和 Internet/Intranet 的网络应用，使得 Java 编程系统以史无前例的速度在世界范围内赢得了人们的认可。现在，在各个领域有很多软件都是采用 Java 语言开发的，Java 语言已经成为软件开发人员的理想工具。

Java 的核心是创新的 Java 编程语言和 SDK 软件开发工具包。通用的 Java 语言完全面向对象，通过构建软件对象来编程。面向对象编程（Object-Oriented Programming，OOP）使得构建、维护、修改和重用软件更加简单。SDK 包含一套丰富的代码库和工具。同时，该语言和 SDK 使得编程更加方便和高效。

本书是 Java 面向对象程序设计的一本经典教材，它将面向对象程序设计的理论、技术、方法和 Java 语言相结合，使用 Java 的最新类库，详细介绍了如何使用 Java 进行面向对象的程序设计、GUI 的程序设计、网络通信应用的程序设计、数据库应用的程序设计和 Web 应用的程序设计。全书内容注重理论、技术和实际应用一体化，内容覆盖面大，实用性强。本书可作为 Java 面向对象程序设计课程的教材，也可作为 Java 的 GUI 程序设计、数据库应用开发、JSP 的 Web 应用开发的入门和提高学习教材。本书在版本升级中，能跟踪和吸收国外一流的 Java 教材特性，融合 Java 科研项目的实际案例。本书的读者对象是各类编程人员、计算机相关专业的本科生和研究生，也可作为 Java 技术和应用的自学者或短训班人员学习教材。为方便人员学习，本书还配有电子教学软件和实例代码，可登录北京邮电大学出版社网站（www. bupt-press. com）下载。本书还配有习题解答与实验指导（张桂珠. Java 面向对象程序设计习题解答与实验. 4 版. 北京：北京邮电大学出版社.）。

全书共有 15 章。第 1 章介绍面向对象编程和设计的基本概念、结构和方法；第 2 章介绍 Java 语言概况，并引入几个 Java 入门程序，说明 Java 程序类型和 Java 开发环境的使用；第 3 章讲述 Java 语言基础知识，包括标识符、运算符、数据类型、表达式、语句、基本算法控制、方法和数组；第 4 章讲述类的定义、对象的创建和使用、包的创建和使用；第 5 章讲述类的继承机制及其程序设计；第 6 章讲述多态性技术，包括抽象类、接口、内部类的定义和通用性的程序设计；第 7 章介绍在编程中常用的 Java 实用包；第 8 章介绍图形和 Java 2D；第 9 章讲述 GUI 组件和用户界面程序设计；第 10 章讲述异常的概念、声明、抛出和处理以及异常的程序设计；第 11 章讲述线程技术和多线程的程序设计；第 12 章讲述输入/输出流处理，应用输入/输出流对顺序文件、随机文件进行读写；第 13 章讲述 Java 网络技术和网络应用程序设计；第 14 章讲述 JDBC 技术和数据库的开发应用；第 15 章讲述 JSP 技术、Struts 2.1 技术和 Web 应用程序开发实例。

本书内容丰富，在使用本书学习或教学时，可分成 4 个部分循序渐进地进行。第一部分

(1～3章)是Java语言的基础知识，学习Java语言的语法和使用；第二部分(4～7章)是Java的面向对象编程技术和应用，训练OOP的程序设计能力；第三部分(8～9章)是Java绘制图形、GUI组件技术和用户界面设计应用，训练开发纯Java应用系统时用户图形界面的设计能力；第四部分(10～15章)是综合训练Java应用系统的开发能力，包括网络通信应用开发能力、数据库的应用开发能力和Web应用开发能力。通过本书的学习，读者对Java的认识和应用能力将会提高到一个新的水平。

全书由张桂珠主编。本书在编写过程中得到了江南大学领导的支持，得到了杨开莜、刘丽、陈爱国、王惠、马晓梅、韩亦强、徐华、韩振、戴月明、张小妹等老师的协助和支持，在此一并致谢。

感谢读者选择使用本书，欢迎各位读者对本书提出批评和修改建议。我们将不胜感激，并在再版时予以考虑。

作者的E-mail为：zhangguizhu@163.com

通信地址为：江苏省无锡市江南大学物联网工程学院计算机系　张桂珠　收

邮政编码：214122

编　者

目 录

第1章 面向对象程序设计概述

本章首先通过比较，介绍传统的面向过程与面向对象程序设计（Object-Oriented Programming，OOP）这两种方法的特点，接着介绍了面向对象编程和设计的基本概念、结构和方法，它们都是Java面向对象程序设计中最具有技术性和挑战性的内容。随着本书的深入，OOP的概念和技术将落实在今后的面向对象编程的应用实例中。本章提供了面向对象编程和设计的核心内容的初步阐述。

1.1 面向对象与面向过程程序设计

1. 面向过程程序设计方法

一个程序往往由多个模块构成，这些模块可以被单独设计、编码和测试，然后被组装成一个完整的程序。在面向过程的语言中，如C、Fortran或Pascal，这些模块就是过程或函数（function），过程或函数组成了面向过程语言编制的程序单位。一个过程或一个函数就是一个语句序列，例如求两个整数的最大值，用C语言的函数表达如下：

```
int max(int num1,int num2) {
if (num1>num2)
    return num1;
else
   return num2;
}
```

面向过程的程序是由赋值语句、条件语句、循环语句和过程调用语句构成的语句序列。

面向过程的程序设计方法 采用自顶向下的功能分解法（top-down，functional decomposition），即一个要解决的问题被分解成若干个子问题，每个子问题又被划分成若干个子子问题。这种自顶向下的功能分解一直持续下去，直到子问题足够简单，可以在相应的子过程中解决。而程序结构也按功能划分为若干个模块，这些模块形成一个树型结构。每个模块用一过程或函数实现，如C语言编制的程序是由main函数加若干个子函数组成，在main中调用子函数。图1-1说明了自顶向下的功能分解法与面向过程的程序结构的关系。

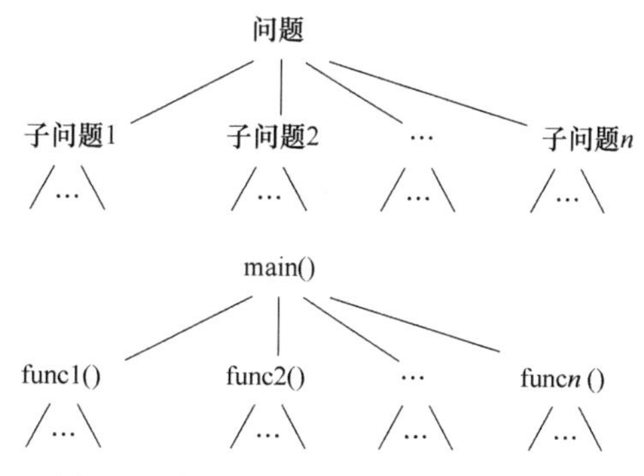

图1-1 自顶向下的功能分解与程序结构

自顶向下设计方法存在的问题：自顶向下设计方法把数据和处理数据的过程分离为相对

独立的实体，当数据结构改变时，所有相关的处理过程都要进行相应的修改，每一种相对于老问题的新方法都要重新设计程序，程序的可复用性差。因此，面向过程的程序设计方法的开发和维护都很困难。

2. 面向对象程序设计方法

面向对象程序设计是代替传统的面向过程程序设计的另一种程序设计方法，是为了解决过程型程序设计和自顶向下设计中的主要问题。程序的模块单位变成了类，类是比函数更大的单位，它将操作数据的函数和数据封装在一起而形成的。在 Java 语言中，像 max 这样的函数被叫做方法，方法变成了类的成员。

面向对象的语言有很多，如 Smalltalk、Ada、Object Pascal、C＋＋、Java，其中 C＋＋和Java 是目前使用最多的语言。而 Java 与 C＋＋语言相比，它去掉了 C＋＋语言的复杂性和二义性的成分，增加了安全性和可移植性的成分。Java 比 C＋＋也简单得多，而且文件尺寸小。Java 去除了头文件、预处理程序、指针运算、多重继承、运算符重载、结构体、共用体和模板。另外，Java 还可以自动执行无用内存单元收集，从而不需要显式进行内存管理。

面向对象的程序设计方法　首先，将数据及对数据的操作行为放在一起，作为一个相互依存、不可分割的整体——对象。再对相同类型的对象进行分类、抽象后，得出共同的特性而形成了类。类中的大多数数据只能用本类的方法进行处理。类通过一个简单的外部接口与外界发生联系。面向对象其实是现实世界模型的自然延伸。现实世界中任何实体都可以看做是对象。对象之间通过消息相互作用。另外，现实世界中任何实体都可归属于某类事物，任何对象都是某一类事物的实例。

面向对象的程序模块间的关系简单，程序的独立性高、数据安全。面向对象方法的显著特性有：封装性、抽象性、继承性和多态性，使软件具有可重用性，开发和维护的代价小。

1.2　类与对象

面向对象的设计，将客观事物(或实体)看做具有属性和行为(或称服务)的对象(object)，通过抽象找出同一类型对象的共同属性(静态特征)和行为(动态特征)，进而形成类(class)的概念。类是相同类型对象的集合的描述。例如，类 Human 就是现实世界中人(对象)的集合，我、你、他都是 Human 的对象。分析类 Human 的所有对象——人，得到这些对象共同的数据属性和行为：

- 数据属性：编号、姓名、年龄等；
- 行为：吃饭、走路、跳舞等。

定义 Human 类，如下所示：

```
class Human {
  int no;
  String name;
  int age; …
  void eat() {…}
  …
}
```

其中，Human 被称为类名。no、name、age 被称为变量或域(fields)，eat()被称为方法(methods)，它们都是类的成员。在面向对象的编程语言中，类是一个数据类型，对象是类的实例

(instance)。在Java中,具有类类型的变量被称为对象引用(object reference)。例如(如图1-2所示):

```
Human p1 = new Human(1,"张三",20,…);
//p1 被称为对象引用变量,new Human(1,"张三",20,…) 被称为 Human 的对象
```

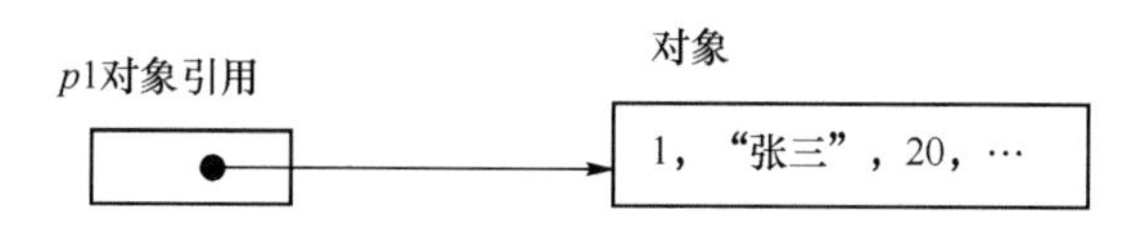

图 1-2　一个对象引用和一个对象

面向对象系统中的对象,是用来描述现实世界中实际存在的事物的实体,它是用来构成系统的一个单位。对象由一组属性和一组行为构成。具体地,对象是具有唯一对象名和固定对外接口的一组属性和操作的集合,用来模拟组成或影响现实世界问题的一个或一组因素。其中对象名是区别于其他对象的标志;对外接口是对象在约定好的运行框架和消息传递机制中与外界通信的通道;对象的属性表示了它所处于的状态;而对象的操作则是用来改变对象状态达到特定的功能。

1.3　封装与信息隐藏

封装(Encapsulation)是面向对象方法的重要原则,就是把对象的属性和操作(或服务)结合为一个独立的整体,并尽可能隐藏对象的内部实现细节。这里有两个含义:其一,把对象的全部属性和全部服务结合在一起,形成一个不可分割的独立单位;其二,"信息隐蔽",尽可能隐蔽对象的内部细节,对外界形成一个边界,只保留有限的对外接口,使之与外部发生联系。

数据封装的作用为:①对象的数据封装特性彻底消除了传统结构方法中数据与操作分离所带来的种种问题,提高了程序的可复用性和可维护性,降低了程序员保持数据与操作相容的负担;②对象的数据封装特性还可以把对象的私有数据和公共数据分离开,保护了私有数据,减少了可能的模块间干扰,达到降低程序复杂性、提高可控性的目的。

Java 语言中,定义类时通过大括号{ }封装了类的成员——域(变量)和方法,并使用 private 和 public 等关键字来控制对类的成员的访问,其中 private 修饰的成员是隐藏的,而 public 修饰的成员则定义了类对外的公共接口。类作为一个抽象的数据类型,允许用户从底层实现细节中抽象出来,提供给用户的是在公共接口上的上层操作,这是抽象性的含义。

1.4　继　承

继承(Inheritance)支持着软件代码的复用,是提高软件开发效率的重要因素之一。继承是在已有类(父类或超类)的基础上派生出新的类(子类),新的类能够吸收已有类的属性和行为,并扩展新的能力。图 1-3 说明了 Java 中的继承的一个例子:给定一个 Window 类,通过继承扩展它而得到一个子类 MenuWin。继承机制中,往往从一组类中抽象出公共属性放在父类。例如,给定类 Car、Motocycle 和 Truck,

```
class Window {
    …// Window 的成员定义
}
class MenuWin extends Window {
    …//MenuWin 新增加成员的定义
```

图 1-3　Java 继承的基本语法

把它们的公共属性放在一个称为 Vehicle 的公共父类中。

继承分为单继承和多继承。单继承是指一个子类最多只能有一个父类。多继承是一个子类可有两个以上的父类。由于多继承会带来二义性，在实际应用中应尽量使用单继承。Java 语言中的类只支持单继承，而接口支持多继承。

如何设计继承并完成继承层次是面向对象设计和编程的核心问题。继承是多态性的前提条件。

1.5 多态性

多态性(Polymorphism)是指在超类中定义的属性或行为，被子类继承之后，可以具有不同的数据类型或表现出不同的行为。这使得同一个属性或行为在超类及其各个子类中具有不同的语义。

例如，定义一个几何图形类 Shape，它具有“画图”行为，用 draw()表示。但具体画什么图并不确定；定义 Shape 类的一些子类，如 Circle(圆)和 Rectangle(矩形)。在子类中“画图”的具体行为可重新定义为：圆类中 draw()画圆，矩形类中 draw()画矩形。定义 Shape s，s 作为引用变量可指向 Circle 圆类的对象，也可指向 Rectangle 矩形类的对象。

通过执行下面的代码：

```
s.draw();
```

s 调用 draw()方法，s 指向对象不同会画出不同的图形(圆或矩形)。

多态性也是泛指在程序中同一个符号在不同的情况下具有不同解释的现象。

面向对象的设计特性为抽象性、封装性、多态性和继承性，必须在实际的面向对象系统设计中体现。将面向对象设计方法应用于程序的开发工具和开发过程中，不仅可以加快开发的速度，还可极大地增强程序的可维护性和可扩展性，提高代码重用率。

1.6 面向对象的建模和 UML

统一建模语言(Unified Modeling Language，UML)是一种流行的建模语言，由对象管理组(Object Management Organization，OMO)发布。UML 是一种图形化语言，允许系统构造人员(软件设计师、系统工程师、程序员等)用一种通用表示法描述系统的需求以及面向对象的分析和设计结果。作为一个建模语言，UML 由一个用于表达模型的词汇表和一个定义怎样组合词汇的语法规则构成，即

UML＝UML 词汇表＋UML 建模的语法规则

UML 词汇表＝UML 事物＋关系＋模型图

UML 事物(Thing)就是被模拟的实体或对象。事物可能是包、类、接口等。事物之间的语义上的联系用关系表示，UML 中共有 4 种关系：依赖关系、关联关系、泛化关系和实现关系。从软件的体系结构出发，UML 把软件模型分成了 4 个视图：用例视图、逻辑视图、实现视图和分布视图。在本书的有关章节中，案例分析和设计的结果用 UML 图形化表示，使读者对 UML 有一个感性化认识。有关 UML 更完整的文档请参阅 Web 站点(www.omg.org)。

1.7 小　结

本章比较了面向过程程序设计与面向对象程序设计方法各自的特点。

类是面向对象编程和设计关键性的概念，是一组对象集合的描述。对象是类的一个特例，引用变量的类型是某个类的变量，其值是对象的地址（或称为指向一个对象）。封装性、抽象性、继承性和多态性必须在以后的程序设计中体现。UML 是面向对象的建模语言。

习　题

1.1　什么是面向过程的程序设计的特点？

1.2　什么是面向对象的程序设计的特点？

1.3　什么是类和对象？

1.4　什么叫引用变量？

1.5　什么叫 OOP 的封装性、抽象性、继承性、多态性？

第2章 Java语言概述和入门程序

本章介绍了Java语言的历史、发展和特点，并对开发Java程序所需的Java的类库概念和Java开发环境进行了介绍。接着讨论了Java程序的基本类型，通过引入两个Java应用程序实例和两个Java小应用程序实例，说明了Java程序结构的特点。通过实例介绍了如何使用JDK 1.6开发包、集成开发环境JCreator和MyEclipse编写、编译和运行Java程序过程。这一章是后续各章中调试运行程序的基础。

2.1 Java历史及发展

Java是由James Gosling、Patrick Naughton等人于1991年在Sun Microsystems公司设计出来的，开发第一个版本花了18个月的时间。Java语言最初名叫“Oak”，是为家用消费电子产品(如电冰箱、电视机等)进行编程控制的语言。后来发现“Oak”已经是Sun公司另外一种语言的注册商标，于1995年更名为“Java”，即太平洋上一个盛产咖啡的岛屿的名字。从1992的秋天Oak问世，到1995春天公开发布Java语言，许多人都对Java的设计和改进做出了贡献。

现在，每天都有上百万人在用Java开发各种各样的软件，Java已用于开发大型企业的应用程序、增强WWW服务器的功能、动画游戏的设计(包括图形图像的调用)、为消费类设备(如手机、传呼机和移动设备等)提供应用程序。Java是一种跨平台、适合于分布式计算环境的面向对象编程语言，能够把整个因特网(Internet)作为一个统一的运行平台，大大地拓展了客户机/服务器(Client/Server)模式应用程序的外延和内涵。

2.2 Java语言特点

Java与C++语言相比，去掉了C++语言的复杂性和二义性的成分，增加了安全性和可移植性的成分。Java语言具有如下特点：简单性、面向对象、平台无关性、可移植性、解释性、高性能、动态性、可靠性和安全性、多线程、分布式处理等。下面将介绍Java语言的主要特点。

1. 简单性

Java系统精简，但功能齐备；语言风格类似于C++，但比C++容易掌握，且摒弃了C++中容易引发程序错误的地方，如指针操作和内存管理；提供丰富的类库。

2. 面向对象

Java语言是纯面向对象的语言，程序的结构由一个以上的类和(或)接口组成。程序的设计集中于类与对象、继承与接口上。通过继承机制，子类可以使用父类所提供的属性与方法，实现了软件代码的复用。

面向对象的设计特性：封装性、抽象性、继承性和多态性。

3. 平台无关性

Java是与平台无关的语言，是指用Java写的应用程序不用修改就可在不同的软硬件平台上运行。平台无关有两种：源代码级和目标代码级。C和C++具有一定程度的源代码级平台无关，表明用C或C++写的应用程序不用修改只需重新编译就可以在不同平台上运行。

Java主要靠Java虚拟机(Java Virtual Machine，JVM)在目标代码级实现平台无关性。JVM是一种抽象机器，它附着在具体操作系统之上，本身具有一套虚拟机的机器指令，并有自己的栈、寄存器组等。但JVM通常是在软件上而不是在硬件上实现。目前，SUN系统公司已经设计实现了Java芯片，主要使用在网络计算机(NC)上。另外，Java芯片的出现也会使Java更容易嵌入到家用电器中。

4. 可移植性

Java应用程序可以在配备了Java解释器和运行环境的任何计算机系统上运行，而与体系结构无关。

5. 解释性

JVM是Java平台无关的基础，在JVM上，有一个Java解释器用来解释Java编译器编译后的程序。Java编程人员在编写完软件后，通过Java编译器将Java源程序编译为JVM的字节码(Bytecode)。任何一台机器只要配备了Java解释器，就可以运行这个程序，而不管这种字节码是在何种平台上生成的。Java程序的运行环境如图2-1所示。另外，Java采用的是基于IEEE标准的数据类型。通过JVM保证数据类型的一致性，也确保了Java的平台无关性。

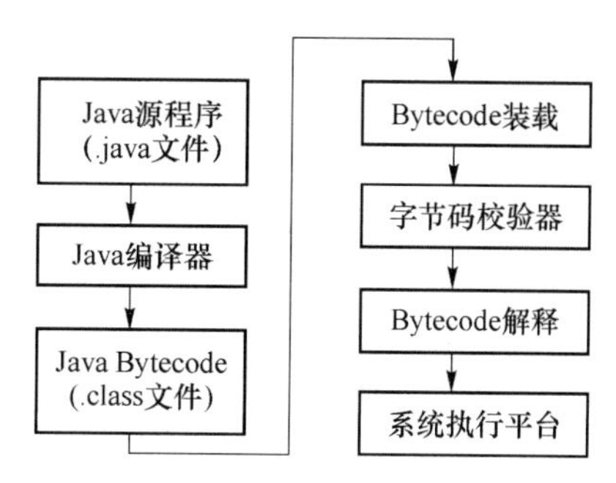

图2-1 Java程序运行环境

6. 高性能

Java系统提供了JIT(Just In Time)编译器，JIT能产生编译好的本地机器代码，以提高Java代码的执行速度。

7. 动态性

允许程序动态地装入运行过程中所需要的类。

8. 可靠性和安全性

可靠性和安全性主要表现在下列几个方面。

(1) Java编译器对所有的表达式和参数都要进行类型相容性的检查，以保证类型是兼容的。任何类型的不匹配都将被报告为错误而不是警告。在编译器完成编译以前，错误必须被改正过来 。

(2) Java不支持指针，这杜绝了内存的非法访问。

(3) Java的自动单元收集防止了内存丢失等动态内存分配导致的问题。

(4) Java解释器运行时实施检查，可以发现数组和字符串访问的越界；Java提供了异常处理机制，以便从错误处理任务恢复。

(5) 由于 Java 主要用于网络应用程序的开发，因此对安全性有较高的要求。Java 通过自己的安全机制防止了病毒程序的产生和下载程序对本地系统的威胁破坏。当 Java 字节码进入解释器时，首先必须经过字节码校验器的检查，然后，Java 解释器将决定程序中类的内存布局，随后，类装载器负责把来自网络的类装载到单独的内存区域，避免应用程序之间相互干扰破坏，最后，客户端用户还可以限制从网络上装载的类只能访问某些文件系统。

上述几种机制结合起来，使得 Java 成为安全的编程语言。

9. 多线程

Java 提供的多线程机制使应用程序能够并发执行，提供的同步机制保证了对共享数据的正确操作。通过使用多线程，程序设计者可以分别用不同的线程完成特定的行为，而不需要采用全局的事件循环机制，这样就很容易实现网络上的实时交互行为。

10. 分布式处理

分布式包括数据分布和操作分布。数据分布是指数据可以分散在网络的不同主机上，操作分布是指把一个计算分散在不同主机上处理。

Java 支持 WWW 客户机/服务器计算模式，因此，它支持以上两种分布性。对于前者，Java 提供了一个叫做 URL 的对象，利用这个对象，可以访问 Internet 上的所有网络资源。对于后者，Java 提供的基于流套接字的网络编程接口，为实现网络的应用和 Web 应用提供了方便。

2.3 Java 类库

程序员在应用 Java 语言开发应用程序时，除了定义本身需要的类外，还会利用 Java 系统中提供的类库，帮助搭建自己的应用。Java 类库也称为 Java API(Application Programming Interface，API)，为程序员提供了方便的编程接口。

2.3.1 Java 中的包

Java 为编写应用程序提供了丰富的预定义类库，这些预定义类库按相关类的范畴进行了分组，这些分组被称为包(package)。

一个包是一个已命名的类的集合，一个包还可以包含其他的子包。Java 所有的包被称为 Java 应用程序接口(Java API)。Java API 包分为：

- 核心包，以 java 开头的包。
- 可选包，以 javax 开头的包。

核心包和可选包一般被包含在 Java 开发工具包(Java Development Kits，JDK)中。JDK 是整个 Java 的核心，包括了 Java 运行环境、一组 Java 工具和 Java 基础类库(rt.jar)。

以下为 JDK 中的一些重要的包。

- java.lang：提供支持 Java 的基础类。
- java.util：提供数据结构和常用方法的类。
- java.awt 和 javax.swing：提供图形用户界面编程的类库集合。
- java.applet：为创建小应用程序提供必要的元件。

- java.io：提供输入、输出流支持。
- java.net：提供支持联网的类。
- java.sql：提供数据库操作的类。

2.3.2 JSE、JEE、JME

JDK 针对不同的的应用，提供不同的服务，分成了 JSE、JEE、JME 三种类型的版本。

- JSE(Java Standard Edition) 是 Java 的标准版，用于标准的 PC 应用开发。
- JEE(Java Enterprise Edition) 是 Java 的企业版，用于企业级的应用服务开发。
- JME(Java Micro Edtion)是 Java 的微型版，用于移动设备、嵌入式设备上的 Java 应用程序。

对于初学者，都是从 JSE 入手的。本书的实验环境使用的版本也是 JSE。

2.4 Java 开发环境

Java 开发环境大体上分成两种方式。一种方式是使用 JDK 工具集，它是一种 DOS 命令行的使用方式。另一种方式是使用集成开发环境 IDE(Integrated Development Environment)，Java 的 IDE 有很多种，如 Eclipse 、MyEclipse、JCreator、NetBean、JBuilder、Visual J++、JPad 等。具体选用哪种开发工具要视项目的具体情况而定。对于初学者而言，集成开发环境的界面不能太复杂。适合于初学者学习的，应该是界面简单一些，最好是不需要对开发环境做很多的设置的软件。MyEclipse 是在 Eclipse 上集成了功能强大的一组插件、支持几乎所有的 Web 应用的开发环境。MyEclipse 安装版本中，已经集成了 JDK 的安装，可以不需要单独装 JDK。当然也可以在 MyEclipse 安装完成后，通过配置更新新的 JDK 版本。

JCreator 是在 JDK 基础上为开发者提供了一个简单的图形化的 Java 开发平台，其占用资源少、运行速度快，但功能有限。安装 JCreator 前，机器上必须预先安装 JDK。安装好后，才能安装 JCreator。

这里推荐读者尽可能使用 MyEclipse 作为学习 Java 的集成开发环境。

本节中将主要讨论 JDK 和集成开发环境 MyEclipse 和 JCreator 的安装，并通过程序例子分别介绍它们的使用方法。

2.4.1 JDK 的安装与使用

JDK 可从 java.sun.com 网站免费下载，在这里下载适合于我们计算机操作系统的 JDK。如 Windows32 操作系统下，我们下载 JDK 的开发工具“jdk-7u51-windows-i586.exe”；而 Windows64 操作系统下，下载“jdk-7u51-windows-x64.exe”。在 Windows32 操作系统下安装 JDK 时，直接运行“jdk-7u51-windows-i586.exe”，按照安装向导进行安装。图 2-2 是 JDK 安装向导的首界面，图 2-3 选择安装 JDK 的目标位置，默认目录是 C:\Program Files\Java\。图 2-4 是安装完成界面。

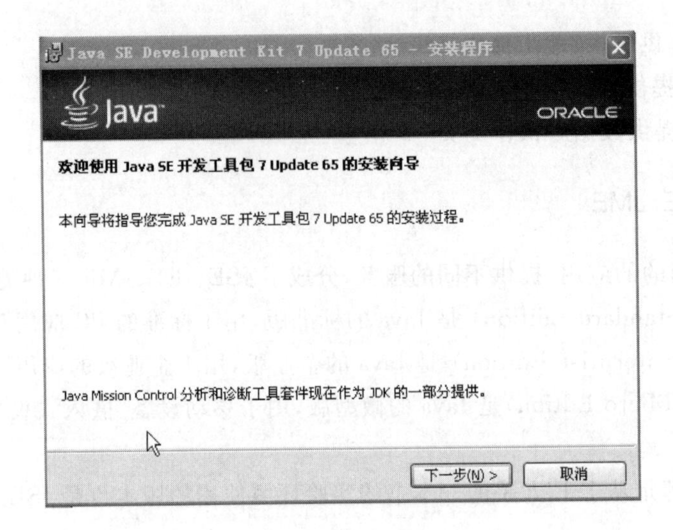

图 2-2　JDK 安装向导界面

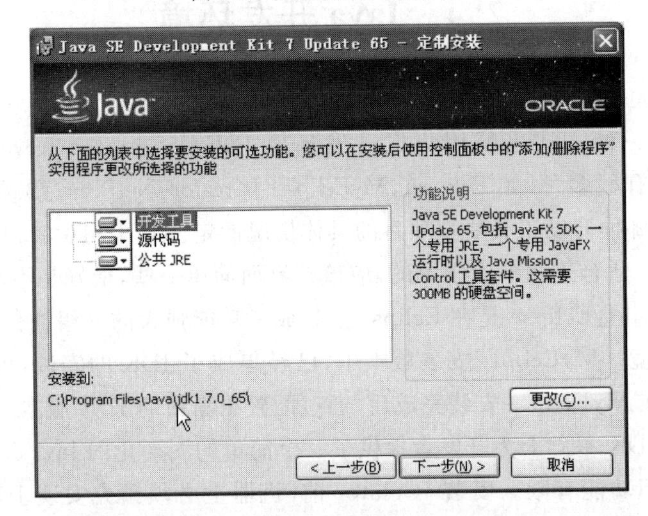

图 2-3　选择安装目标位置

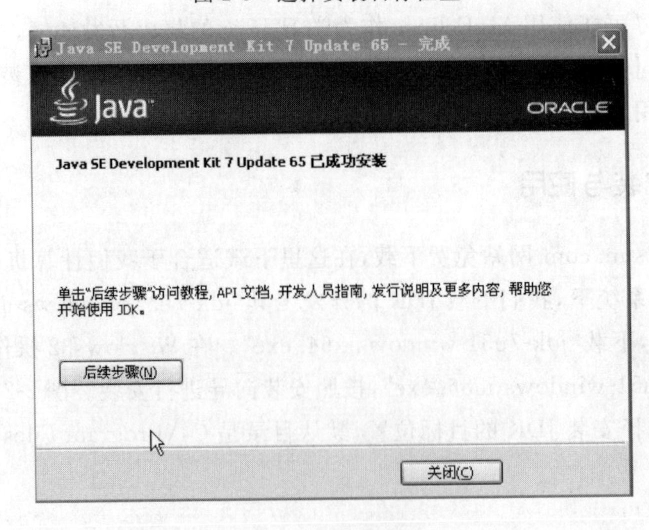

图 2-4　完成安装界面

安装完 JDK 后，必须设置两个环境变量 PATH 和 CLASSPATH。PATH 环境变量指定 Java 工具包的路径，CLASSPATH 环境变量指定了类的路径。假如 JDK 安装在 C:\Program Files\Java 目录下，而用户的 Java 类文件放在 D:\javaExamples 目录下，则设置环境变量的过程和方法如下。

(1) 右击“我的电脑”，打开“属性”，选择“高级”里面的“环境变量”。

(2) 在系统变量里找到“path”，单击“编辑”，添加变量值“C:\Program Files\Java\ jdk1.7.0_65\bin”，界面如图 2-5 所示，单击“确定”。

(3) 在系统变量中单击“新建”，变量名写上“classpath”，值为“D:\javaExamples”。图 2-6 是设置环境 classpath 变量界面 。

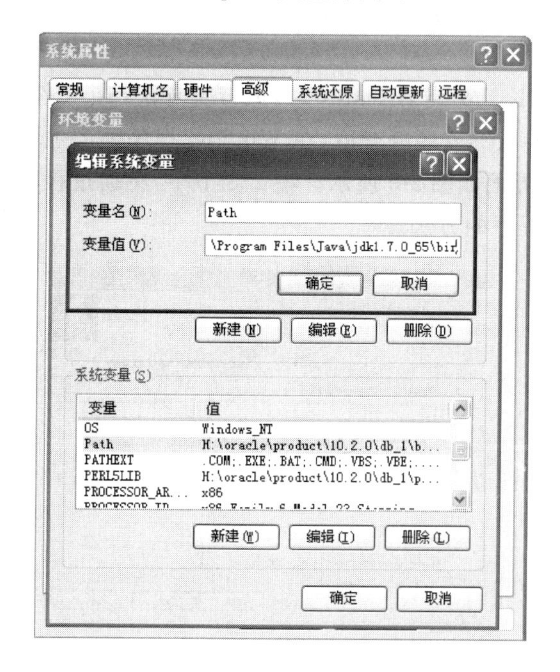

图 2-5 设置环境 path 变量界面

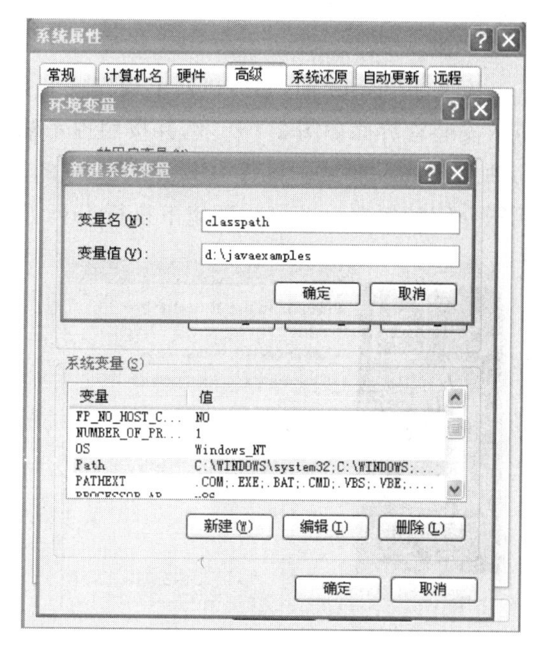

图 2-6 设置环境 classpath 变量界面

安装 JDK 后产生主要的目录结构如下。

(1) \bin 目录:Java 开发工具，包括 Java 编译器、解释器、applet 解释器等。

javac.exe:Java 编译器，用来将 Java 程序编译成 Bytecode。

java.exe:Java 解释器，执行已经转换成 Bytecode 的 Java 应用程序。

jdb.exe:Java 调试器，用来调试 Java 程序。

javap.exe:反编译，将类文件还原回方法和变量。

jconsole: Java 进行系统调试和监控的工具。

javadoc.exe:文档生成器，创建 HTML 文件。

appletviwer.exe:applet 解释器，用来解释已经转换成 Bytecode 的 Java 小应用程序。

(2) \demo 目录:一些实例程序。

(3) \lib 目录:Java 开发类库。

(4) \jre 目录: Java 运行环境，包括 Java 虚拟机、运行类库等。

从 java.sun.com 网站，我们可以免费下载 JDK 的开发帮助文档“jdk-6-doc.zip”。解开“jdk-6-doc.zip”，得到 JDK 的开发帮助文档。

在 2.5 节中,将通过程序例子介绍如何在 JDK 的环境下执行 Java 应用程序的方法。

2.4.2 JCreator 的安装和使用

JCreator 为用户提供了相当强大的功能,例如:项目管理功能,可个性化设置语法高亮属性、行数、类浏览器、标签文档、多功能编译器,向导功能以及完全可自定义的用户界面。通过 JCreator,不用激活主文档即可直接编译或运行 Java 程序。JCreator 能自动找到包含主函数的文件或包含 applet 的 Html 文件,然后它会运行适当的工具。JCreator 的设计接近 Windows 界面风格,用户对它的界面比较熟悉。其最大特点是与机器中所装的 JDK 完美结合,是其他任何一款 IDE 所不能比拟的。JCreator 的下载地址: http://www.jcreator.com/Download.htm。

安装 JCreator 前,必须机器上预先安装好 JDK。安装 JCreator 时,运行 setup.exe 文件,进入安装首界面如图 2-7 所示,并按照提示输入参数。安装完毕后,在 Window 程序组产生程序项 JCreator PRO。运行 JCreator PRO 的显示界面如图 2-9 所示。在 2.5 节中,将通过程序例子介绍如何在 JCreator 环境下执行 Java 应用程序的方法。

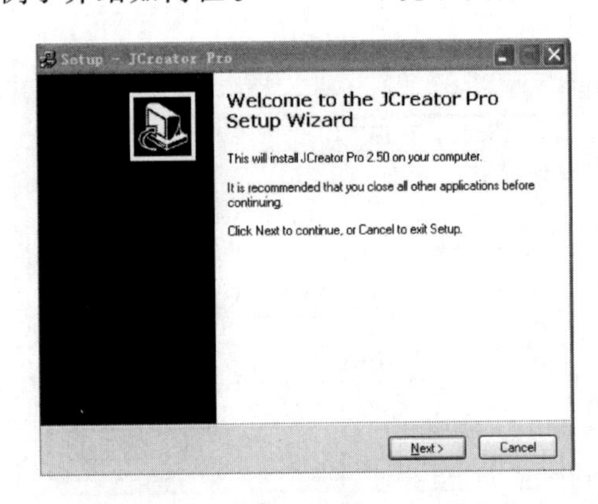

图 2-7 JCreator 安装首界面

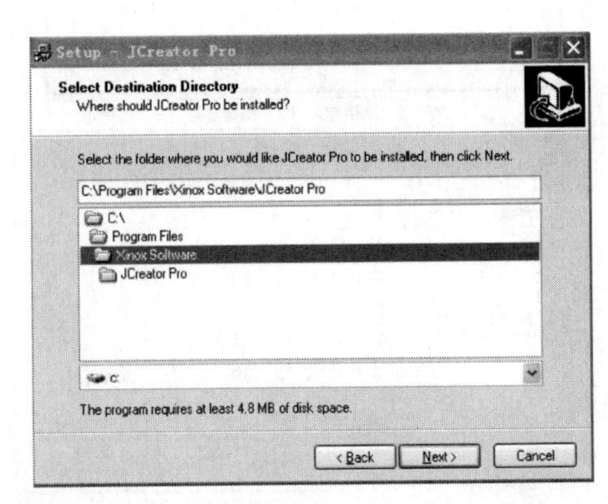

图 2-8 选择 JCreator 安装目录

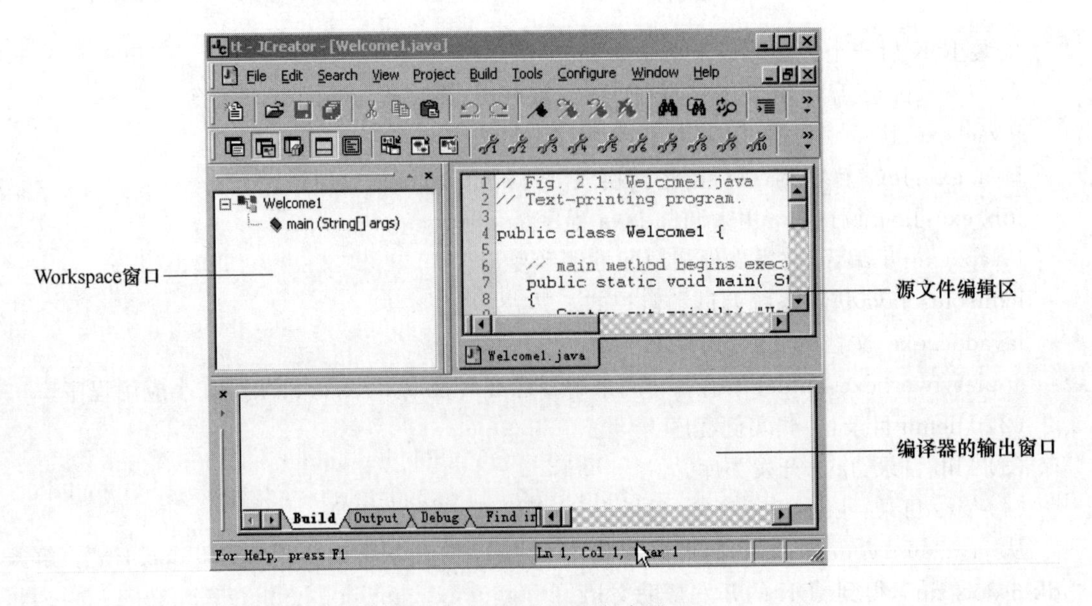

图 2-9 JCreator 运行主界面

2.4.3 MyEclipse 8.5 的安装和简单使用

Eclipse 是一个免费的开放的 IDE，它允许安装第三方开发的插件来扩展和增强自身的功能。而 MyEclipse 是以 Eclipse 为核心，并集成开发者常用到的一些有用插件，如 struts、Hibernate、Spring 等。MyEclipse 提供了一种高级编程环境，包括了完备的编码、调试、测试和发布功能，完全支持 HTML、Struts、JSF、CSS、Javascript、SQL、Hibernate 编程。MyEclipse 的唯一不足是收费软件，但读者可以从网络上下载相应的版本或试用版本。本书使用的是 myeclipse-8.5.0-win32.exe 版本。

安装 MyEclipse 时，直接运行文件 myeclipse-8.5.0-win32.exe。首先进入解压缩界面，如图 2-10 所示，并按照提示输入参数，安装完毕后，在 Windows 桌面的程序组产生 MyEclipse 程序项。

单击“MyEclipse 8.5”，进入启动 MyEclipse 的工作空间选择界面，如图 2-11 所示，可以在 workspace 中选择好项目所存放的路径，如“c:\JavaWorkSpace”，如图 2-11 所示，单击“OK”按钮后，将进入 MyEclipse 的窗口主界面，如图 2-12 所示。第一次启动后还会显示一个欢迎界面(Welcome)，单击上面的图标关闭欢迎界面即可。

MyEclipse 是以项目(Project)为单位管理一组相关的文件，一个 Java 程序文件必须包含在某一项目中才能被执行。而一组 Projects 又包含在一个工作空间(Workspace)中。在 2.5 节中，将通过程序例子介绍如何在 MyEclipse 环境下执行 Java 应用程序的过程。

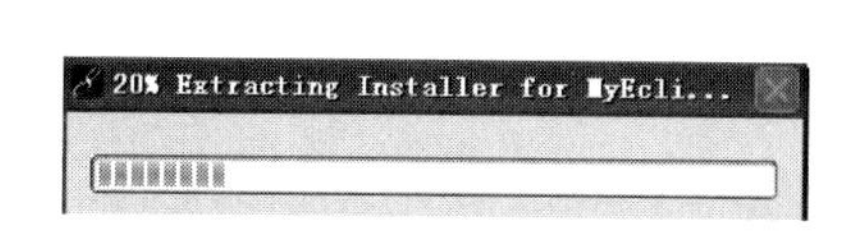

图 2-10 MyEclipse 安装的解压缩界面

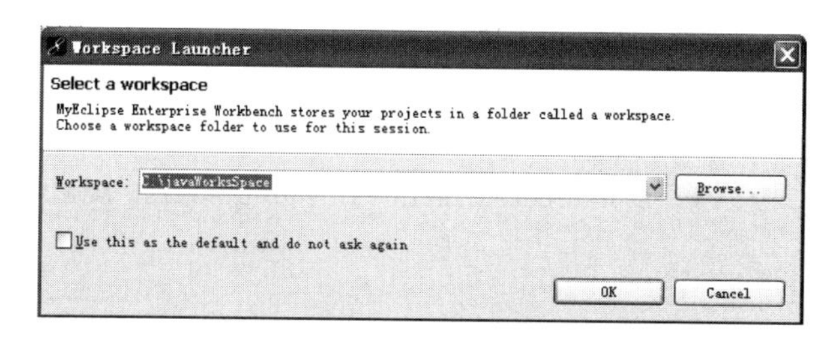

图 2-11 启动 MyEclipse 的工作空间选择界面

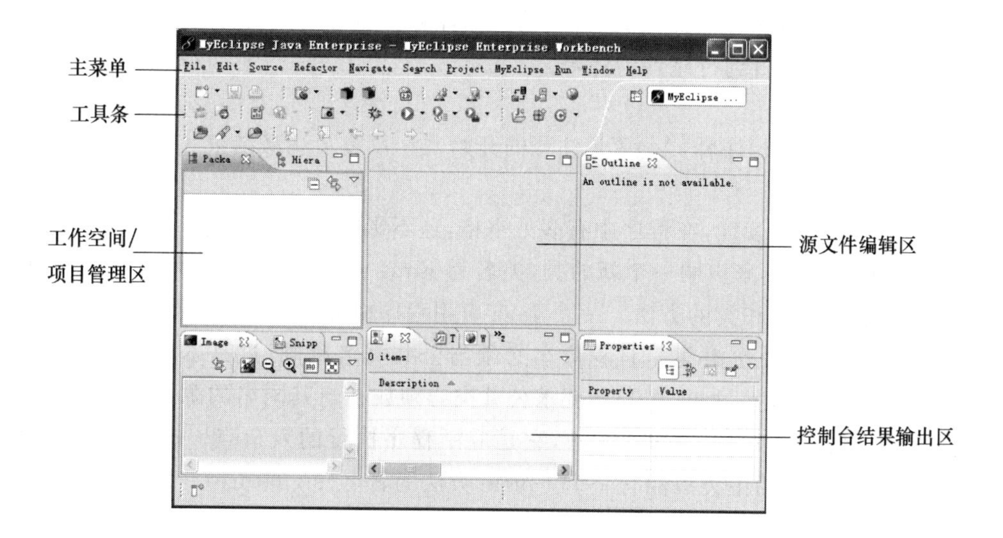

图 2-12 MyEclipse 的主界面

2.5 Java 程序类型、简单例子以及开发环境的使用

Java 程序有四种基本类型:应用程序(application)、小应用程序(applet)、servlet 和 bean。这四种类型程序的区别很小,Java 的每一个程序都至少有一个类,并且 Java 语言的基本编程结构对于所有类型的程序都适用,且每一种类型的程序都要运行在 Java 虚拟机 JVM 上。下面将重点介绍前面两种基本类型的程序。

2.5.1 应用程序与开发环境 JDK 1.7、JCreator 和 MyEclipse 的使用举例

应用程序(applications)是一个可以独立运行的程序,它只需要 Java 虚拟机就能够运行。相反,其他三种类型的程序(applet、servlet 和 bean)则不同,它们需要一个主机程序(host program)才能运行。一般来说,applet 的主机程序是一个 Web 浏览器,servlet 的主机程序是一个 Web 服务器,bean 的主机程序可以是其他几种程序类型的任何一种:应用程序、applet 或 servlet,也可以是另外一个 bean。

一个 Java application 中,一定要有一个类包含 main 方法。下面我们来看 Java 应用程序的例子。

例 2-1 第一个简单的 Java application,并使用 JDK 对程序编写、编译和运行。

FirstApp. java 文件的代码如下:

```
//程序命名为:FirstApp. java
public class FirstApp
  {
    public static void main(String args[ ])
      {
        System. out. println("This is the first Java Application!");
      }
  }
```

程序中的语句解释如下:

(1) 注释行。在本程序中第一行"//"后的内容为注释,进行编译时,这一行所有内容会被忽略。注释不影响程序的执行结果,编译器将忽略注释。Java 有两种注释形式:

(a) 从"//"开始一直到行尾均为注释。

```
// 在一行的注释
```

(b) 可用于跨越多行注释,"/ * "是注释的开始,"* /"表示注释结束。

```
/* 一行或多行的注释 */
```

为了增强程序的可读性,在程序中有多处空格,这不影响程序的执行。

(2) 用关键字 class 来声明一个新的类,类名为 FirstApp。类定义由大括号{}括起来,类中封装了类的变量(域)和类的方法。一个 Java 应用程序由 $n(n>0)$ 个类组成,但这 n 个类中只能有一个是 public 类(公共类),且程序名必须与公共类名相同,这是本应用程序文件起名为 FirstApp. java 的原因。Java 解释器要求公共类必须放在与其同名的源文件中。

(3) 在该类中定义了一个 main 方法,它是应用程序执行的入口点。main 方法所在的类叫主类,一个应用程序的主类只能有一个。main 方法的署名(signature)一定是:

```
static void main(String args[ ])
```

main 方法的署名中的符号解释如下。

- public:指明为公共方法。public 方法可以被类的对象使用。
- static:指明方法是一个静态方法,静态方法是类的方法,而不是对象的方法,静态方法可以通过类名直接调用。
- void:表示 main 方法执行后不会返回任何值。
- 括号中的 String args[]是定义传递给 main 方法的参数:参数名为 args,类型为 String 的数组。

(4) 在 main 方法中,只有一条语句:

```
System.out.println("This is my first java pplication!");
```

其功能是在命令窗口上输出一行文本。

(5) Java 语句结束符号。每一条 Java 语句都由分号";"结束。Java 的一条语句可跨越多行。例如,跨越两行的语句:

```
total = a + b + c +
        d + e + f + g;
```

与语句:

```
totals = a + b + c + d + e + f + g;
```

是等价的。

使用 JDK 的命令行方式执行上述程序。程序的执行分三步:编写、编译和运行。

第一步:利用文本编辑工具,如记事本,编写生成 Java 源程序,命名为 FirstApp. java(注意:这里的文件名的后缀名一定为. java),并将该文件放入 D:\javaExamples 目录中。

第二步:编译。调用 Java 编译器执行 javac. exe,将源文件 FirstApp. java 编译生成类文件 FirstApp. class。方法:单击"开始"->"运行",输入"cmd",进入 MS-DOS 命令行界面。在 MS-DOS 命令行提示符下输入下列命令执行编译:

```
d:                        //改变盘符
cd javaExamples           //改变当前目录
javac FirstApp.java       //编译
```

第三步:运行。调用 Java 解释器 java. exe 对类文件 FirstApp. class 解释执行,输出程序的执行结果。在 MS-DOS 命令行提示符"D:\java\application>"下,输入下列命令执行:

```
java  FirstApp
```

运行时,程序将在命令窗口中输出下面一行信息,如图 2-13 所示。

```
This is the first Java Application!
```

图 2-13　例 2-1 程序运行输出结果

例 2-2　用格式化方法 printf 输出文本,并在 JCreator 环境下执行应用程序。

Second. java 文件的代码如下:

```
public class Second {
    public static void main(String[] args) {
      System.out.printf("%s\n%s\n%s\n","*","* *","* * *" ,"* * * *");
    }
}
```

程序运行时输出结果如图 2-14 所示。

```
You print
Two lines
```

图 2-14　例 2-2 程序运行输出结果

用 JCreator 完成上述程序的执行,分三步:程序书写、编译和运行。

第一步:利用 JCreator,编写生成 Java 源程序,命名为 Second. java。

建立过程:进入 JCreator 主界面,如图 2-15 所示。选择主菜单“File—>New”,将出现输入文件名界面,如图 2-16 所示,输入 Java 的源文件名 Second 和文件所在的目录 D:\javaExamples(若 D 盘根目录下无目录名 javaExamples,请先创建它),将出现文件内容编辑界面,如图 2-17 所示。在 Java 源程序窗口中输入文件内容。

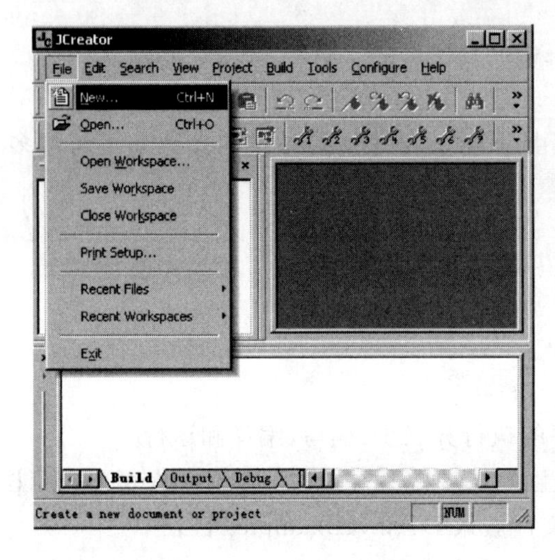

图 2-15　建立文件界面

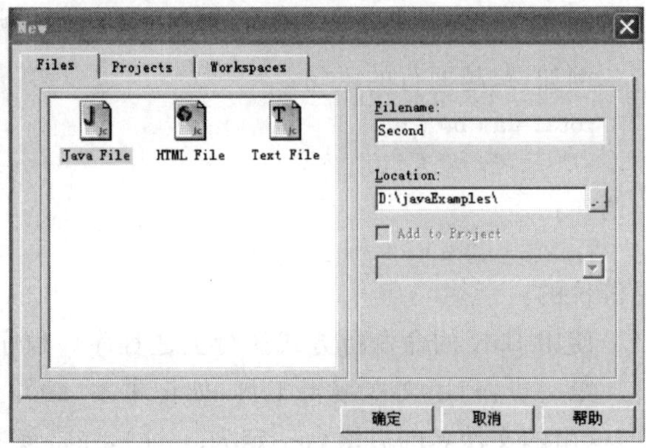

图 2-16　输入文件名界面

第二步:程序编译。选择主菜单“Build”的子菜单“Compile File”,如图 2-18 所示。编译器将源文件 Second. java 编译生成类 Second. class 放在 D:\javaExamples 目录下。

第三步:程序运行。选择主菜单“Build”的子菜单“Execute File”,如图 2-18 所示,对类文件 Second. class 解释执行,并输出如图 2-14 所示的结果。

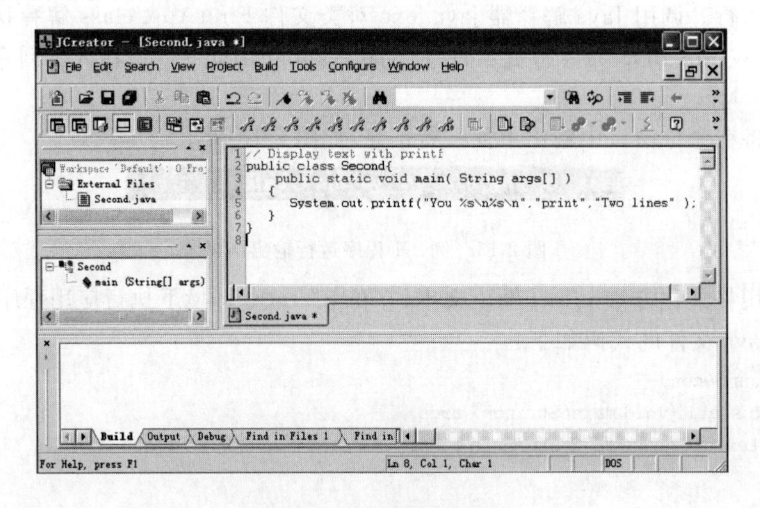

图 2-17　文件编辑的界面

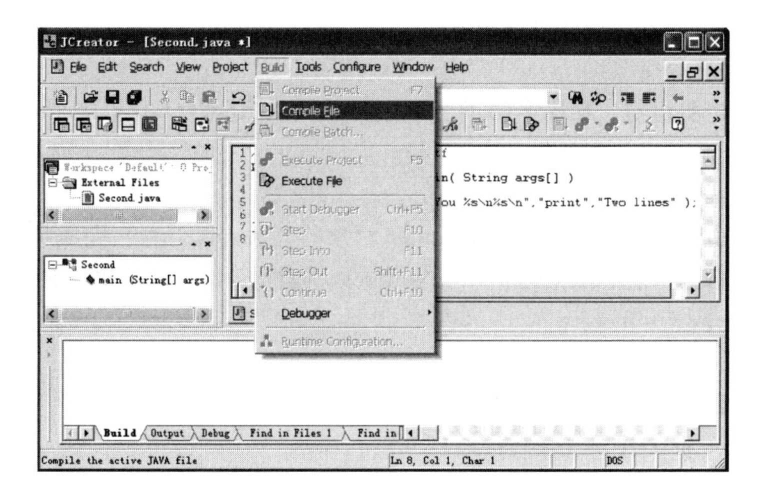

图 2-18 文件编译和运行界面

例 2-3 第三个 Java application，并在 MyEclipse 环境下执行应用程序。

程序要求用户从键盘输入两个整数，计算这两个整数之和，并显示结果。Addition.java 源程序代码如下：

```
//Addition.java
import java.util.Scanner;
public class Addition {
    public static void main( String args[] )
    {
        Scanner  input = new Scanner(System.in);
        //声明三个整型变量
        int firstNumber;
        int secondNumber;
        int sum;
        // read in first number
        System.out.print( "输入第一个数：" );
        firstNumber = input.nextInt();
        // read in second number
        System.out.print( "输入第二个数：");
        secondNumber = input.nextInt();
        // addtion numbers
        sum = firstNumber + secondNumber;
        System.out.printf( "相加后的结果是：%d\n",sum);
    }
}
```

程序中的语句说明如下。

(1) `import java.util.Scanner;`

此句是 import 声明语句，导入 java.util 包中的类 Scanner。此行告说编译程序将使用 javax.swing 包中预定义的类 Scanner。应用 Scanner 类可以帮助程序读入数据。

(2) `Scanner  input = new Scanner(System.in);`

此句是一个赋值语句，即将等号"="右边的表达式的值，赋给等号左边的变量，此等号被称为赋值号。在这条语句中，赋值号左边声明的是一个变量，其类型为 Scanner，变量名为 input。赋值号右边的"new Scanner(System.in)"创建了一个标准输入对象。通过赋值语句，将

Scanner 类型的对象赋给了应用变量 input。

(3) int firstNumber;

此句是变量声明语句。程序声明了三个整型的变量 firstNumber、secondNumber 和 sum 用以存放被减数、减数以及相加的和。

(4) firstNumber = input.nextInt();

此语句通过引用变量 input，调用 Scanner 对象的方法 nextInt()，完成接受从键盘输入的第一个整数，并将此整数值赋给变量 firstNumber。同样接着的语句提示用户输入第二个整数，赋给变量 secondNumber。

(5) sum = firstNumber + secondNumber;

此句是将两数相加的结果赋给变量 sum。

(6) System.out.printf("相加后的结果是：%d\n",sum);

程序将相加的结果用 printf 格式化输出。"%d"是格式说明符，表示对应位置被替换的值是十进制整形值，这里是 sum 变量的值。

在 MyEclipse 环境下，完成上述程序的执行，需要四个步骤：创建项目，给项目添加类文件，输入、编译类文件，运行类文件。

(1) 创建 Java Project 项目

启动 MyEclipse 程序，进入 MyEclipse 的工作空间选择界面如图 2-11 所示，在 Workspace 中选择项目所存放的路径，如"c:\JavaWorkSpace"，然后单击"OK"按钮后，进入 MyEclipse 的主界面，如图 2-19 所示，单击"File—>New—>Java Project"，出现新建项目界面如图 2-20 所示。在 Project Name 中输入项目的名字，如 ch2，选择"Next"，选择"Finish"按钮后，将进入已经创建好项目的界面，如图 2-21 所示。

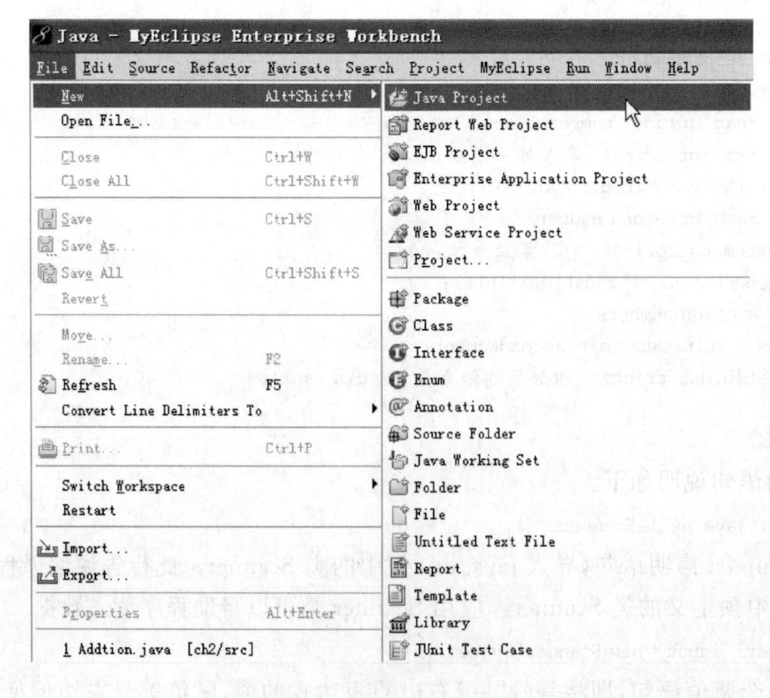

图 2-19 新建项目界面 1

图 2-20　新建项目界面 2

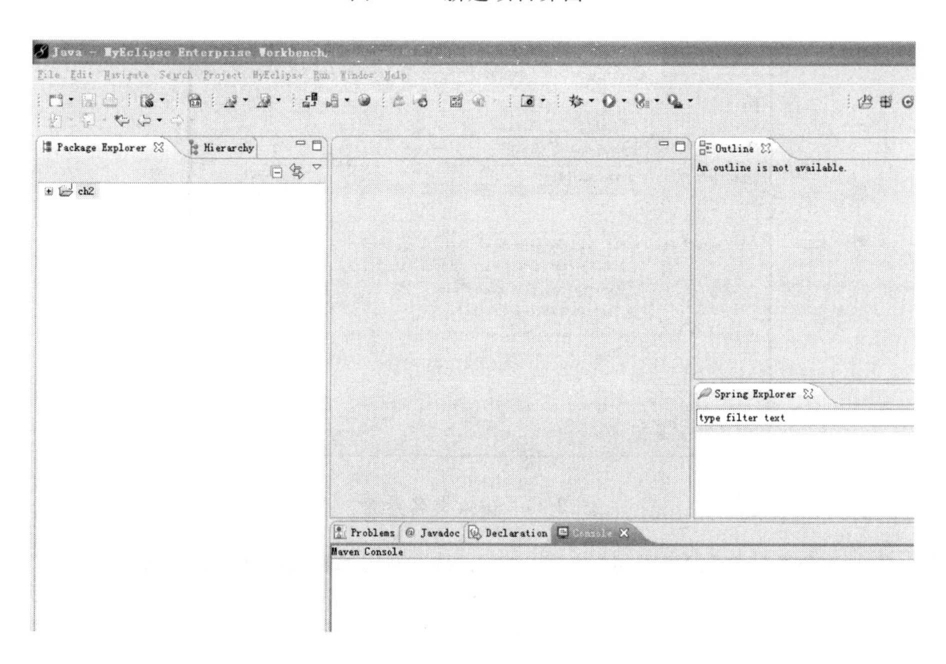

图 2-21　新建项目完成界面

(2) 给项目添加类文件

在"src"(Source 的速写)上右击,在弹出的下拉式菜单上选择"new—>class",如图 2-22 所示,进入输入类名的界面,如图 2-23 所示,在 Name 中输入类名 Addition,选择"Finish"按钮。进入类的源代码输入、编译界面,如图 2-24 所示。MyEclipse 集成开发环境为新建的类名 Addition 生成了存放的源文件 Addition.java ,并将类自动放在默认包(Default package)中。

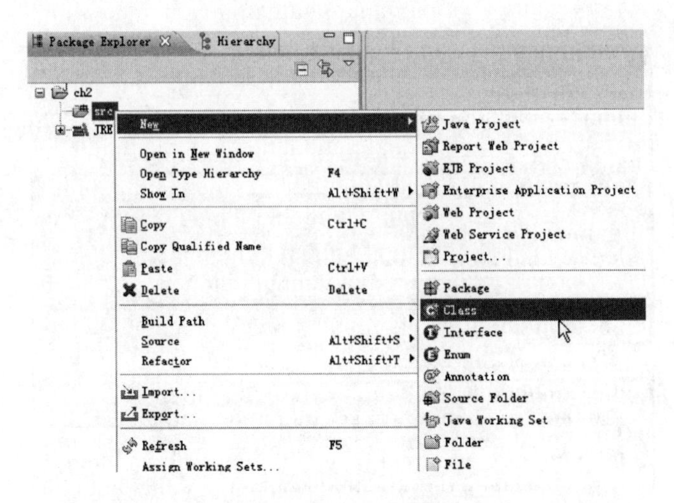

图 2-22　给项目添加类文件

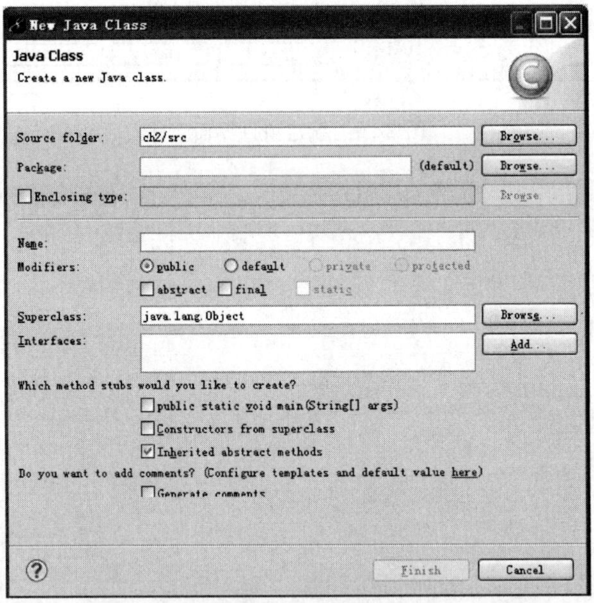

图 2-23　输入类名界面

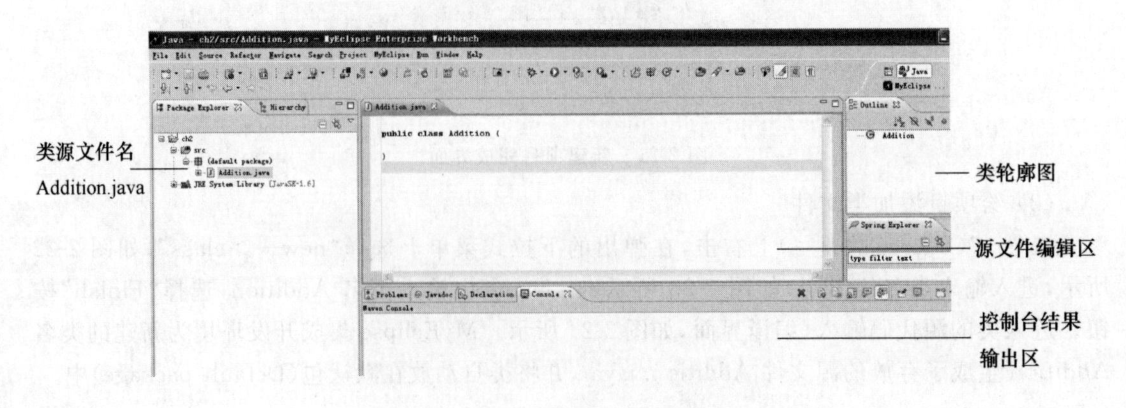

图 2-24　类的源文件输入、编译界面

(3) 输入源代码,并编译

在图 2-21 的源文件编辑区,输入上面的例 2-3 程序的源代码,如图 2-22 所示。MyEclipse 集成环境会在你输入源代码时,自动编译输入的每一行语句,如果发现语句有错,会在出错的语句前面打上红色的叉"×",提醒你修改相应的行,直到整个源程序编译正确完成(即每一条语句前都没有出现红叉)。源程序被编译完成后,系统会将类的源文件 Addition.java 翻译成二进制代码的目标类文件 Addtion.class,并放在当前项目的缺省包中。

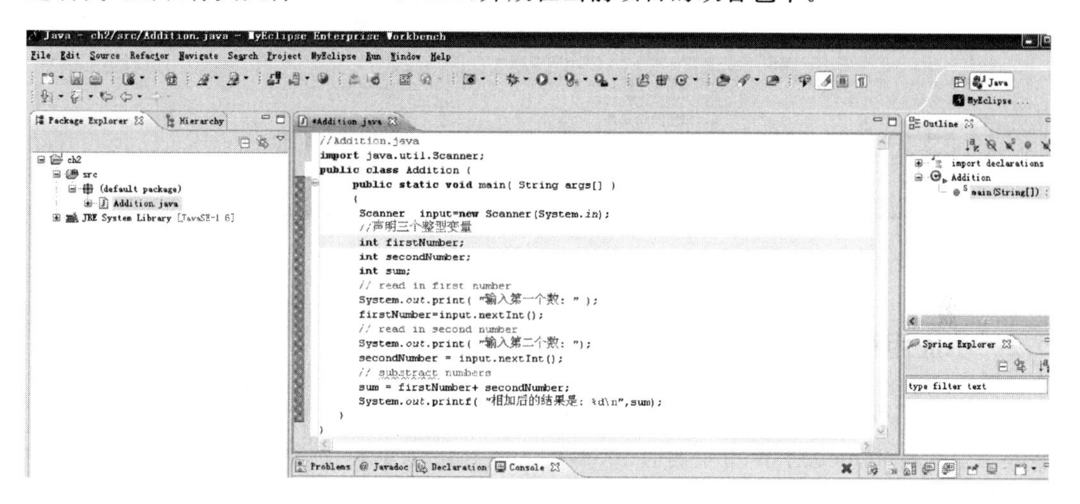

图 2-25　类的源文件输入、编译完成界面

(4) 运行 Java 主类

在图 2-22 的源文件编辑区内的任意位置上右击,将弹出下拉式菜单,如图 2-23(a)所示,选择"Run As—>Java application",将运行 Java 应用程序(主类 Addition.class),运行结果将显示在控制台结果输出区,如图 2-24 所示。首先显示第一行"输入第一个数:",输入 10;接着显示第二行"输入第二个数:",输入 30;接着显示第三行"相加的结果是:40"。

运行 Java 应用程序的另一方法:单击 MyEclipse 主界面工具栏上的按钮"Run",如图 2-23(b)所示,再选择下拉式菜单中的"Addition",将运行 Addition.class 主类文件。

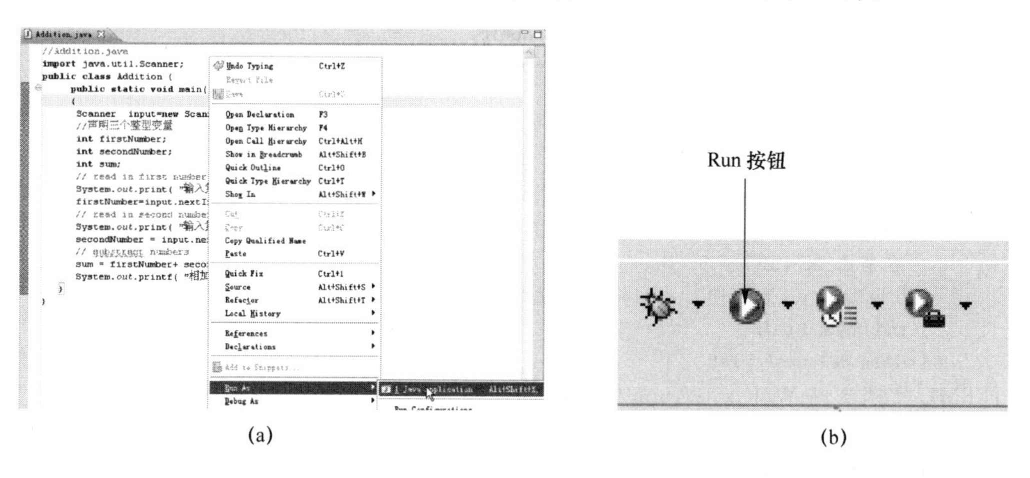

图 2-26　运行 Java 应用程序

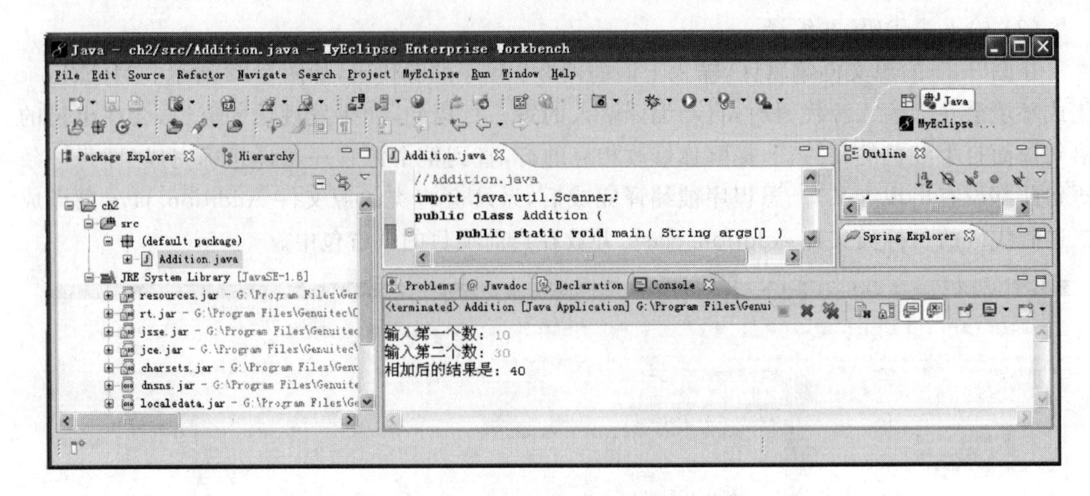

图 2-27　例 2-3 程序运行的输出结果

2.5.2　小应用程序 applet

Java 小应用程序(applet)是一种嵌入在 HTML 文档(即 Web 页)中的 Java 程序。当浏览器装入一个含有 applet 的 Web 页时，applet 被下载到该浏览器中，并开始执行。执行 applet 的浏览器一般称为 applet 容器。Java 2 软件开发工具包含有 appletviewer. exe，它是一个 applet容器，一般用于对嵌入 Web 页的 applet 进行测试。在 JCreator 集成环境中，会调用 appletviewer 容器，运行含有 applet 的 HTML 文档。

例 2-4　第一个 Java applet：绘制字符串。

第一个 applet 的功能是在 applet 显示区域上绘制字符串"欢迎进入 applet 程序设计"。文件名为 WelcomeApplet. java，经过编译后将生成 WelcomeApplet. class。由于 Java 小应用程序是不能直接运行的，必须嵌入到网页中。因此我们将小应用程序 WelcomeApplet. class 嵌入在 HTML 文件 WelcomeApplet. html 中。小应用程序 WelcomeApplet. java 的代码如下：

```
import java.awt.Graphics;        // 导入类 Graphics
import javax.swing.JApplet;      // 导入类 JApplet
public class WelcomeApplet extends JApplet {
    // draw text on applet background
    public void paint( Graphics g )
    {  // 调用基类方法 paint，刷新 applet 显示区域
      super.paint( g );
      // 在坐标(25 , 25)画 String
      g.drawString( "欢迎进入 applet 程序设计", 25, 25 );
    } // end method paint
} // end class WelcomeApplet
```

HTML 文档文件 WelcomeApplet. html 的代码如下：

```
<html>
<applet code = "WelcomeApplet.class" width = "300" height = "45">
</applet>
</html>
```

分别使用支持 Java 的微软 Internet Explorer 浏览器和小应用程序查看器 AppletViewer

这两个 applet 容器，来执行 WelcomeApplet.html 的显示结果如图 2-28 所示。

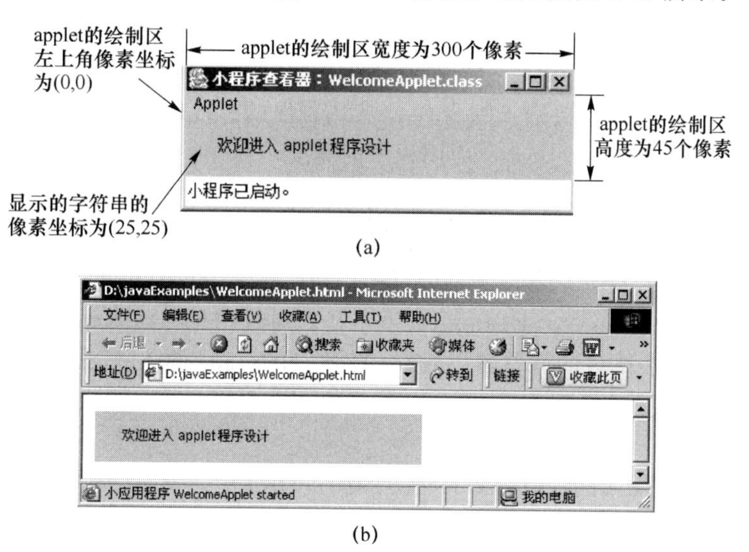

图 2-28　绘制一个字符串的 applet

（1） `import java.awt.Graphics;`

此句导入 javax.swing 包中的 Graphics 类。Graphics 类使 Java applet 能绘制各种图形如直线、矩形、椭圆以及字符串等。

（2） `import javax.swing.JApplet;`

此句导入 javax.swing 包中的 JApplet 类。创建一个 JApplet 类必须从 JApplet 类继承，并且此 applet 类必须声明为 public 类型。Japplet 类是从旧类 applet（在 java.applet 包中）派生出来的新类。

（3） `public class WelcomeApplet extends JApplet;`

此句是 applet 类的声明。它定义了一个 applet 类名为 WelcomeApplet，关键字 extends 表示继承，即表示 WelcomeApplet 是超类 JApplet 的子类，它拥有 JApplet 类的所有属性和方法，并且在 WelcomeApplet 类可以重写这些方法并扩展新的功能。

（4） `public void paint( Graphics g);`

此句是 applet 的 paint 方法的声明。paint 方法由 applet 容器调用执行。一个 applet 类中声明的方法有 5 个：init、start、paint、stop 和 destroy（此次序也是 applet 容器调用方法的执行次序），这些方法的含义请参见电子课件中 Java 文档中的 JApplet 类的说明。

利用 paint 方法可在 applet 上绘制出用户想要的各种图形。paint 方法的参数类型为 Graphics，它使用 Graphics 对象的引用 g，通过引用变量 g 调用 Graphics 对象的方法，以绘制各种图形以及字符串等。Graphics 类的对象的创建也是由 applet 容器完成的。

（5） `g.drawString( "欢迎进入 applet 程序设计", 25, 25 );`

此句是通过引用变量调用 Graphics 对象的 drawString 方法，在 applet 上指定的像素位置绘制一个字符串。drawString 方法有三个参数：第一个参数是要绘制的字符串；第二个和第三个参数是要绘制的字符串的左下角在 applet 上的平面坐标(25,25)，它们也是以像素为单位。

（6） `<applet code = "WelcomeApplet.class" width = "300" height = "45">`

此句是在 HTML 文件中嵌入小应用程序 WelcomeApplet. class，并定义该 applet 在 applet 容器中显示区域：以像素为单位的宽度和高度。在 HTML 文件中嵌入 Java 小应用程序应使用<APPLET> </APPLET>标志，这个标志包含几个参数：

- CODE=""：用来指出要嵌入的 Java 小应用程序的名称；
- WIDTH=x：用来指定 Java 小应用程序的宽度；
- HEIGHT=x：用来指定 Java 小应用程序的高度。

(7) WelcomeApplet 的编译和运行

对 WelcomeApplet. java 编译生成类文件。假设 WelcomeApplet. class 和 WelcomeApplet. html 都放在文件夹"D:\javaExamples"下。我们通过两种方法来运行 Java 小应用程序。第一种方法是使用支持 Java 的浏览器（如微软的 IE、Netscape 的 Navigator）。

使用 IE 浏览器，打开 IE 浏览器，在地址栏上输入命令：

```
d\javaExamples\WelcomeApplet.html
```

将看到小应用程序的输出结果，如图 2-9（下方的图）所示。

另一种方法是在 JCreator 集成环境中，调用 appletviewer 容器，运行含有 applet 的 HTML 文档。其过程是：

(1) 在 JCreator 中分别创建生成 Java 程序 WelcomeApplet. java 和 HTML 文件 WelcomeApplet. html；

(2) 选中 WelcomeApplet. java 对其进行编译；

(3) 选中 WelcomeApplet. html，运行它（执行菜单命令"Build—>Excute File"），将看到小应用程序的输出结果，如图 2-9（上方的图）所示。

例 2-5 第二个 Java applet：绘制字符串和图形。

第二个 applet 的功能是在 applet 显示区域上绘制字符串"欢迎进入 applet 程序设计!"，并且在此字符串的上方和下方绘制两条直线。类 WelcomeLines 从超类 JApplet 继承。在 paint 方法中，画两条直线 drawLine，画一个字符串。画直线 drawLine 方法需要四个参数，表示直线在 applet 上的起点坐标(x1,y1)和终点坐标(x2,y2)，坐标值均以像素为单位。小应用程序 WelcomeLines. java 的代码如下：

```
import java.awt.Graphics;        // import class Graphics
import javax.swing.JApplet;      // import class JApplet
public class WelcomeLines extends JApplet {
    public void paint( Graphics g )
    {  super.paint( g );          // call superclass version of method paint
                                 // draw horizontal line from (15, 10) to (210, 10)
       g.drawLine( 15, 10, 180, 10 );
                                 // draw horizontal line from (15, 30) to (210, 30)
       g.drawLine( 15, 30, 180, 30 );
                                 // draw String between lines at location (25, 25)
       g.drawString( "欢迎进入  applet  程序设计!", 25, 25 );
    }
}
```

网页文件 WelcomeLines. html 的代码如下：

```
<html>
<APPLET CODE = "WelcomeLines.class" WIDTH = "300" HEIGHT = "40">
</APPLET>
</html>
```

在 JCreator 中，运行 WelcomeLines. html，将看到小应用程序的输出结果如图 2-29 所示。

图 2-29　绘制一个字符串和两条直线的 applet

2.5.3　简单输入和输出

1. 应用 System. out. printf( )方法进行格式化的输出

printf()方法中常用的格式说明符如表 2-1 所示。

表 2-1　printf 方法中常用的格式说明符

格式说明符	含义	示例
%s	输出字符串，对应的参数类型为 String	"This is a string"
%c	输出字符，对应的参数类型为 char	'a'
%d	输出整型值，对应的参数类型为整型	123
%f	输出浮点型值，对应的参数类型为 float 或 double	12. 34
%b	输出布尔型值，对应的参数类型为 boolean	true

2. 应用 Scanner 类进行输入

Scanner 类在包 java. util 中，使用时必须用 import 语句导入。此类中常用的数据输入实例方法如表 2-2 所示。

表 2-2　Scanner 常用的输入数据方法

方法	含义
int nextInt( )	输入 int 整型数据
*** next *** ( )	输入 *** 类型的数据，其中 *** 为基本数据类型 byte、short、int、long、float、double、boolean
String nextLine()	输入一行字符串
String next()	输入一个单词

3. 利用对话框 JOptionPane 进行输入和输出

JOptionPane 类在包 javax. swing 中，使用时必须用 import 语句导入。JOptionPane 类提供静态方法 showInputDialog()和 showMessageDialog()分别用于输入和输出。静态方法直接通过加类名前缀调用。这两个方法头的声明分别如下：

```
staticString showInputDialog(Object message)
staticString showInputDialog(Component parentComponent,Object message, String title, int mes-
sageType)。
```

例 2-6　利用对话框从键盘输入两数，进行相减后输出结果。

Subtraction. java 源程序代码如下：

```
import javax.swing.JOptionPane;   //导入类 JOptionPane
public class Subtraction {
```

```
    public static void main( String args[] )
    { String firstNumber;           //定义第一个字符串变量
      String secondNumber;          //定义第二个字符串变量
      int number1;                  //定义三个整型变量
      int number2;
      int sub;
      // read in first number from user as a string
      firstNumber = JOptionPane.showInputDialog("输入第一个数");
      // read in second number from user as a string
      secondNumber = JOptionPane.showInputDialog("输入第二个数");
      // 将字符串型转换成整型
      number1 = Integer.parseInt( firstNumber );
      number2 = Integer.parseInt( secondNumber );
      sub = number1 - number2; // 两数相减
      // display result
      JOptionPane.showMessageDialog( null, "结果是 " + sub, "结果", JOptionPane.PLAIN_
          MESSAGE );
      System.exit( 0 );
    }
}
```

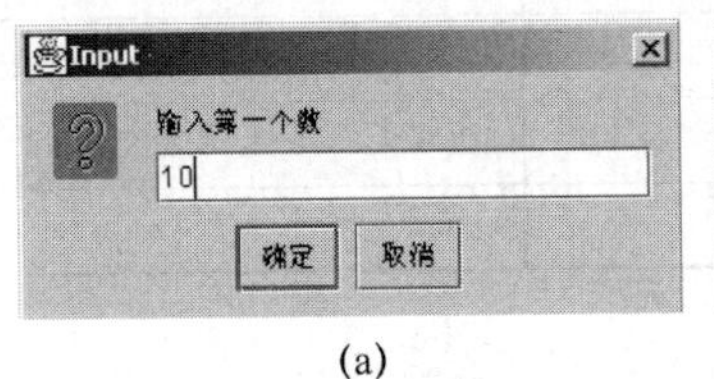

(a)

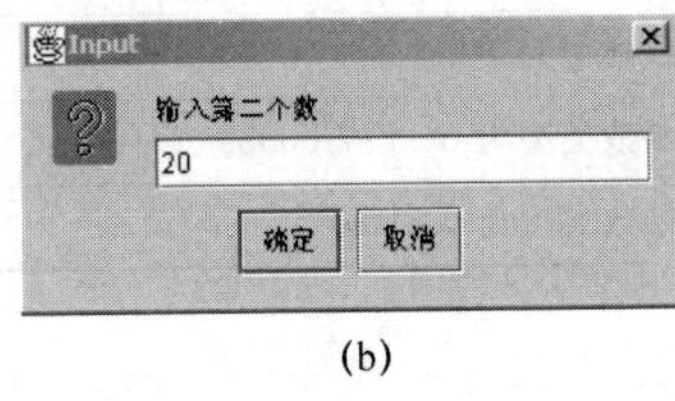

(b)

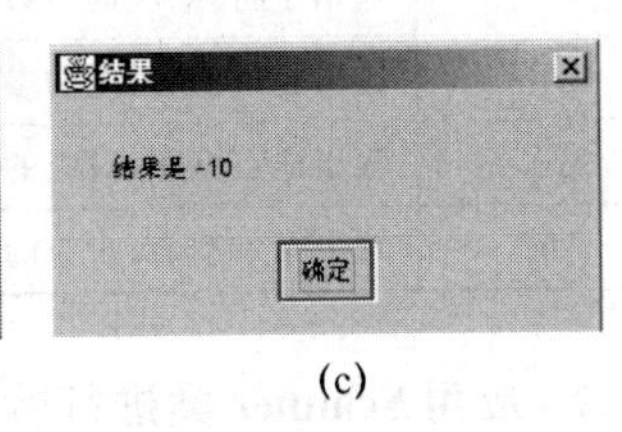

(c)

图 2-30　显示两数相减的程序运行结果

图 2-30 所示为程序的运行结果，其中图 2-30(a)和图 2-30(b)是提示用户输入两个数的输入界面，输入数后按“确定”按钮，图 2-30(c)显示两数相减结果的对话框。程序解释如下。

(1) `String firstNumber;`
`String secondNumber;`

这两句是两条变量声明语句，声明两个字符串型的变量名为 firstNumber 和 secondNumber，在 Java 中字符串型用类 String 表示，类 String 在包 java.lang 中，由编译器自动导入。

(2) `firstNumber = JOptionPane.showInputDialog("输入第一个数");`
`secondNumber = JOptionPane.showInputDialog("输入第二个数");`

这两句是使用对话框输入方法 showInputDialog，两次提示用户各输入一个数，用户应在文本字段中输入数值。输入数值后，单击“确定”按钮，输入的内容作为字符串对象分别赋值给字符串变量 firstNumber 和 secondNumber。

(3) `number1 = Integer.parseInt( firstNumber );`
`number2 = Integer.parseInt( secondNumber );`

这两句是将两个 String 值转换成计算用的整型 int。类 Integer 的静态方法 parseInt 把 String 型的参数转换成一个整型值。两个整型值通过赋值语句赋给数据类型为 int 的两个变量 number1 和 number2。

(4) `JOptionPane.showMessageDialog`
`( null, "结果是 " + sub, "结果", JOptionPane.PLAIN_MESSAGE );`

在表达式中运算符“+”是字符串连接运算。字符串连接是将两个字符串合并以形成一个大的字符串。在字符串合并前，数值型变量 sub 的内容为－10，被 Java 编译器转换成 String 型“－10”，再将两字符串值“结果是 ”和“10”合并生成字符串“结果是 10”。第三个参数是定义对话框的标题。第四个参数是定义对话框的消息类型。消息类型除了 PLAIN_MESSAGE 类型外，还有另外四种消息类型 ERROR_MESSAGE、INFORMATION_MESSAGE、WARNING_MESSAGE 和 QUESTION_MESSAGE，详细内容参见 Java 开发文档中的 JOptionPane 类的说明。

2.6 小　结

通过本章的学习，读者应了解 Java 语言的基本特点和 Java 程序的基本类型，并懂得 Java 应用程序、小应用程序的基本结构，能够使用 JDK 开发包和 JCreator 集成开发环境，完成编写、编译和运行 Java 应用程序和小应用程序。

本章通过例子引入了 JavaAPI 的几个类的使用。这些类是：System 类，System. out 称为输出设备，允许在执行应用程序的 MS-DOS 命令窗口中显示字符集（调用这个对象的方法 println、print、printf 的输出数据的区别是：println 输出文本后换行，print 输出文本，printf 是格式化的输出文本）；java. util 包的类 Scanner 提供了一系列输入数据的方法，其中 nextInt() 方法用于输入整数；JOpitionPane 类提供了对话框的输入和输出；Graphics 类能绘制各种图形。

在学习 Java 时，实际上学习内容分成两部分。第一部分是 Java 语言本身，以便编写自己的类；第二部分学会扩展 Java 类库的类。本书通过例子讨论了 Java 语言中的大量类库的使用。

习　题

2.1　什么是 Java 语言的基本特点？

2.2　什么是 Java 程序的基本类型，各有何特点？

2.3　执行如下代码后会打印什么？

```
System.out.print("* /n*  *\n***\n**** /n*****");
```

2.4　执行如下代码后会打印什么？请编制一个完整的程序验证其结果。

```
int a, b; a = 20; b = 10;
System.out.printf ("a = %d, b = %d",a,b);
System.out.println("*\n*  *  *");
System.out.printf("%s %s", "string1", "string2);
```

2.5　编写一个程序，输入用户的姓名和性别，输出姓名和性别，并且它们之间加一个空格。

2.6　编写一个应用程序，要求输入两个整数，并显示这两个数的和、差。

2.7　编写一个 applet，要求画一个矩形，并在矩形的上方显示“这是一个矩形”的字符串。

第3章 Java程序设计基础

任何程序设计语言,都是由语言的基本语法和一组 API(Application Interface)接口库组成的。本章主要介绍编写 Java 程序必须了解的若干语言基础知识,包括 Java 程序的组成结构、基本数据类型、变量、常量、表达式、流程控制语句、方法和数组等。算法由一系列控制结构组成,顺序结构、选择结构和循环结构是程序设计中的基本结构,也是构成复杂算法的基础。掌握这些基础知识,是书写正确的 Java 程序的前提条件。

3.1 Java 程序的组成

第 2 章中已经介绍了几个简单的 Java 程序的例子,通过它们可以了解 Java 程序的一般构成规则,下面仍以一个程序为例,对 Java 程序的结构进行小结。

例 3-1 Java 程序的结构。

```
//FileStructure.java
package myPackage;                          //包声明语句
import javax.swing.JOptionPane;             //导入类 JOptionPane
public class FileStructure {                //类定义
  public static void main( String args[] )  //方法定义
  {  JOptionPane.showMessageDialog( null,"欢迎你学习 Java 程序基础!");
        System.exit( 0 ); // 终止应用程序
  }
}
```

此程序编译后生成的类 FileStructure.class 放入包 myPackage 中,程序运行结果是用对话框输出一行信息。分析此程序代码结构,可得到 Java 源程序代码,由 3 个要素组成:

(1) 一个包声明(package statement,可选);

(2) 任意数量导入语句(import statement, 可选);

(3) 任意数量的类定义和接口定义。

该三要素必须以上述顺序出现。如果使用包声明,则包声明必须出现在类和导入语句之前,包声明和使用将在第 4 章详细讨论;任何导入语句出现在所有类定义之前;每个 Java 的源文件(编译单元)可包含多个类或接口,但是要注意:每个源文件最多只能有一个类或接口是 public 型,且当一个源文件中包含一个 public 类或接口时,Java 源文件名必须与该类名或接口名相同。

一个类或接口中包含一组数据成员和语句,比语句更小的语言单位是表达式、变量、常量和关键字等,Java 的语句就是由它们构成的。

词法记号是程序中最小的有意义的单词。Java 中的词法记号有:关键字、标识符、文字、操作符、分隔符、空白符。下面分别介绍它们。

1. 关键字

关键字(keyword)是 Java 预定义的单词,这些单词在程序中具有不同的使用目的。下面列出 Java 中的关键字:

abstract break byte boolean catch case class char continue default double do else extends false final float for finally final if import implements int interface instanceof long length native new null package private public protected return switch synchronized short static super try true this throw throws threadsafe transient void while

关于这些关键字的意义和用法,将在以后章节介绍。

2. 标识符

标识符是程序员声明的单词,它命名程序正文中的一些实体,如类名、变量名、方法名,对象名等。Java 标识符的构成规则如下:

- 以字母、下画线(_)或美元符($)开始;
- 可以由以大写字母、小写字母、下画线、$ 或数字 0~9 组成;
- 不能是 Java 的关键字;
- 大写字母和小写字母代表不同的标识符。

标识符的长度是任意的。例如:

- 合法的标识符 identifier、userName、User_Name、_sys_value、$ change;
- 非法的标识符 2mail、room#、class。

注意,Java 中的命名最好遵照如下约定:

- 类名应为名词,含有大、小写字母,每个字的首字母大写。如 HelloWorld、Customer、MergeSort。
- 方法名应是动词,含有大、小写字母,首字母小写,其余各字的首字母大写。尽量不要在方法名中使用下画线。如 getName、setAddress、search。
- 简单类型常量的名字应该全部为大写字母,字与字之间用下画线分隔,如 PI、RED。
- 所有的实例、类和全局变量都使用混合大小写,首字符为小写。用大写字符作字间的分隔符。变量名中不要使用下画线。如 balance、orders。

3. 文字

文字是在程序中直接使用符号表示的数据,包括数值、字符、字符串和布尔型文字,如 123、34.56、'A'、"Hello"、ture 等,在本章 3.2.2 节将详细各种文字。

4. 运算符(操作符)

运算符是用于各种运算的符号,如"+"、"-"、"*"、"/"等。在本章后续章节将详细介绍各种运算符。

5. 分隔符

分隔符用于分隔各个词法记号或程序正文,如"( )"、"{ }"、","、":"、";"等,这些分隔符不表示任何实际的操作,仅用于构造程序,具体用法将在以后的章节中介绍。

6. 空白符

空白符用于指示单词的开始和结束位置。空白符是空格(Space)、制表符(Tab 键产生的字符)、换行符(Enter 键所产生的字符)和注释的总称。

3.2 基本数据类型、常量与变量

3.2.1 基本数据类型

Java语言的数据类型分为基本数据类型和引用数据类型(或称引用类型)。基本数据类型如表3-1所示,是Java语言本身定义的数据类型,在程序中可直接使用。基本类型有8种,分为4类:布尔型、字符型、整型和浮点型。

表3-1 Java的基本数据类型

数据类型	关键字	占用位数	取值范围
布尔型	boolean	8	true,false
字符型	char	16	“\u0000”～“\uFFFF?”
字节型	byte	8	－128～127
短整型	short	16	－32 768～32 767
整型	int	32	－2 147 483 648～2 147 483 647
长整型	long	64	-2^{-63}～$2^{63}-1$
浮点型	float	32	1. 401 298 464 324 817 07e－45～3. 402 823 466 385 288 60e＋38
双精度型	double	64	4. 940 656 458 412 465 44e－324～1. 797 693 134 862 315 70e＋308d

boolean是用来表示布尔型数据的数据类型。boolean型的变量或常量的取值只有true和false两个。其中,true代表“真”, false代表“假”。

Java的字符类型char与其他语言相比有较大的改进。C、Fortran等语言的字符是采用ASCII编码(见光盘附录A所示),每个数据占用8 bit的长度,总共可以表示256个不同的字符。ASCII编码是国际标准的编码方式,但是也有其一定的局限性,最典型的体现在处理以汉字为代表的东方文字方面。汉字的字符集大,仅用8位编码是不够的,所以传统的处理方法是用两个8位的字符数据来表示一个汉字。为了简化问题,Java的字符类型采用了一种新的国际标准编码方案——Unicode编码。每个Unicode码占用16 bit,包含的信息量比ASCII码多了一倍,无论东方字符还是西方字符,都可以统一用一个字符表达。由于采用Unicode编码方案,为Java程序在基于不同语言的平台间实现平滑移植铺平了道路。

整数类型包括字节型(byte)、短整型(short)、整型(int)、长整型(long),它们都是有符号整数,并按照长度分类。byte类型范围很小,为－128～＋127。int类型较常用,基本满足需要的数值范围,再大的整数用long类型。

浮点类型包括浮点型(float)和双精度型(double),它们代表有小数精度要求的数值,也就是大家常听说的实型(real)数据,当计算的表达式有精度要求时被使用。例如,计算平方根或计算正弦、余弦等。这些计算结果的精度要求使用浮点型。双精度浮点型比单精度浮点型的精度更高,表示数据的范围更大,见表3-1。

Java的各种基本数据类型占用固定的内存长度,与具体的软硬件平台环境无关。使用数据类型时要注意:①由于各种不同的类型所能表示的数据范围不同,要根据需要选择合适的数据类型;②Java的类中的域变量根据数据类型都会对应一个默认的值。

Java的引用数据类型是用户根据自己的需要定义的数据类型，包括数组、类和接口，引用数据类型将在后续章节详细讨论。

3.2.2 常量

常量是指在程序运行的整个过程中其值始终不可改变的量，也就是直接使用文字符号表示的值，也通常称为常数。例如，123，5.6，'B'都是文字常量。Java中的常量有5种类型：整型常量、浮点型常量、布尔常量、字符常量和字符串常量。下面分别讨论它们。

1. 整型常量

整型常量即以文字形式表示的整数，包括正整数、负整数和零。

Java的整型常量的表示形式有：十进制、八进制和十六进制。

- 十进制的整型常量的一般形式与数学中的表示形式一样：

[±]若干个0～9的数字

正数的符号可省略，如：100，－50，0。

- 八进制的整型常量用以0开头的数值表示：

[±]0若干个0～7的数字

例如，0123表示十进制数83，－011表示十进制数－9。

- 十六进制的整型常量用0x或0X开头的数值表示：

[±]0x若干个0～9的数字和A～F的字母(大小写均可以)

例如，0x2F代表十进制的数字47，0x123表示十进制数291，－0X12表示十进制数－18。

整型常量按照所占用的内存长度又可分为一般整型(int)和长整型(long)常量，其中int型常量占用32位，如123，－34；long型常量占用64位，其尾部有一个大写的L或小写的l，如－386L，017777l。

2. 浮点型(实型)常量

浮点型常量表示的是可以含有小数部分的数值常量。根据占用内存长度的不同，可以分为单精度浮点常量和双精度浮点常量两种。单精度浮点常量占用32位内存，用带后缀F、f的数值表示，如19.4F，3.0513E3f，8701.52f；双精度浮点常量占用64位内存，用带后缀D、d的数值或不加后缀的数值表示，如2.433E－5D，700041.273d，3.1415。

与其他高级语言类似，浮点常量也有两种不同的表示形式：一般形式和指数形式。

一般形式由数字和小数点组成，如0.123，1.23，123.0。

指数形式，如：0.123e3表示 0.123×10^{3}，－35.4E－5表示 -35.4×10^{-5}，其中e或E之前必须有数字，且e或E后面的指数必须为整数。

3. 字符常量

字符常量是用一对单引号括起来的单个字符，这个字符可以是拉丁字母表中的字符，如'a'，'Z'，'8'，'#'；也可以是转义字符，如'\n'。Java的一个字符用两个字节的Unicode字符表示，其值是16位无符号整数，范围为0～65 535。下面给出常用字符对应的ASCII编码值：

'A'～'Z'：65，66，67，…，90 (按字母次序递增1)

'a'～'z'：97，98，99，…，122 (按字母次序递增1)

'0'～'9'：48，49，…，57 (按数字次序递增1)

要注意：字符'1'和整数1是不同的概念，因为它们在计算机内存存放的数值分别是49

和1。

转义字符(Escape sequences)是一些有特殊含义、很难用一般方式表达的字符,如回车、换行等。为了表达清楚这些特殊字符,Java 中引入了转义字符的定义,即用反斜线(\)开始,详见表 3-2。表中'\ddd'是用八进制表示一个字符常量。例如,'\101'就是用八进制表示一个字符常量,它与'\u0041'表示的是同一个字符'A'。需要补充说明的是,八进制表示法只能表示'\000'~'\377'范围内的字符,即不能表示全部的 Unicode 字符,而只能表示其中 ASCII 字符集的部分。

表 3-2　常用转义字符

转义字符	含　义	对应 Unicode 码
'\b'	退格	'\u0008'
'\t'	水平制表符 tab	'\u0009'
'\n'	换行	'\u000a'
'\f'	表格符	'\u000c'
'\r'	回车	'\u000d'
'\"'	双引号	'\u0022'
'\''	单引号	'\u0027'
'\\'	反斜线	'\u005c'
'\ddd'	3 位八进制数表示的字符	
'\uxxxx'	4 位十六进制数表示的字符	

4. 字符串常量

字符串常量是用双引号括起的若干个字符(可以是 0 个)。字符串中可以包括转义符,标志字符串开始和结束的双引号必须在源代码的同一行上。下面是几个字符串常量的例子:"Hello","My\nJava","How are you?1234",""。

在 Java 中可以使用连接操作符(+)把两个或更多的字符串常量串接在一起,组成一个更长的字符串。例如,"How do you do?"+"\n"的结果是"How do you do? \n"。

5. 布尔常量

在流程控制中经常用到布尔常量。布尔常量包括 true 和 false,分别代表真和假。

Java 是一种严格的类型语言,它不允许数值类型和布尔类型之间进行转换。

3.2.3 变量

变量是在程序的运行过程中数值可变的数据,通常用来记录运算中间结果或保存数据。Java 中的变量必须先声明后使用,声明变量时要指明变量的数据类型和变量的名称。

1. 变量的声明语句格式

```
数据类型 变量名 1,变量名 2,…,变量名 n;
```

变量名必须是一个合法的标识符,满足前面所讲的 Java 标识符的规则。另外,Java 对变量名区分大小写,变量名应具有一定的含义,以增加程序的可读性。例如:

```
int num,total;
double d1;
```

定义一个变量意味着给变量分配存储空间,用以存放对应类型的数据。上面的声明语句给变量 num 和 total 各分配 4 B 的空间；给变量 *d*1 分配 8 B 的空间。

变量声明后才能使用，下面的赋值语句

```
num = 1;
```

是给变量 num 赋值 1，实际上是给变量 num 的存储空间写值 1。

2. 变量的初始化

声明变量时还可以指定变量的初始数值。下面的语句

```
boolean flag = true;
```

声明了一个布尔型的变量，名字 flag，该变量的初值是 true。

又如：

```
char a = 'a',b;
double d1, d2 = 0.0;
```

前面的一些例子中，变量的初始化，仅将文字常量作为其初始值。Java 语言也允许在变量声明时使用任何有效的表达式来动态地初始化变量。例如，下面的程序代码用于在给定直角三角形两个直角边长度的情况下，求其斜边长度。

```
class DynInit {
    public static void main(String args[])
    {  double a = 3.0, b = 4.0;
           // c is dynamically initialized:
       double c = Math.sqrt(a * a + b * b);
       System.out.println("Hypotenuse is:" + c);
    }
}
```

在这里，定义了 3 个变量 *a*、*b*、*c*。前两个变量 *a* 和 *b* 被初始化为常量，而直角三角形的斜边 *c* 则被动态地初始化(使用勾股定理)。该程序用了 Java 中 Math 类的一个静态的方法 sqrt( )，计算它的参数的平方根。这里关键的一点是初始化表达式可以使用任何有效的元素，包括方法调用、其他变量或常量。

3.2.4 符号常量

在程序中除可以直接用文字表示常量外，也可以为常量起一个名字，这就是符号常量。符号常量在使用之前先声明。符号常量的声明语句格式为：

```
final 数据类型 常量名 = 缺省值;
```

例如：`final int COUNT = 1000;`

注意，符号常量声明时一定要赋初值，而在程序的中间不能改变其初值。例如：

```
final double PI = 3.14;
PI = 3.1415 //错误
```

例 3-2 变量声明和赋值。

```
//UseVariables.java
public class UseVariables {
    public static void main(String args[]){
        final int PRICE = 30;          // 声明整型常量
        long l = 12341;                // 声明长整型变量并赋值
        int num,total;                 // 声明整型变量
        float r,v,h = 3.5f;            // 声明浮点型变量,并给 h 赋值
        double w = 3.1415;             // 声明双精度型变量并赋值
```

```
        boolean truth = true;   // 声明布尔类型变量并赋值
        boolean false1;         //声明布尔类型变量
        char c;                 //声明字符类型变量
        c ='A';                 // 给字符类型变量赋值。变量使用之前,要先声明
        num = 10;               // 给整型变量赋值
        total = num * PRICE;
        r = 2.5f;               // 给浮点型变量赋值
        v = 3.14159f * r * r * h;
        false1 = 6 > 7;         // 给布尔类型变量赋值
        String s = "I am a student"; //给字符串引用变量赋值
        System.out.println("final int PRICE = " + PRICE);
        System.out.println("long l = " + l);
        System.out.println("int num = " + num + "\ntotal = " + total);
        System.out.println("boolean truth = " + truth);
        System.out.println("boolean false1 = " + false1);
        System.out.println("char c = " + c);
        System.out.println("float r = " + r);
        System.out.println("float v = " + v);
        System.out.println("String s = " + s);
    }
}
```

程序运行结果如图 3-1 所示。

```
final int PRICE=30
long l=1234
int num=10
total=300
boolean truth=true
boolean false1=false
char c=A
float r=2.5
float v=68.72228
String s=I am a student
```

图 3-1　例 3-2 程序运行结果

3.3 运算符与表达式

在程序中,表达式是用于计算求值的基本单位,可以简单地将表达式理解为计算的公式。它是由运算符(如"+"、"-"、"*"、"/")、运算量(也叫操作数,可以是常量、变量等)和括号组成的式子。符合语法规则的表达式可以被编译系统理解、执行或计算,表达式的值就是对它运算后所得的结果。

运算符指明对操作数的运算方式。组成表达式的 Java 操作符有很多种。运算符按其要求的操作数数目来分,可以有单目运算符(如求负"-"),双目运算符(如加"+"、减"-")和三目运算符(如"?"、":"),它们分别对应于 1 个、2 个和 3 个操作数。运算符按其功能来分,有算术运算符、赋值运算符、关系运算符、逻辑运算符、位运算符和其他运算符。下面将按功能详细讨论它们。

3.3.1 算术运算符与算术表达式

算术运算符用于对整型数据或实型数据的运算。根据需要的操作数个数不同,算术运算符分为双目运算符和单目运算符两类。

1. 双目运算符

双目运算符定义如表 3-3 所示，这里需要注意两点。

(1) 只有整型(int, long, short)的数据才能够进行取余运算(%)，float 和 double 不能取余。

(2) 两个整型的数据做除法时，结果是截取商的整数部分，而小数部分被截断。如果希望保留小数部分，则应该对除法运算的操作数做强制类型转换。例如 2/4 的结果是 0，而 2.0/4 的结果是 0.5。

表 3-3　双目算术运算符

运算符	运　算	举　例	功　能
+	加	a+b	求 a 与 b 相加的和
−	减	a−b	求 a 与 b 相减的差
*	乘	a*b	求 a 与 b 相乘的积
/	除	a/b	求 a 除以 b 的商
%	取余	a%b	求 a 除以 b 所得的余数

2. 单目运算符

单目运算符的操作数只有一个，算术运算中有 3 个单目运算符：求负(−)、自增 1(++)和自减 1(−−)。其定义如表 3-4 所示。

单目运算符中的自增 1 和自减 1，其运算符的位置可以在操作数的前面，称为前缀运算符，也可以在操作数的后面，称为后缀运算符。例如：

```
++a;        //前缀 ++，将 a 变量的值加 1
a++ ;       //后缀 ++，将 a 变量的值加 1
```

表 3-4　单目算术运算符

运算符	运　算	举　例	功　能
++	自增 1	a++或++a	a=a+1
−−	自减 1	a−−或−−a	a=a−1
−	求负数	−a	a=−a

当单目运算符单独作为一条语句出现时，如−−a 和 a−−，它们表达的功能是一样的，都是将变量 a 的值减 1。而当单目运算符位于一个复杂表达式的内部，前缀运算和后缀运算是有区别的。前缀运算表示变量在参加表达式值计算之前，先将前缀运算的变量加(减)1，然后再将此变量的值参加表达式的计算，即用加(减)1 的变量值参加表达式的计算。而后缀运算表示先用变量的原先值参加表达式的计算，然后再将后缀运算的变量值加(减)1，即用未加(减)1 的变量值参加表达式的计算。下面给出两个例子说明。

例如：

```
int    x = 1 ;
int    y = ( ++ x ) * 3;        //前缀 ++
```

语句执行后的结果：x =2, y =6。

例如：

```
int    x = 1;
int    y = (x++ ) * 3;        //后缀 ++
```

语句执行后的结果：y =3, x =2。

可见，自增 1 和自减 1 的位置不同，却会改变整个表达式的值。并且自增和自减的操作对象只能是变量，而不能用于常量或表达式，如 12＋＋或(x＋y)＋＋都是不合法的。

例 3-3 算术运算符的应用。

程序是一个 application：利用对话框，接受用户从键盘输入的两个字符串数据；将这两个字符串数据转化为整型数值，并赋值给两个整型变量 a 和 b；以 a 和 b 为操作数进行算术运算，并在对话框中输出运算结果。

Java 的源文件是 ArithmeticOperator.java 代码如下：

```
import javax.swing.JOptionPane;
public class ArithmeticOperator {
  public static void main(String args[])
   {  String input1;  String input2;
      int a,b;int plus,minus;
      input1 = JOptionPane.showInputDialog("输入第一个数");
      input2 = JOptionPane.showInputDialog("输入第二个数");
      a = Integer.parseInt(input1);
      b = Integer.parseInt(input2);
      plus = a + b;minus = a - b;
      String s1 = a + " + " + b + " = " + plus + "\n";
      s1 + = a + " - " + b + " = " + minus + "\n";
      s1 + = a + " * " + b + " = " + (a * b) + "\n";
      s1 + = a + "/" + b + " = " + (a/b) + "\n";
      s1 + = a + " % " + b + " = " + (a % b) + "\n";
      JOptionPane.showMessageDialog(null,s1,"算术运算结果",JOptionPane.PLAIN_MESSAGE);
  }
}
```

程序运行时，可输入任意两个整数，观察程序运行的不同结果。图 3-2 是程序的运行输出的部分结果。

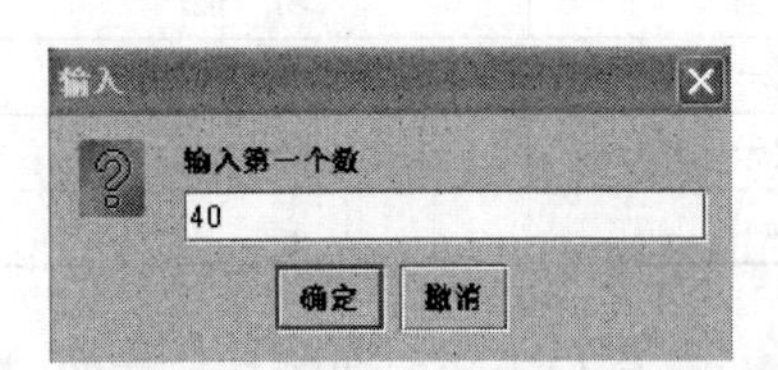

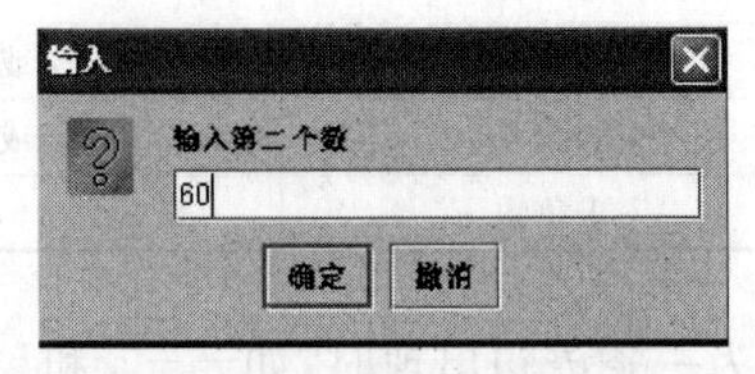

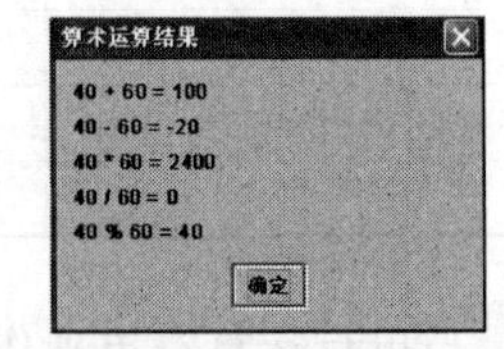

图 3-2　例 3-3 程序运行的部分输出结果

3.3.2 赋值运算符与赋值表达式

Java 中提供了几个赋值运算符(如“＝”、“＋＝”、“－＝”、“＊＝”等)，其中，“＝”是最简单的赋值运算符。带有赋值运算符的表达式叫做赋值表达式。例如，i＝i＋5 就是一个赋值表达式。赋值表达式的含义是等号右边表达式的值赋给等号左边变量。赋值表达式值的类型为等号左边变量的类型，运算的结合性为自右向左。下面是一些简单赋值运算的例子：

```
i = 1;      //表达式值为 1
i = j = k = 1;
```

表达式值为 1，i，j，k 的值均为 1。这个表达式从右向左运算，在 k 被赋值 1 后，表达式 k＝1 的值为 1，接着 j 被赋值 1，最后 i 被赋值 1。

```
i = 2 + (j = 4);          //表达式值为 6, j 的值为 4,i 的值为 6
i = (j = 10) * (k = 2);   //表达式值为 20, j 的值为 10,k 的值为 2, i 的值为 20
```

除了“＝”外，Java 中还提供了其他 11 个赋值运算符：在“＝”之前加上其他运算符＜op＞，就可以构成复合赋值运算符＜op＞＝，它们的含义如表 3-5 所示。

表 3-5　复合赋值运算符

运算符	举　例	运算符	举　例
＋＝	x＋＝a 等价于 x＝x＋a	&＝	x&＝a 等价于 x＝x&a
－＝	x－＝a 等价于 x＝x－a	|＝	x|＝a 等价于 x＝x|a
＊＝	x＊＝a 等价于 x＝x＊a	^＝	x^＝a 等价于 x＝x^a
/＝	x/＝a 等价于 x＝x/a	＜＜＝	x＜＜＝a 等价于 x＝x＜＜a
%＝	x%＝a 等价于 x＝x%a	＞＞＝	x＞＞＝a 等价于 x＝x＞＞a
		＞＞＞＝	x＞＞＞＝a 等价于 x＝x＞＞＞a

3.3.3　关系运算符与关系表达式

关系运算是比较两个数据之间的大小关系的运算，常用的关系运算符如表 3-6 所示。

表 3-6　关系运算符

运算符	运　算	举　例
＝＝	等于	a＝＝b
！＝	不等于	a！＝b
＞	大于	a＞b
＜	小于	a＜b
＞＝	大于等于	a＞＝b
＜＝	小于等于	a＜＝b

关系运算的结果是布尔型的量，即 true 或 false。例如：

```
int x = 5,y = 7;
boolean b = (x == y);
```

计算关系表达式 x ＝ ＝ y 的值为 false，则 b 的初始值是 false。另外，需注意区分等于号“＝＝”和赋值号“＝”，不要混淆。

例 3-4　关系运算符的应用。

Java 的源文件是 RelationOperator. java 代码如下：

```
import javax.swing.JOptionPane;
public class RelationOperator {
    public static void main( String args[] )
    {   String input1;
        String input2;
        int a,b;
        boolean ee;
        input1 = JOptionPane.showInputDialog("输入第一个数");
        input2 = JOptionPane.showInputDialog("输入第二个数");
        a = Integer.parseInt( input1 );
        b = Integer.parseInt( input2 );
        ee = (a!= b);
        String s1 = a + ">" + b + " = " + (a>b) + "\n";
        s1 + = a + "<" + b + " = " + (a<b) + "\n";
        s1 + = a + "> = " + b + " = " + (a> = b) + "\n";
```

```
        s1 + = a+"<= "+b+"= "+(a<= b) +"\n";
        s1 + = a+"= = "+b+"= "+(a= = b) +"\n";
        s1 + = a+"!= "+b+"= "+ee+"\n";
        JOptionPane.showMessageDialog(null,s1,"比较关系运算结果",JOptionPane.PLAIN_MESSAGE);
    }
}
```

程序运行时,可输入任意两个整数,观察程序运行的不同结果。图 3-3 是程序运行的部分结果。

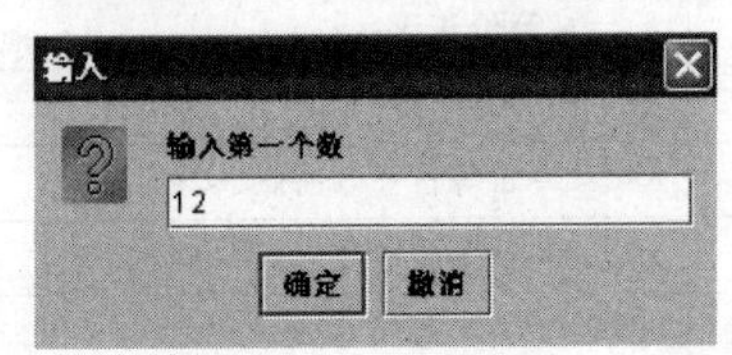

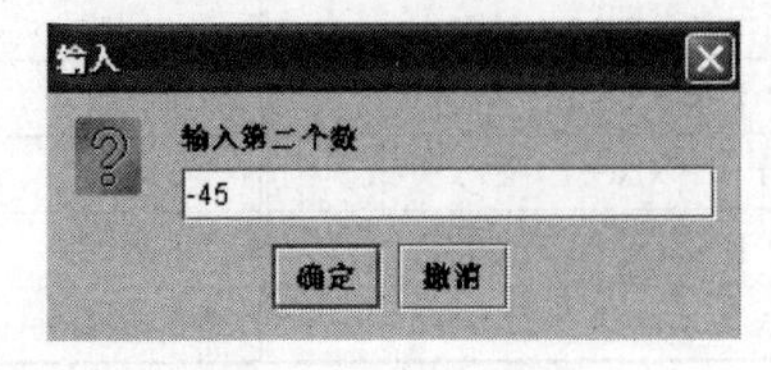

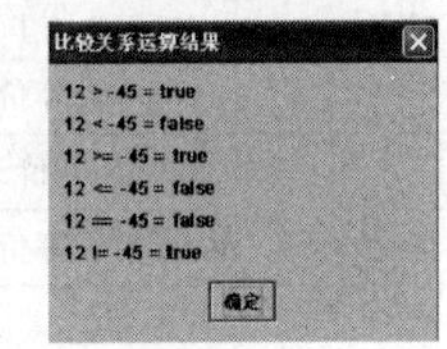

图 3-3　例 3-4 程序运行部分结果

3.3.4　逻辑运算符与逻辑表达式

逻辑运算是针对布尔型数据进行的运算,运算的结果仍然是布尔型量。常用的逻辑运算符如表 3-7 所示。

表 3-7　逻辑运算符

运算符	运　算	举　例	运算规则
!	逻辑取反	! x	x 真时为假,x 假时为真
‖	逻辑或	x‖y	x,y 都假时结果才为假
&&	逻辑与	x&&y	x,y 都真时结果才为真
^	布尔逻辑异或	x^y	x,y 同真假时结果为假
&	布尔逻辑与	x&y	x,y 都真时结果才为真
|	布尔逻辑或	x|y	x,y 都假时结果才为假

"&"、"|"和"^"称为布尔逻辑运算符,因为在利用它们做与、或、异或运算时,运算符左右两边的表达式总会被运算执行,然后再对两表达式的结果进行与、或运算;而在利用"&&"和"||"做逻辑运算时,运算符右边的表达式有可能被忽略而不加执行。例如:

```
int x = 3,y = 5;
boolean b = x>y&&x++  == y--;
```

在计算布尔型的变量 b 的值时,先计算 && 左边的关系表达式 $x>y$,得结果为假,根据逻辑与运算的规则,只有参加与运算的两个表达式都为真时,最后的结果才为真;所以不论 && 右边的表达式结果如何,整个式子的值都为假,右边的表达式就不予计算执行了;最后 3 个变量的取值分别是:x 为 3,y 为 5,b 为 false。

如果把上题中的逻辑与(&&)换为布尔逻辑与(&),则与号 & 两边的两个表达式都会被运算,最后 3 个变量的取值分别是:x 为 4,y 为 4,b 为 false。

同样,对于逻辑或(‖),若左边的表达式的运算结果为真,则整个或运算的结果一定为真,右边的表达式就不会再运算执行了。

逻辑表达式由逻辑运算符、关系表达式、逻辑值构成。例如:

```
x == y && a>b
! (a > b)
x > y && a > b ‖ ! ( x > a )
true
```

都是合法的逻辑表达式。

例 3-5 逻辑运算符的应用。

Java 的源文件是 RelationOperator. java 代码如下：

```
import javax.swing.JOptionPane;
public class LogicalOperator {
    public static void main( String args[] )
    {  String input1;
       String input2;
       boolean a,b;
       boolean ee;
       input1 = JOptionPane.showInputDialog("输入第一个布尔值");
       input2 = JOptionPane.showInputDialog("输入第二个布尔值");
       a = Boolean.valueOf(input1).booleanValue();
       b = Boolean.valueOf(input2).booleanValue();
       String s1 = "";
       s1 + = a + "&&" + b + " = " + (a&&b) + "\n";
       s1 + = a + " || " + b + " = " + (a || b) + "\n";
       s1 + = "!" + a + " = " + (! a) + "\n";
       s1 + = a + "&" + b + " = " + (a&b) + "\n";
       s1 + = a + "|" + b + " = " + (a|b) + "\n";
       s1 + = a + "^" + b + " = " + (a^b);
       JOptionPane.showMessageDialog(null,s1,"逻辑运算结果",JOptionPane.PLAIN_MESSAGE);
    }
}
```

程序运行时，可输入任意两个逻辑值，观察程序运行的不同结果。图 3-4 是程序运行的部分结果。

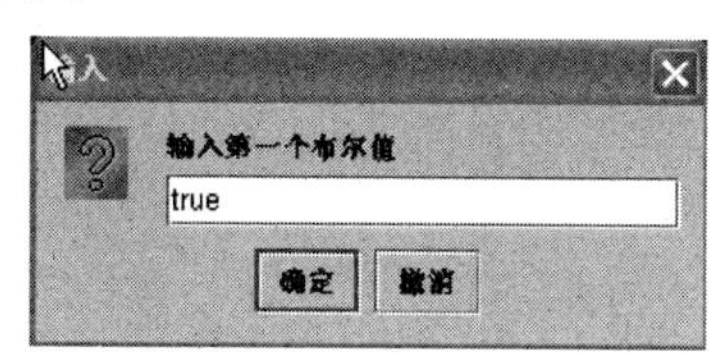

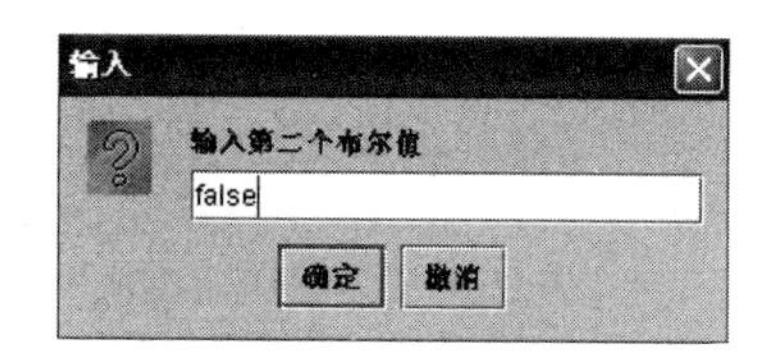

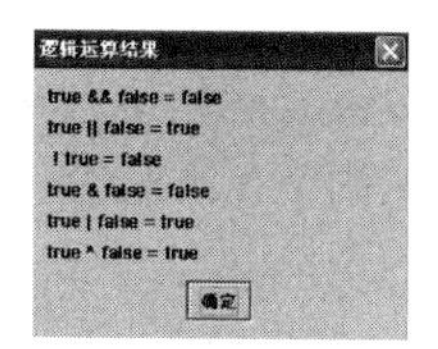

图 3-4 例 3-5 程序运行的部分结果

3.3.5 位运算符

位运算是对操作数以二进制比特位为单位进行的操作和运算，位运算的操作数只能为整型和字符型数据，结果都是整型量。几种位运算符和相应的运算规则如表 3-8 所示。

1. 按位取反(～)

"～"是一元运算法，对数据的每个二进制位取反，即把 1 变为 0，把 0 变为 1。例如：

025：00000000 00010101

～025：11111111 11101010

2. 按位与(&)

参与运算的两个值，如果两个相应位都为 1，则该位的结果为 1，否则为 0。即：0&0=0，0&1=0，1&0=0，1&1=1。例如：

3：	00000011
5：	(&) 00000101
3&5：	00000001

按位与可以用来对操作数中的若干位置0，或者取操作数中的若干指定位。请看下面的例子。

(1) 下列语句将 byte 型的 x 的第4位和第1位置0，假设 x 的各位从低位到高位编号依次为第0位、第1位、…、第7位。

```
x = x&0xde;
```

(2) 下列语句要取 byte 型的 x 的低4位。

```
x = x&0x0f;
```

表 3-8　位运算符

运算符	运算	例	运算规则
~	位反	~x	将 x 各二进位取反
&	位与	x&a	求 x 和 a 各二进位与
|	位或	x&a	求 x 和 a 各二进位异或
^	位异或	x^a	求 x 和 a 各二进位异或
>>	右移	x>>a	x 各二进位右移 a 位
<<	左移	x<<a	x 各二进位左移 a 位
>>>	不带符号的右移	x>>>a	x 各二进位右移 a 位，左边的空位一律填零

3. 按位或(|)

参与运算的两个值，只要两个相应位中有一个为1，则该位的结果为1。即：0|0=0，0|1=1，1|0=1，1|1=1。例如：

```
      3:          00000011
      5:       (|)00000101
    ------------------------
    3|5:          00000111
```

按位或操作可以用来把操作数的某些特定的位置1(其他位保持不变)，例如：将 int 型的变量 a 的第3、4位置1：

```
a = a|ox18
```

4. 按位异或(^)

参与运算的两个值，如果两个相应位不同，则结果为1，否则为0。即：0^0=0，1^0=1，0^1=1，1^1=0。例如：

```
      071:          00111001
      052:       (^)00101010
    --------------------------
  071^052:          00010011
```

按位异或操作可以用来使某些特定的位翻转。如果是某位与0异或，结果是该位的原值；如果是某位与1异或，则结果与该位原来的值相反。例如：要使11010110的第4、5位翻转，可以将数与00011000进行按位异或运算。

5. 移位(<<、>>、>>>)

移位运算都是二元运算符，是将某一变量所包含的各二进位按指定的方向移动到指定的位数，表3-9是3个移位运算符的例子。

(1) 左移运算(<<)：将一个数的各个二进位全部左移若干位。左移后，低位补0，高位舍弃。在不产生溢出的情况下，一个无符号数左移一位相当于乘2，而且用左移来实现乘法比乘法运算速度要快。

表 3-9　移位运算的例子

十进制表示	二进制补码表示	x<<2	x>>2	x>>>2
30	00011110	01111000	00000111	00000111
－17	11101111	10111100	11111011	00111011

(2) 右移运算(>>)：将一个数的各二进制位全部右移若干位。右移后，低位舍弃；高位若是无符号数，补 0；若是有符号数，补“符号位”。右移一位相当于除 2 取商，而且用右移实现除法比除法运算速度要快。

(3) 无符号右移运算符(>>>)：用来将一个数的各二进制位无符号地右移若干位，与运算符 >> 相同，移出的低位被舍弃，但不同的是最高位补 0。

6. 不同长度的数据进行位运算

如果两个数据长度不同，如 a & b，a 为 byte 型，b 为 int 型，对它们进行位运算时，系统首先会将 a 的左侧 24 位填满，若 a 为正，则填满 0；若 a 为负，则填满 1。

例 3-6　位运算符的应用。

```
public class BitOperator {
    public static void main( String args[] )
    {   byte a = 025,b = 4;
        System.out.println("~" + a + " = " + (~a));
        System.out.println(a + "&" + b + " = " + (a&b));
        System.out.println(a + "|" + 6 + " = " + (a|6));
        System.out.println(a + "^" + b + " = " + (a^b));
        a = 59;
        System.out.println(a + "<<" + 2 + " = " + (a<<2));
        System.out.println(a + ">>" + 2 + " = " + (a>>1));
        a = - 77;
        System.out.println(a + "<<" + 1 + " = " + (byte)(a<<1));
        System.out.println(a + ">>>" + 1 + " = " + (byte)(a>>>1));
    }
}
```

程序运行的结果如图 3-5 所示。

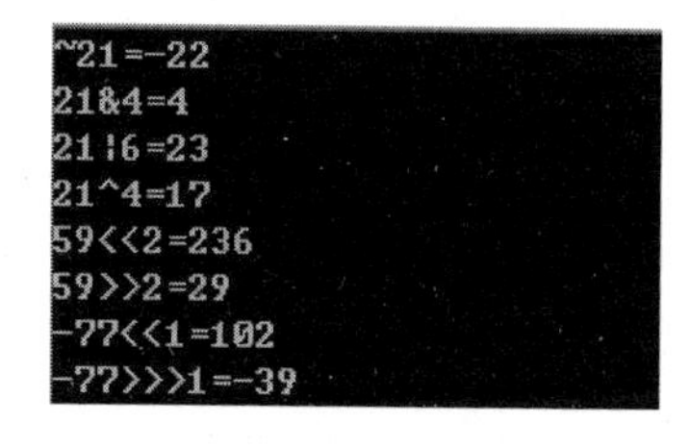

图 3-5　例 3-6 程序运行结果

3.3.6　其他运算符

1. 条件运算符与条件表达式

条件运算符“?:”是 Java 语言中仅有的一个三目运算符，它能实现简单的选择功能。条件表达式的形式为

```
e1? e2:e3
```

其中 e1 必须是 boolean 类型。计算规则为：先计算表达式 e1 的值，若 e1 为真，则整个三目运算的结果为表达式 e2 的值；若 e1 为假，则整个运算结果为表达式 e3 的值。请看下面的例子：

```
int x = 4,y = 8,z = 2;
int k = x<3? y:z;              // x<3 为 false,所以 k 取 z 的值,结果为 2
int y = x>0? x: - x;           // y 为 x 的绝对值
int max = a>b? a:b;            // max 取 a,b 中的大值
```

2. 括号与方括号

括号运算符"()"在某些情况下起到改变表达式运算先后顺序的作用；在另一些情况下代表方法或函数的调用。它的优先级在所有的运算符中是最高的。

方括号运算符"[ ]"是数组运算符，它的优先级也很高，其具体使用方法将在后面章节介绍。

3. 对象运算符

对象运算符 instanceof 用来测定一个对象是否属于某一个指定类或其子类的实例，如果是，则返回 true；否则返回 false。例如：

```
boolean b = MyObject instanceof TextField;
```

3.3.7 运算符的优先级与结合性

表达式的运算次序取决于表达式中各种运算符的优先级。优先级高的运算符先运算，优先级低的运算符后运算，同一行中的运算符的优先级相同。运算符的结合性决定了并列的相同运算符的先后执行顺序。

Java 语言规定的运算符优先级如表 3-10 所示，表中排在上面的运算符有较高的优先级，例如，关系运算符的优先级高于逻辑运算符，x>y&&! z 相当于(x>y)&&(! z)。如对于左结合的"+"，x+y+z 等价于(x+y)+z，对于右结合的"!"，!! x 等价于！(! x)。

表 3-10 Java 运算符的优先级与结合性

<table>
<tr><th>优先级</th><th>描 述</th><th>运算符</th><th>结合性</th></tr>
<tr><td>1</td><td>最高优先级</td><td>[]、()</td><td>左→右</td></tr>
<tr><td>2</td><td>单目运算</td><td>+(正号)、-(负号)、++、--、~、!
强制类型转换符</td><td>右→左</td></tr>
<tr><td>3</td><td>算术乘除运算</td><td>*、/、%</td><td>左→右</td></tr>
<tr><td>4</td><td>算术加减运算</td><td>+、-</td><td>左→右</td></tr>
<tr><td>5</td><td>移位运算</td><td>>>、<<、>>></td><td>左→右</td></tr>
<tr><td>6</td><td>关系运算</td><td><、<=、>、>=</td><td>左→右</td></tr>
<tr><td>7</td><td>相等关系运算</td><td>==、!=</td><td>左→右</td></tr>
<tr><td>8</td><td>按位与，布尔逻辑与</td><td>&</td><td>左→右</td></tr>
<tr><td>9</td><td>按位异或</td><td>^</td><td>左→右</td></tr>
<tr><td>10</td><td>按位或，布尔逻辑或</td><td>|</td><td>左→右</td></tr>
<tr><td>11</td><td>逻辑与</td><td>&&</td><td>左→右</td></tr>
<tr><td>12</td><td>逻辑或</td><td>||</td><td>左→右</td></tr>
<tr><td>13</td><td>三目条件运算</td><td>?:</td><td>右→左</td></tr>
<tr><td>14</td><td>赋值运算</td><td>=、+=、-=、*=、/=、%=、<<=、>>=</td><td>右→左</td></tr>
</table>

另外，使用圆括号可以提高括在其中的运算的优先级。例如，考虑下列表达式：

```
a>>b+3
```

该表达式首先将 3 加到变量 b,得到一个中间结果,然后将变量 a 右移该中间结果位。该表达式可用添加圆括号的办法改写如下:

```
a>>(b+3)
```

3.3.8 混合运算时数据类型的转换

当表达式中出现了多种类型数据的混合运算时,需要进行类型转换。Java 的类型转换有较严格的规定:从占用内存较少的短数据类型转化成占用内存较多的长数据类型时,可以不做显式的类型转换声明(隐含转换);而从较长的数据类型转换成较短的数据类型时,则必须做强制类型转换。

1. 隐含转换

算术运算符、赋值运算符、关系运算符、逻辑运算符和位运算符,这些二元运算符要求两个操作数的类型一致。如果类型不一致,编译系统会自动对数据进行类型转换。Java 的类型转换是一种加宽转换。隐含转换的规则

低——————————→高

byte short char int float double

2. 强制类型转换

将表达式的类型强制性地转换成某一数据类型。强制类型转换的格式为

(数据类型)表达式

通过强制类型转换,可将数据范围宽的数据转换成范围低的数据,但这可能会导致溢出或精度的下降。

例 3-7 数据类型转换的例子(完整的程序见光盘)。

```
int iVal = 258;
long lVal = 1000;
short sVal = 12;
byte bVal  =  2;
char cVal  =  ´a´;
float fVal  =  5.67f;
double dVal  =  .1234;
lVal = iVal; dVal  =  fVal;        //自动转换,lVal 的值是 258。dVal 的值是 5.67
bval = (byte) iVal;                /* 当值 258 被强制转换为 byte 变量时,其结果是 258 除以 256(256 是
                                      byte 类型的变化范围)的余数 2; */
iVal = (int)lVal;                  //强制转换
iVal = cVal + 1;                   /* 自动转换。将´a´对应的 16 位的 Unicode 值 97 转换为 32 位的 int 型
                                      ,与 1 相加,iVal 的值是 98 */
lVal = (long) fVal;                //变量 fVal 被强制转换为 long 型,舍弃了小数部分。lVal 的值是
                                     5dVal  =  (fVal  *  bVal)  +  (iVal / cVal)  -  (dVal  *  sVal);
```

计算时,在第一个子表达式 fVal * bVal 中,变量 bVal 被提升为 float 类型,该子表达式的结果当然是 float 类型。接下来,在子表达式 iVal/cVal,中,变量 cVal 被提升为 int 类型,该子表达式的结果当然是 int 类型。然后,子表达式 dVal * sVal 中的变量 sVal 被提升为 double 类型,该子表达式的结果当然也是 double 类型。最后,考虑 3 个中间值,float 类型、int 类型和 double 类型。float 类型加 int 类型的结果是 float 类型。然后 float 类型减去提升为 double 类型的 d * s 变量,该表达式的最后结果是 double 类型。又如:

```
bVal  =  bVal  *  2;            //Compiling Error. Cannot assign an int to a byte
```

```
bVal = (byte) (bVal * 2); // 将 int 型表达式强制转换成 byte
```

当分析表达式时,Java 自动提升 byte 型(short 型也一样)的操作数 bVal 到 int 型,这意味着子表达式 a * b 使用整型而不是字节型来执行。

3.3.9 语句和块

在 Java 语言中,一个基本的语句以分号结尾。语句是程序的基本组成单位,包括:声明语句、表达式语句、选择语句、循环语句、跳转语句、空语句、复合语句和方法调用语句几类。下面先介绍一些简单的语句,其他复杂一些的语句将在后续章节介绍。

1. 表达式语句

一个合法的表达式加上分号,就成为一个表达式语句。例如:

```
a + b;
a + b * c + f;
```

2. 空语句

空语句是什么都不做的语句,其形式为

```
;    //这是一条空语句
```

3. 复合语句(语句块)

将多条语句用一对大括号"{}"括起来,就构成复合语句,也被称为语句块。例如:

```
{  int x,y,z;
   x = 5;y = 6;z = x + y;
}
```

在任何可使用单个语句的地方都可以使用复合语句。

3.4 算法的基本控制结构

学习了数据类型、表达式、赋值语句后,就能编写程序以完成一些简单的功能,这些程序只是一些顺序执行的 Java 语句序列。但实际上,现实世界中要解决的问题往往不会这么简单,解决问题的方法也不是仅仅用顺序步骤就可以描述的。例如,根据输入的 x 值,求输出的 y 值。

$$y=\begin{cases}-1 & (x<0)\\ 0 & (x=0)\\ 1 & (x>0)\end{cases}$$

要解决这个问题,必须进行判断,这需要用选择条件型控制结构。

又如,要统计出 n 个数的和,当 n 很大时($n=100,1000$),就必须重复地进行累加,这需要用循环控制结构。

算法的基本控制结构有 3 种:顺序结构、选择结构、循环结构,如图 3-6 所示。这 3 种基本结构就构成了程序的控制流程。

- 顺序结构:3 种结构中最简单的一种,即语句按照书写的顺序依次执行。
- 选择结构(又称为分支结构):将根据布尔表达式的值来判断应选择执行哪一个流程的分支。
- 循环结构:在一定条件下反复执行一段语句的流程结构。

在 Java 中专门负责实现分支结构的是条件分支语句;负责实现循环结构的是循环语句。

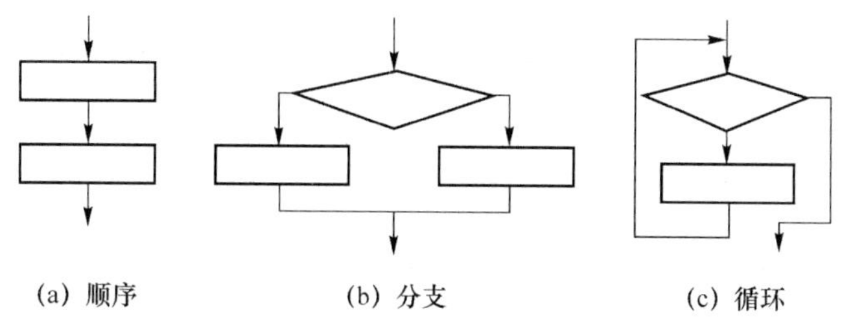

图 3-6　算法的 3 种基本结构

3.4.1　分支语句

分支语句提供了一种控制机制，使得程序根据相应的条件去执行对应的语句。Java 中的分支语句有两个，一个是负责实现双分支的 if 语句；另一个是负责实现多分支的 switch 语句。

1. if 语句

(1) if 语句的一般形式：

```
if ( 布尔表达式 )
    语句 1;          // if 分支
else
    语句 2;          // else 分支
```

执行顺序：首先计算布尔表达式的值，若表达式值为 true，则执行语句 1，否则执行语句 2，如图 3-7 (a)所示。语句 1 和语句 2 可以是复合语句。

例如：

```
if (x>y) z = x;
else z = y;
```

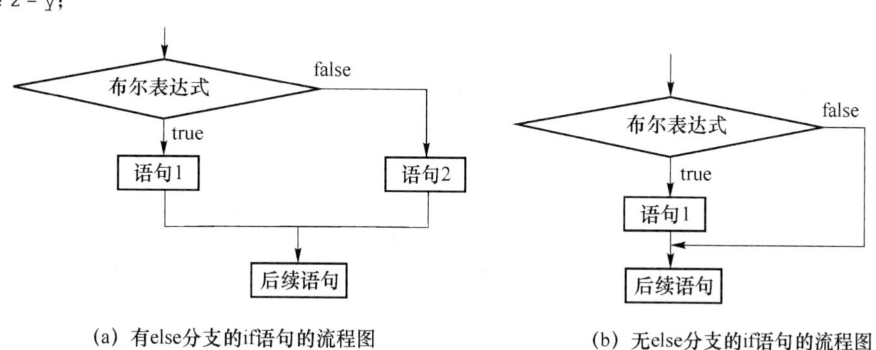

图 3-7　if 语句流程图

例 3-8　输入一个年份，判断是否为闰年。

分析：判断闰年条件是，年份能被 4 整除而不能被 100 整除，或者被 400 整除。程序的流程是：先输入年份到变量 year 中，接着判断是否为闰年，最后输出判断结果。

LeapYear.java 的代码：

```
import javax.swing.JOptionPane;
class LeapYear {
  public static void main(String args[])
  {
  int year;
  boolean isLeapYear;
  String sYear  =  JOptionPane.showInputDialog("Enter the year:");
```

```
    year = Integer.parseInt( sYear );
    isLeapYear = ((year % 4 == 0 && year % 100 != 0)||(year % 400 == 0));
    if (isLeapYear)
      sYear = year +"is a leap year";
    else
      sYear =  year +"is not a leap year";
    JOptionPane.showMessageDialog(null, sYear,"结果",JOptionPane.PLAIN_MESSAGE);
    }
}
```

例 3-8 程序运行结果如图 3-8 所示。

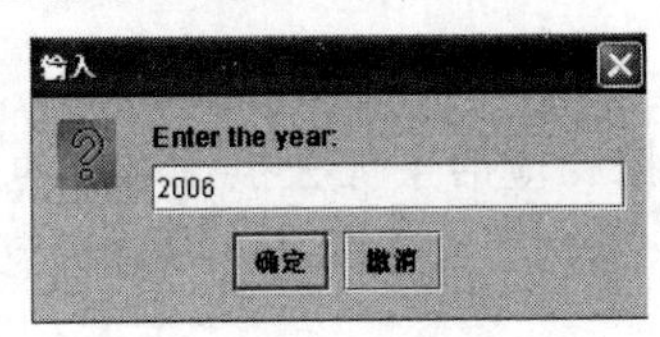

图 3-8　例 3-8 程序运行结果

(2) if 语句中的 else 分支可为空。当 else 分支为空时,成为如下形式:

```
if ( 布尔表达式 )
    语句 1;        // if 分支
```

在这种情况下,当计算条件表达式的值为假时,不需做任何操作,执行流程如图 3-7(b)所示。

例如:实现求某数的绝对值功能的程序片段。

```
if(x< = 0)  x = - x;
```

(3) if 语句的嵌套

if 语句中内嵌的语句 1 或语句 2 又是 if 语句的情况称为 if 语句的嵌套。Java 规定,else 总是与离它最近的 if 配对。如果需要,可以通过使用花括号"{ }"来改变这种配对关系。

例如:
```
if(x>0)  y = 1;
    else if(x = = 0)  y = 0;
    else if(x<0&&x> - 10)  y = - 1;
```

例 3-9　从键盘输入 3 个整数,求 3 个数之中最大数。

Java 的源文件是 FindMax.java 代码如下:

```
import javax.swing.JOptionPane;
public class FindMax {
    public static void main( String args[] )
    {  String input1,input2,input3;
       int a,b,c,max;
       input1 = JOptionPane.showInputDialog("输入第一个数");
       input2 = JOptionPane.showInputDialog("输入第二个数");
       input3 = JOptionPane.showInputDialog("输入第三个数");
       a = Integer.parseInt( input1 );
       b = Integer.parseInt( input2 );
       c = Integer.parseInt( input3 );
       if(a>b)
           if (a>c)
               max = a;            //a>b a>c
           else
               max = c;            //b<a< = c
```

```
        else
            if (b>c)
                max = b;            //a< = b c<b
            else
                max = c;            //a< = b b< = c
        JOptionPane.showMessageDialog( null, a+","+b+","+c+"中的最大数是:"+max,"最大数结果", JOptionPane.PLAIN_MESSAGE );
    }
}
```

例 3-9 程序运行结果如图 3-9 所示。

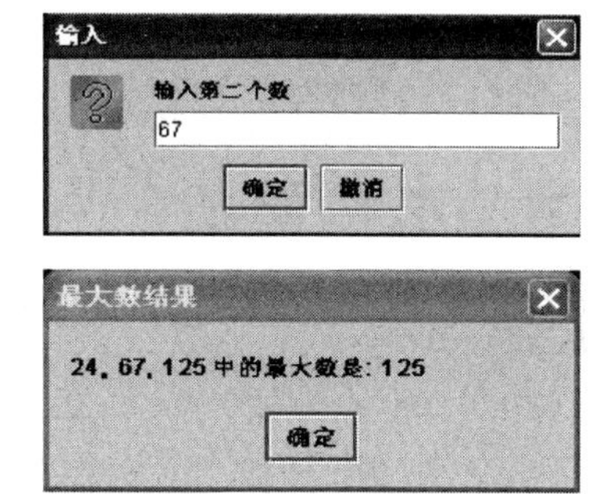

图 3-9　例 3-9 程序运行结果

2. switch 语句

switch 语句是多分支的开关语句，它的一般格式如下：

```
switch(表达式) {
    case 常量判断值 1:语句 1          // case 分支 1
    case 常量判断值 2:语句 2          // case 分支 2
    …
    case 常量判断值 n:语句 n          // case 分支 n
    [default: 缺省处理语句]           // default 分支
}
```

switch 语句的执行顺序：首先计算表达式的值，将它先与第一个 case 分支的判断值相比较，若相同，则程序的流程转入第一个 case 分支的语句块；否则，再将表达式的值与第二个 case 分支相比较……依此类推。如果表达式的值与任何一个 case 分支都不相同，则转而执行最后的 default 分支，在 default 分支不存在的情况下，则跳出整个 switch 语句。

switch 语句需要注意的是：

(1) 表达式、判断值都是 int 型或 char 型。

(2) default 分支可选。

(3) switch 语句的每一个 case 判断，只是指明流程分支的入口点，而不是指定分支的出口点，分支的出口点需要用相应的跳转语句(break)来标明。请看下面的例子：

```
switch (MyGrade)
{   case 'A':      MyScore = 5 ;
    case 'B':      MyScore = 4 ;
    case 'C':      MyScore = 3 ;
    default:       MyScore = 0 ;
}
```

假设变量 MyGrade 的值为“A”，执行完 switch 语句后，变量 MyScore 的值被赋成 0，而不是 5。因为 case 判断只负责指明分支的入口点,表达式的值与第一个 case 分支的判断值相匹配后,程序的流程进入第一个分支,将 MyScore 的值置为 5。由于没有专门的分支出口,所以流程将继续沿着下面的分支逐个执行,MyScore 的值被依次置为 4、3,最后变成 0。如果希望程序能正常完成分支的选择,则需要为每一个分支加上 break 语句。

修改后的例子代码如下:

```
switch (MyGrade)
{     case 'A':       MyScore = 5 ;
                      break;
      case 'B':       MyScore = 4;
                      break ;
      case 'C':       MyScore = 3;
                      break;
      default:        MyScore = 0;
  }
```

switch 语句的各 case 分支中带有 break 语句的执行流程,如图 3-10 所示。

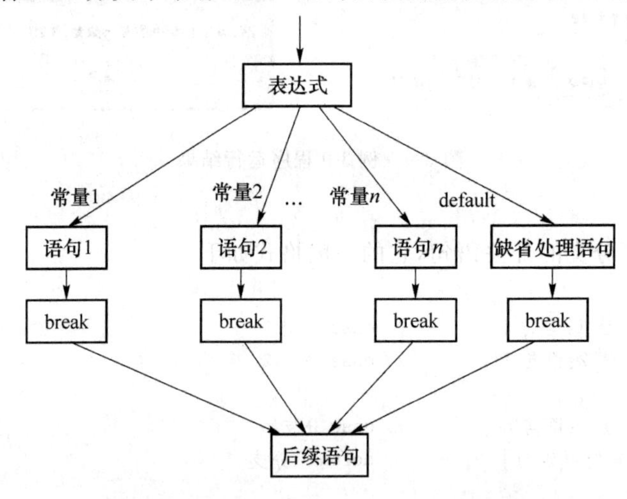

图 3-10　带有 break 的 switch 语句的流程图

(4) case 分支中可包含多个语句,且不用{ }。

(5) 在一些特殊的情况下,多个不同的 case 值要执行一组相同的操作。下面的例子是上例的进一步修改,仅划分及格与不及格。

```
switch (MyGrade)
{     case 'A' :
      case 'B':
      case 'C':     MyScore = 1 ;        //及格
                    break ;
      default :     MyScore = 0 ;        //不及格
}
```

(6) 使用 if-else 语句可以实现 switch 语句所有的功能。但通常使用 switch 语句更简练,且可读性强,程序的执行效率也高。

(7) if-else 语句可以基于一个范围内的值或一个条件来进行不同的操作,但 switch 语句中的每个 case 子句都必须对应一个单值。

例 3-10　输入一个 0～100 的整数,实现学生成绩的百分制到等级制的转换。

SwitchDemo.java 程序代码如下:

```
class SwitchDemo{
  public static void main(String[ ] args){
    String input = args[0];
    int score = Integer.parseInt(input);
    char grade;
    switch (score/10)          // 两个整型数相除的结果还是整型
    {
      case 10:
      case 9:grade = 'A';      //值为 10 和 9 时的操作是相同的
             break;
      case 8: grade = 'B';
              break;
      case 7: grade = 'C';
              break;
      case 6: grade = 'D';
              break;
      default: grade = 'F';
    }
    System.out.println("grade is:" + grade);
  }
}
```

此程序用 JDK 命令方式运行。要注意几点。

(1) SwitchDemo.class 类必须放在系统环境变量 classpath 设置的路径中，如 d:\javaExamples。

(2) 使用 Java Application 命令行参数：Java Application 是用命令行来启动执行的，命令行参数是向 Java Application 传入数据的常用而有效的手段。本例调用 Java 解释器 java.exe 对类文件 SwitchDemo.class 解释执行。命令行上跟在类名 SwitchDemo 之后的是 n 个 String 型的运行参数($n \geq 0$)，运行时将这 n 个 String 型参数放入 main( )方法的参数中，即数组元素 args[0],args[1],args[2],…,args[$n-1$]中。本例中 $n=1$，输入的分数是字符串"88"放入 args[0]。因此程序运行的结果如图 3-11 所示。

图 3-11　例 3-10 用 JDK 命令方式运行输出结果

MyEclipse 环境中设置运行参数：Run as→Run configuration－＞arguments→program arguments，如图 3-12 所示。

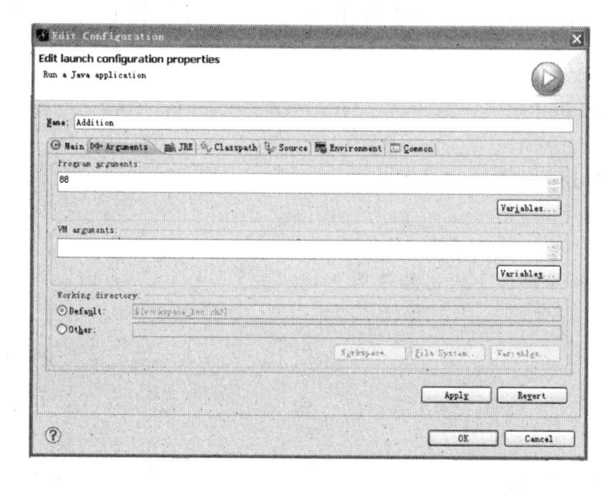

图 3-12　MyEclipse 环境中设置运行参数

3.4.2 循环语句

循环结构是在一定条件下反复执行某段程序的流程结构，被反复执行的程序被称为循环体。例如：求累加和。循环结构是程序中非常重要和基本的一种结构，它是由循环语句来实现的。Java 的循环语句共有 3 种：while 语句、do-while 语句和 for 语句。它们的执行流程结构如图 3-13 所示。

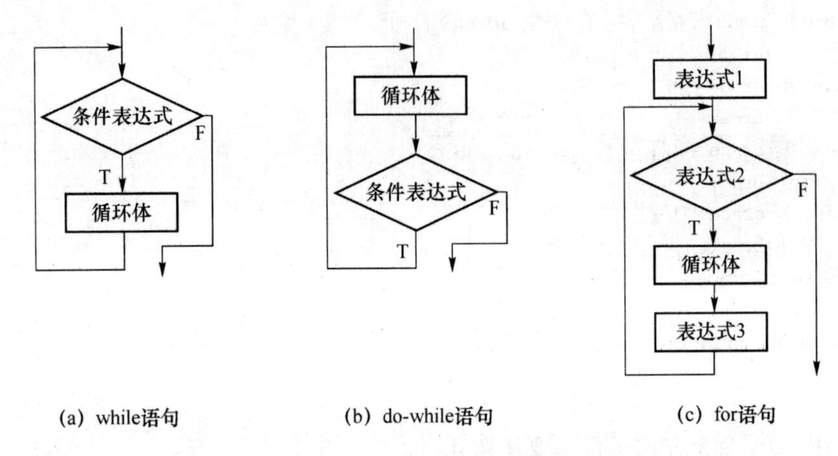

图 3-13　Java 的 3 种循环语句执行流程

1. while 语句

while 语句的一般语法格式如下：

```
while(条件表达式)
    循环体
```

其中条件表达式的返回值为布尔型，循环体可以是单个语句，也可以是复合语句块。

while 语句的执行过程如图 3-12(a)所示。该语句执行的过程为：首先，计算布尔表达式的值，如果其值为真，就执行循环体，循环体执行完之后再无条件转向条件表达式进行计算与判断；当计算出条件表达式为假时，跳过循环体执行 while 语句的后续语句。

应用 while 语句时应该注意：在循环体中，一般要包含改变循环条件表达式的语句，否则会造成无限循环（死循环）。

例 3-11　用 while 语句求 1＋2＋…＋100 的和。

分析：用累加算法：sum＝sum＋i（i 范围为 100），累加过程是一个循环过程，用 while 语句实现。

WhileTry.java 代码如下：

```
public class WhileTry {
    public static void main(String[ ] args) {
    int i,sum;
    sum = 0;            //累加和变量置 0
    i = 1;              //计数器 i 的初始值为 1
    while (i< = 100)
    {
      sum + = i;
      i++ ;
    }
    System.out.println("sum =" + sum);
```

```
    }
}
```

运行输出结果如下：

```
sum = 5050
```

2. do-while 语句

do-while 语句的语法形式如下：

```
do {
    循环体
} while (布尔表达式);
```

do-while 的流程结构如图 3-12(b)所示。该语句执行的过程为：先执行循环体语句，后判断条件表达式的值。如果条件表达式的值为 true，继续执行循环体；条件表达式的值为 false，则结束循环。

与应用 while 语句一样，应该注意，在 do-while 的循环体中，也要包含改变循环条件表达式的语句，否则便会造成无限循环(死循环)。

例 3-12 用 do-while 语句求 1+2+…+100 的和。

```
public class DoWhileTry {
    public static void main(String[ ] args) {
        int i,sum;
        sum = 0;        //累加和变量置 0
        i = 1;          //计数器 i 的初始值为 1
        do{
            sum + = i;
            i++ ;
        } while (i< = 100);
        System.out.println("sum = " + sum);
        }
}
```

程序运行输出结果同例 3-11。

do-while 语句和 while 语句都是实现循环结构，但两者是有区别的：do-while 语句先执行循环体，然后判断条件表达式的值，再决定是否继续循环；而 while 语句是先判断条件表达式的值，再决定是否执行循环体。所以，用 do-while 语句时，循环体至少执行一次。现在将例 3-11和例 3-12 程序中的 i 初值由 1 修改为 101，比较两个程序的输出结果可知，例 3-11 程序输出结果为 0，而例 3-12 程序输出结果为 101，从而得到 do-while 语句和 while 语句的区别。

例 3-13 输入一个整数，将各位数字反转后输出。

分析：将一个整数的各位数字反转输出，即先输出个位，十位、百位……，采用不断除以 10 取余数的方法，直到商数等于 0 为止。由于无论整数 n 是多少，至少要输出个位位数(即使 $n=0$)，因此用 do-while 语句。算法分析过程描述如下：

```
输入 n;
rightDigit = n % 10;  ⎫
输出 rightDigit;       ⎬   重复此过程直到 n == 0 为止
n = n/10;             ⎭
```

ReverseDigits .java 的代码如下：

```
import javax.swing.JOptionPane;
class ReverseDigits{
```

```
    public static void main (String [] args) {
        int n,rightDigit;
        String str = JOptionPane.showInputDialog("Input a integer number:");
        String output = "";
        n = Integer.parseInt(str);
        do {
            rightDigit = n % 10;
            output = output + rightDigit;
            n = n/10;
        }
        while (n!= 0);
        JOptionPane.showMessageDialog(null,"The reverse digits is:" + output,"结果", JOption-
Pane.PLAIN_MESSAGE);
    }
}
```

例 3-13 程序运行结果如图 3-14 所示。

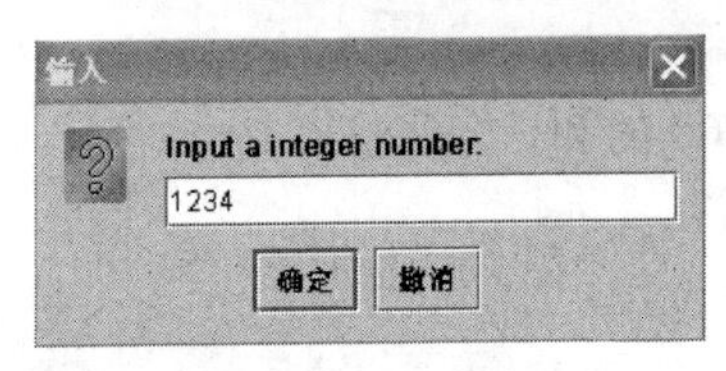

图 3-14　例 3-13 程序运行结果

3. for 语句

for 语句是 Java 语言 3 个循环语句中功能较强，使用较广泛的一个，它的流程结构可参见图 3-12(c)。for 语句的一般语法格式如下：

```
for(表达式 1;表达式 2;表达式 3)
    循环体
```

其中表达式 2 是返回布尔值的条件表达式，用来判断循环是否继续；表达式 1 完成初始化循环变量和其他变量的工作；表达式 3 用于改变循环控制变量的值。3 个表达式之间用分号隔开。

for 语句的执行过程是：

(1) 首先计算条件表达式 1 的值，完成必要的初始化工作；

(2) 再判断表达式 2 的值，若值为 false，则跳出整个 for 语句(循环结束)，若表达式 2 的值为 true，则执行循环体，执行完循环体后再计算表达式 3 的值(表达式 3 往往用于修改循环条件)这样一轮循环就结束了；

(3) 跳转到(2)重新开始下一轮循环。

例 3-14　用 for 语句求 1＋2＋…＋100 的和。

```
public class ForTry {
  public static void main(String[ ] args) {
    int i;              //声明循环控制变量
    int sum = 0;        //累加和变量置 0
    for(i = 1;i< = 100;i ++ )
    {
      sum + = i;
    }
    System.out.println("sum = " + sum);
  }
}
```

在上述例子中，称变量 i 为循环控制变量，由它的值决定循环开始和结束。一般来说，在循环次数预知的情况下，用 for 语句比较方便，而 while 语句和 do-while 语句比较适合于循环次数不能预先确定的情况。

关于 for 语句的说明如下。

(1) for 语句的 3 个表达式可以为空(但分号不能省略)，但若表达式 2 也为空，则表示当前循环是一个无限循环，需要在循环体中书写另外的跳转语句终止循环。例如：

```
for (;;) 语句;              //相当于 while (true)语句
for (; i<=100)语句;         //相当于 while (i<=100)语句
```

(2) 有时，在表达式 1 和表达式 3 的位置上可包含多个语句。

请看下面几个程序段。

程序段 1：

```
sum=0;
for(int i=1;i<=100;i++ )        //在 for 语句中声明循环控制变量并赋初值
{
    sum+=i;
}
```

程序段 2：

```
sum=0;
i=1                                 //在 for 语句之前给循环控制变量赋初值
for(;i<100;i++ )sum=sum+i;          //省略表达式 1
```

程序段 3：

```
i=1                                 //在 for 语句之前给循环控制变量赋初值
for(sum=0;i<100;i++ )sum=sum+i;     //表达式 1 与循环控制变量无关
```

程序段 4：

```
for (sum=0, i=1; i<100;)            //省略表达式 3
  {sum=sum+i;i++ ;}                 //在循环体中改变循环控制条件
```

程序段 5：

```
for(i=0,j=10;i<j;i++,j--){……}      //表达式 1 和表达式 3 可以是逗号表达式
```

例 3-15 输入一个整数，求出它的所有因子。

求一个整数 n 的所有因子的算法：对 $1\sim n$ 的全部整数判断，如果能整除 n 的均是 n 的因子，即 n%k==0(k 从 1 变化到 n)。

Fctors.java 的代码如下：

```
class Factors{
  public static void main(String[ ] args){
    int n=Integer.parseInt(args[0]);
    for (int k=1;k<=n;k++ )
      if (n%k==0)
        System.out.print(k+" ");
    System.out.println("");
    }
}
```

用 JDK 工具，程序运行结果如图 3-15 所示。

4. 循环的嵌套

一个循环体内又包含另一个完整的循环结构，称为循环的嵌套。内嵌的循环中还可以嵌

套循环，这就是多重循环。3 种循环语句（while 循环、do-while 循环和 for 循环）可以相互嵌套使用。

图 3-15　例 3-15 程序运行结果

例 3-16　打印出星星菱形图案。

```
public class Star{
    public static void main(String[ ] args) {
      int i,j;
      // 画出上半部分图案
      for(i = 1;i< = 4;i++ ) //画上面 4 行
      {
          for (j = 1;j< = 4 - i;j++ )
             System.out.print(" "); //画空格
          for (j = 1;j< = 2 * i - 1;j++ )
             System.out.print(" *"); //画星号
          System.out.println(); // 换行
      }
    // 画出下半部分图案
    for(i = 3;i> = 1;i-- ) //画下面 3 行
{
        for (j = 1;j< = 4 - i;j++ )
           System.out.print(" "); //画空格
        for (j = 1;j< = 2 * i - 1;j++ )
           System. out. print (" *"); //画
           星号
        System.out.println(); // 换行
      }
    }
}
```

```
      *
    * * *
  * * * * *
* * * * * * *
  * * * * *
    * * *
      *
```

图 3-16　星星菱形图案

程序运行结果如图 3-16 所示。

5. 跳转语句：break 语句和 continue 语句

除了选择语句和循环语句外，Java 还提供了 continue 语句和 break 语句用于改变控制流。

(1) continue 语句

continue 语句只能出现在循环体中，其作用是结束本轮循环，接着开始判断是否执行下一轮循环。continue 语句的一般语法格式为：

```
continue [标号];
```

其中，用“[ ]”括起的标号部分是可选的，即 continue 语句分为带标号和不带标号两种形式。

不带标号的 continue 语句，它的作用是终止当前这一轮的循环，跳过本轮剩余的语句，直接进入当前循环的下一轮。在 while 或 do while 循环中，不带标号的 continue 语句会使流程直接跳转至条件表达式；在 for 循环中，不带标号的 continue 语句会跳转至表达式 3，计算修改循环变量后再判断循环条件。

例 3-17　使用不带标号的 continue 语句的一个程序段（完整程序可到北京邮电大学出版社网站下载电子资源）。

```
int count;
```

```
for (count = 1;count< = 10;count ++ ) {
        if (count == 5)
          continue;
        System.out.print(count + " ");
}
System.out.println("count = " + count);
```

此程序段的运行结果如下：

```
1 2 3 4 6 7 8 9 10 count = 11
```

带标号的 continue 语句，其格式是：

```
continue 标号名;
```

这个标号名应该定义在程序中外层循环语句的前面，用来标志这个循环结构。标号的命名应该符合 Java 标识符的规定。带标号的 continue 语句使程序的流程直接转入标号标明的循环层次。

例 3-18 使用带标号的 continue 语句的一个程序段（完整程序可到北京邮电大学出版社网站下载电子资源）。

查找 1～100 之间的素数。判断 i 是否是素数的过程：如果找到整数 i 的一个因子 j（当 j 从 2 到 $i-1$ 变化时），则说明这个 i 不是素数。程序将跳过本轮循环剩余的语句，直接进入下一轮循环，检查下一个数是否是素数。

```
OutterLoop:
for (int i = 1;i< = 100;i ++ )
{
      for (int j = 2;j<i;j ++ )
      {   if (i % j == 0)
            continue OutterLoop;
      }
      System.out.println ( i );        //屏幕输出素数
}
```

此程序段的运行结果如下：

1 2 3 5 7 11 13 17 19 …

(2)break 语句

break 语句仅出现在 switch 语句或循环体中，其作用是使程序的流程从一个语句块内部跳转出来，如从 switch 语句的分支中跳出，或从循环体内部跳出。break 语句的一般语法格式为：

```
break [标号];
```

break 语句也分为带标号和不带标号两种形式。带标号的 break 语句的使用格式为：

```
break 标号名;
```

这个标号应该标志某一个语句块。执行 break 语句就从这个语句块中跳出来，流程进入该语句块后面的语句。

例 3-19 使用不带标号的 break 语句的一个程序段（完整程序可到北京邮电大学出版社网站下载电子资源）。

修改例 3-17，将 continue 语句替换成 break 语句后，得到的程序段的运行输出结果为：

```
1 2 3 4 count = 5
```

例 3-20 使用带标号的 break 语句的一个程序段（完整程序见可到北京邮电大学出版社网站下载电子资源）。

```
stop:
```

```
for (int i = 1;i< = 10;i ++ )
{     for (int j = 1;j< = 5;j ++ )
      {  if (i = = 5) break stop;
         System.out.print(" * ");
      }
      System.out.println ( "" );
}
```

程序运行结果如图 3-17 所示。

```
* * * * *
* * * * *
* * * * *
* * * * *
```

图 3-17　例 3-20 程序运行的输出结果

3.5　方　法

对象的行为在 Java 中用方法来实现。一个方法(在其他语言中也称做函数或过程)往往完成一个具体的、独立的功能。方法是处理问题过程的一种抽象,方法编写好后,可以被重复地使用。使用时只关心方法的功能和如何使用,而不必关心方法的功能的具体实现,这样可有利于代码的重用,提高程序的开发效率。

在 Java 中所有的方法都被封装在类中,不能单独使用。

3.5.1　方法的声明

方法声明的格式为:

```
返回值类型 方法名([形式参数表])        //方法头
{
    语句序列                           //方法体
}
```

(1) 返回值类型:规定了方法返回给调用者的结果值类型,可以定义为简单数据类型或类的类型。在 Java 中声明一个方法时必须定义返回值类型。如果返回类型为 void,则表示无返回值,在方法体中不必写 return 语句,如果写 return 语句,在 return 之后必须不带返回值。如果方法的返回类型不是 void,则这个方法的方法体中必须包含一个 return 语句,且 return 之后必须带返回值,如 return 0。

(2) 方法名:是任何合法的 Java 标识符。

(3) 形式参数表:形式参数表的内容如下。

类型 1 形参 1,类型 2 形参 2,…,类型 *n* 形参 *n*

指明该方法所需要的若干参数和这些参数的类型。各参数之间用逗号分开。一个方法的形式参数表可为空,表示该方法无参数。

(4) 方法头与方法体:方法声明的第一行是方法头,花括号中的语句序列构成了方法体。方法头定义了方法的功能是什么,以及怎样调用它。方法体定义功能的具体实现过程代码,对调用者而言,只需要关心方法头,而无须关心方法体。

(5) return 语句的一般格式是:

```
return 表达式;
```

return 语句用来使程序流程从方法调用中返回，表达式的值就是调用方法的返回值。如果方法没有返回值，则 return 语句中的表达式可以省略。

例 3-21　求阶乘 $n!$。

```
long Factor(int n) {
    long s;
    for (i = 1;i< = n;i ++ )
        s = s * i;
    return s;
}
```

3.5.2　方法的调用

方法的调用的一般格式：

```
方法名（实参表）
```

其中实参表是用逗号分开的实参。要求实参表和方法声明中的形参表的形参类型和形参个数完全相同。如果方法的调用无返回值(void)，方法调用可作为一条语句单独出现在程序中。

例如，上述求 $n!$ 的方法 fn()的调用形式为：

```
int y = Factor (10);        //求阶乘 10!
y = Factor (5);             //求阶乘 5!
```

在面向对象的语言中，方法的调用还有另外两种方式：

```
对象的引用名.方法名(实参表)          //通过对象调用非静态方法
类名.方法名(实参表)                  //通过类名调用静态方法
```

这两种方法的调用方式将在第 4 章详细讨论。

例 3-22　输出 10 000 之内的所有完全数。

完全数是指等于其所有因子和(包括 1，但不包括这个数本身)的数。例如，6＝1×2×3，6＝1＋2＋3，则 6 是一个完全数。

定义两个方法：

① boolean isPerfect(int x)，用于判断 x 是否完全数；

② void displayPerfect(int x)，用于输出 x 的因子之和的公式。

由于这两个方法的调用语句出现在静态方法 main 中，所以这两个方法声明的开头必须加关键字 static。有关静态方法将在第 4 章详细讨论。

在 main()方法中使用一个 for 循环，检查 1～10 000 之间的所有整数是否是完全数。

PerfectNumber. java 的代码如下：

```
public class PerfectNumber
{
  public static void main(String args[])
  {  for(int n = 1;n<10000;n ++ )
         if(isPerfect(n))
             displayPerfect(n) ;
  }
  static boolean isPerfect(int x)               //判断 x 是否完全数
  {  int y = 0;
     for(int i = 1;i<x;i ++ )
         if(x % i == 0)                         //i 是 x 的因子
```

```
            y+ = i;
        if(y == x)
            return true;
        else
            return false;
        }
        static void displayPerfect(int x)        //输出 x 的因子之和的公式
        {  System.out.print(x + " = ");
           for(int i = 1;i<x;i++ )
                if(x % i == 0) {
                  if (i!= 1) System.out.print(" + ");
                  System.out.print(i);
                  }
           System.out.println();
        }
    }
```

程序运行结果如图 3-18 所示。

```
6 = 1+2+3
28 = 1+2+4+7+14
496 = 1+2+4+8+16+31+62+124+248
8128 = 1+2+4+8+16+32+64+127+254+508+1016+2032+4064
```

图 3-18　例 3-22 的运行结果

3.5.3　方法的参数传递

参数传递方式讨论的是:方法的调用者如何将实际参数值传递给被调用者。Java 的参数传递方式分为按值传递和按引用传递。

1. 按值传递

当方法的参数为简单数据类型时,则将实参的值传递给形参。这种传递不会因为调用方法中对形参值的改变而影响实参的值。例如:

```
void swap( int x,int y) {int hold = x; x = y; y = hold; }        //方法的声明
int a = 1,b = 2; swap (a,b);                   //方法调用执行后,a 的值仍为 1,b 的值为 2
```

2. 按引用传递

当方法的参数为复合数据类型(如类、数组名、接口)时,则对形参的任何访问等同于对实参的访问,即形参被认为是实参的别名。例如:

```
void bubbleSort( int array1[ ]) {…}          // 方法的声明,形参 array1 是数组名
bubbleSort(array2);                          //方法的调用
```

在 bubbleSort 方法执行时,对 array1 的操作被等同地替换成对 array2 的操作。即在调用方法中对形参 array1 值的改变就是对实参 array2 值的改变。

3.5.4　方法的重载

方法的重载(Overloading)定义为:在一个类中定义多个同名的方法,但要求各方法具有不同的参数的类型或参数的个数。调用重载方法时,Java 编译器能通过检查调用的方法的参数类型和个数选择一个恰当的方法。方法重载通常用于创建完成一组任务相似但参数的类型或参数的个数不同的方法。

注意,重载的方法参数必须有所区别:①参数的类型不同;②参数的顺序不同;③参数的个数不同。

重载虽然表面上没有减少编写程序的工作量，但实际上重载使得程序的实现方式变得很简单，只需要记住一个方法名，就可以根据不同的输入类型选择该方法不同的版本。

例 3-23 用重载方法实现计算 int 类型和 double 类型的平方。

```
public class MethodOverload {
  public static void main(String args[])
  {  System.out.println("整型数 7 的平方值:" + square( 7 ));
     System.out.println("浮点型数 7.5 的平方值:" + square( 7.5 )) ;
  }
  // 声明带有 int 型参数的 square 方法
  public static int square( int intValue )
  {
     System.out.println("调用 int 型参数的 square 方法");
     return intValue * intValue;
  }
  // 声明带 double 型参数的 square 方法
  public static double square( double doubleValue )
  {
     System.out.println("调用 double 型参数的 square 方法");
     return doubleValue * doubleValue;
  }
}
```

程序运行结果如图 3-19 所示。

```
C:\PROGRA~1\XINOXS~1\JCREAT~1\GE2001.exe
调用int型参数的square方法
整型数 7 的平方值: 49
调用double型参数的square方法
浮点型数7.5 的平方值: 56.25
```

图 3-19 例 3-23 程序运行结果

3.5.5 嵌套与递归

嵌套是指在一个方法声明的方法体中直接调用了另外一个方法。而递归(recuision)是指在一个方法声明的方法体中直接或间接地调用了自身方法。递归技术体现了在解决实际问题时采用的“依此类推”、“用同样的步骤重复”的基本思想，通过递归使用户能用简单的程序来解决某些复杂的计算问题，但是运算量较大。

例 3-24 求阶乘 $n!$ 的递归例子。

求阶乘 $n!$ 的递归思路如下：

$$n! = \begin{cases} 1 & (\text{当 } n=1 \text{ 时}) \\ n(n-1) & (\text{当 } n \geqslant 2 \text{ 时}) \end{cases}$$

所以，求阶乘 $n!$ 的递归方法的声明如下(完整程序可到北京邮电大学出版社网站下载电子资源)：

```
long Factorial(int n){                //方法声明
    if(n == 1)
        return 1;                     //递归终结点
    else
        return n * Factorial (n - 1); //递归调用自身
}
```

下面来分析调用语句 y=Factorial(5);的执行过程,如图 3-20 所示。

如果一个递归方法只是一味地自己调用自己,则将构成无限循环,永远无法返回,所以任何一个递归方法都必须有一个“递归终结点”,即当同性质的问题被简化得足够简单时,将可以直接获得问题的答案,而不必再调用自身。例如当求 $n!$ 的问题被简化成求 1! 的阶乘的问题时,可以直接获得 1! 的答案为 1,这就是递归终结点。

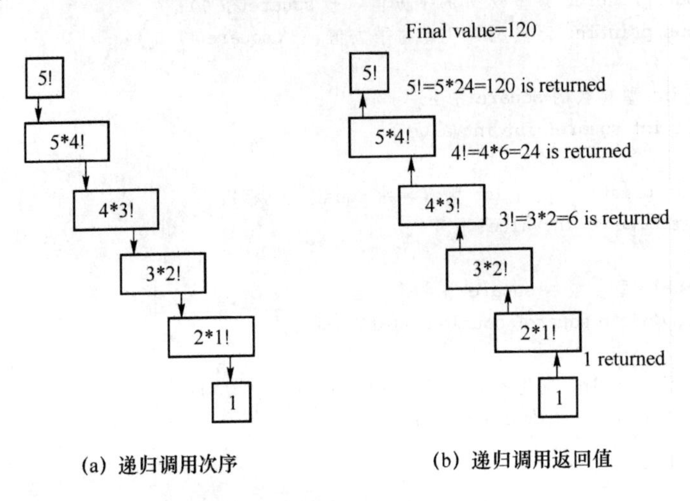

图 3-20 递归计算 5! 的过程

3.5.6 变量的作用域

变量的作用域指明可访问该变量的一段代码。声明一个变量的同时也就指明了变量的作用域。

作用域的基本规则如下:

(1) 参数声明的作用域是声明方法所在的方法体;

(2) 局部变量在方法或方法中的一块代码中声明,它的作用域为它所在的代码块(整个方法或方法中的某块代码);

(3) 在带标号的 break 和 continue 语句中,标号的作用域是带标号结构范围的语句(即带标号语句的主体);

(4) 出现在 for 结构头初始化部分的局部变量,其作用域是 for 结构体和结构体头中的其他表达式;

(5) Catch 语句中的异常处理参数将传递给异常处理代码块,它的作用域就是 Catch 的异常处理块;

(6) 类中的方法和域(数据成员),它们的作用域是整个类。这种作用域使一个类的方法可以用简单的名称调用类中声明的其他方法以及该类继承的方法,并使之能直接访问在该类中声明的域。

如果块嵌套在方法体中,在外层块中出现与内层块中已声明过的变量同名的声明时,编译时会产生语法错误,说明变量已经声明过了。如果方法中的局部变量或参数与域同名,则域将被“屏蔽”,直到块结束。将在第 4 章讨论使用 this 关键字访问被屏蔽的域。

例如:

```
class C1{
    int g;                    //声明实例变量 g
```

```
    void method1(int k){        //声明参数 k
    int i;                      //声明局部变量 i
    i = k + g;                  //求参数 k 与实例变量 g 的和,将结果值赋给局部变量 i
}
void method2(int k){            //声明参数 k
    int g;                      //声明局部变量 g
    int j = 0;                  //声明局部变量 j
    g = j + k;                  //给局部变量 g 赋值
    this.g = j;                 //给实例变量 g 赋值
    i = 10;                     //错误! i 变量未声明
  }
}
```

3.6 数　组

一组相同数据类型的元素按一定顺序线性排列,就构成了数组。数组的主要特点如下:

(1) 数组是相同数据类型的元素的集合;

(2) 数组中的各元素是有先后顺序的,它们在内存中按照这个顺序连续存放在一起;

(3) 每个数组元素用整个数组的名字和它自己在数组中的位置表达(此位置被叫做下标或索引),如 a[0]代表数组 a 的第一个元素,a[1]代表数组 a 的第二个元素,依此类推。

例如:int a[]=new int [5];

是一个包含 5 个整型元素的数组,在内存中的顺序存储结构如图 3-21 所示。

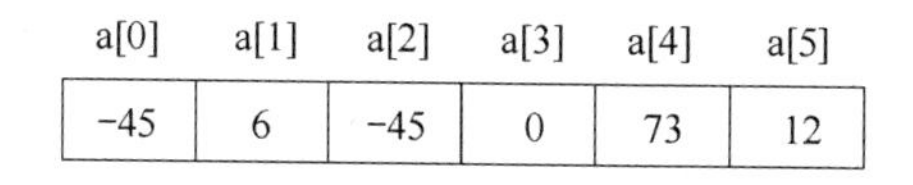

a[0]	a[1]	a[2]	a[3]	a[4]	a[5]
-45	6	-45	0	73	12

图 3-21　顺序存储结构

Java 中的数组是对象,因此属于引用类型,数组对象需要使用 new 关键字来创建。数组有一维数组和多维数组之分,下面分别详细介绍。

3.6.1　一维数组

Java 语言中定义数组的操作需要下面 3 个步骤。

1. 声明数组

声明数组是指声明数组的名称和数组所包含的元素的数据类型。声明数组的语法格式有两种:

```
数组元素类型 数组名[ ];
数组元素类型 [ ]数组名;
```

其中,方括号[ ]是数组的标志,它可以出现在数组名的后面,也可以出现在数组元素类型名的后面,两种定义方法没有什么差别。下面的例子声明了一个整型的数组:

```
int intArray[ ];
```

等同于

```
int [ ] intArray;        //声明数组变量
```

在 Java 中数组元素的数据类型,可以是简单数据类型,也可以是某一类的类型(这时每一个数组元素被称做引用变量)。例如语句

```
String stringArray[ ];        //声明数组变量
```

声明了每一个数组元素的类型是字符串的数组 stringArray。

2. 创建数组空间

声明数组仅仅指定了数组的名字和数组元素的类型，并没有为数组中的每一个元素开辟内存空间。在创建数组空间时，需要使用 Java 中的 new 关键字，并且指定数组元素的个数（数组的长度），以确定所开辟的内存空间的大小。创建数组空间的语法格式为：

```
数组名 = new 数组元素类型 [ 数组元素的个数 ]；
```

为上面声明的两个数组变量创建空间的语句：

```
intArray = new int [10];
```

为整型数组 intArray 分配 10 个整数元素所占据的内存空间，并使每个元素初值为 0。

```
stringArray = new String[5];          //创建数组
```

为字符串型的数组 stringArray 分配 5 个字符串元素所占据的内存空间，并使每个元素初值为 null。

声明数组可以与创建数组空间的工作合在一起，用一条语句完成。其格式如下：

```
数组元素类型 数组名[ ]= new 类型 [数组元素的个数]；
```

例如：

```
int intArray[] = new int[10];          //声明并创建数组
String stringArray [ ] = new String[5];
```

用 new 关键字为一个数组分配内存空间后，系统将为每个数组元素都赋予一个初值，这个初值取决于数组的类型。所有数值型数组元素的初值为 0，字符型数组元素的初值为一个不可见的 ISO 控制符（"\u0000"），布尔型数组元素的初值为 false，类的类型的数组在创建时，元素的初值为 null。在实际应用中，用户应根据具体情况来对数组元素重新进行赋值。

数组对象一旦创建之后，在程序整个执行期间，就不能再改变数组元素的个数。

3. 数组的初始化

数组初始化（initializtion）就是为数组元素指定初始值。如果不希望数组的初值为默认值，此时，就需要用赋值语句来对数组进行初始化。数组的初始化方式有如下两种。

(1) 在声明数组的同时，进行初始化。例如：

```
int a[] = {1,2,3,4,5};          //通过初始化列表给数组元素指定初始值
```

声明创建了包含 5 个元素的数组 *a*，并为每个数组元素赋初值，即 a[0]=1，a[1]=2，a [2]=3，a[3]=4，a[4]=5。这种声明不需要用 new 创建数组对象。当编译器遇到一个含有初始化列表的数组声明时，会通过计算列表中的初始值的个数来确定数组元素的个数。

(2) 在声明创建数组之后，用赋值语句为每个元素赋值。例如：

```
String stringArrar[];                  //定义一个 String 类型的数组变量
stringArray = new String[3];           //给数组 stringArray 分配 3 个元素空间
                                       //初始化每个引用值为 null
stringArray[0] = new String("how");    //初始化各数组元素
stringArray[1] = new String("are");
stringArray[2] = new String("you");
```

4. 数组元素的使用

声明并创建的数组对象后，就可以在程序中像使用任何变量一样来使用数组元素。一维数组元素的使用方式为：

```
数组名[下标]
```

其中：

- 下标必须是整型或者可以转化成整型的量。下标的取值范围从 0 开始到数组的长度减 1。例如长度为 10 的数组,其元素下标为 0～9。
- 所有的数组都有一个属性 length,这个属性存储了数组元素的个数,利用它可以方便地完成许多操作。
- Java 系统能自动检查是否有数组下标越界的情况,例如数组 intArray 的长度为 10,包含 10 个元素,下标分别为 0～9,如果在程序中使用 intArray[10],就会发生数组下标越界,此时 Java 系统会自动终止当前的流程,并产生一个名为 ArrayIndexOutOfBoundsException 的异常,通知使用者出现了数组下标越界,避免这种情况的一个有效方法是利用上面提到的 length 属性作为数组下标的上界。

例 3-25 求数组元素之和。

```
public class SumArray {
    public static void main( String args[] )
    {
        int array[] = { 1, 2, 3, 4, 5, 6, 7, 8, 9, 10 };
        int total = 0;
        for( int counter = 0; counter < array.length; counter ++ )
            total + = array[ counter ];         // add each element's value to total
        System.out.println("Total of array elements:" + total);
        System.exit( 0 );
    }
}
```

程序输出结果为:

```
55
```

例 3-26 用冒泡排序法对数组进行排序。

冒泡排序法是常见的排序算法,其思路是:将相邻的两个数据进行比较,把较大的数据交换到后面。如果纵向看一个数组,那么交换过程中较小的数据就好像水中的气泡不断浮起。为了提高中的冒泡法的可重用性,把它设计为方法 bubbleSort(int array[]),方法的参数是待排序的一维数组,以增序(从小到大)方式排序。

BubbleSort.java 程序代码如下:

```
public class BubbleSort {
    public static void main( String args[] ) {
        int array[] = { 12, 6, 4, -8, 10, 11, 89, 68, 45, 37 };
        System.out.println ("排序前:");                //输出排序前结果
        for(int counter = 0; counter < array.length; counter ++ )
            System.out.print(array[ counter ] + " ");
        System.out.println ();
        bubbleSort( array );                          // 调用排序方法
        System.out.println("\n\n 排序后:");            //输出排序后结果
        for(int counter = 0; counter < array.length; counter ++ )
            System.out.print(array[ counter ] + " ");
        System.out.println ();
}
static public void bubbleSort( int array2[] )          //增序的冒泡排序法的实现
{// 外循环控制排序的趟数
  for ( int pass = 1; pass < = array2.length - 1; pass ++ ) {
      //内循环控制一趟的比较排序
        for(int element = 0;element< = array2.length - pass - 1;element ++ )
```

```
            for ( int element = 0; element < array2.length - 1; element++ ) {
                if ( array2[ element ] > array2[ element + 1 ] ) {      //比较,交换
                    int hold = array2[ element ];
                    array2[ element ] = array2[element + 1 ];
                    array2[ element + 1 ] = hold;
                }
            }
        } // end loop to control passes
    } // end method bubbleSort
} // end class BubbleSort
```

程序运行结果如图 3-22 所示。

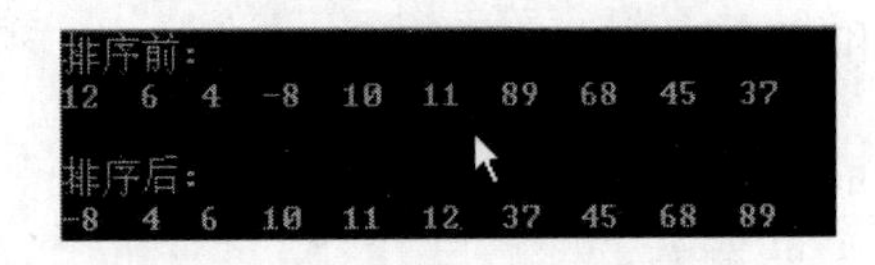

图 3-22 例 3-26 程序运行结果

3.6.2 增强的 for 循环语句

从 JDK 1.5 版本起,Java 新增了增强的 for 循环语句,实现通过一个循环对列表或数组的元素逐个遍历。增强的 for 循环语句的格式如下:

```
for (ElementType element:arrayName) {  };
```

其中,element 是 for 循环的局部变量,数据类型 ElementType 必须和数组 arrayName 的元素类型一致。用变量 element 代表数组遍历中的下一项。

例如,下面的代码完成逐个打印整形数组的所有元素。

```
int[]numArray = {1,2,3,4,5,6};
for(int conut = 0;count<numArray.length;count ++ ){
    System.out.print(numArray[count]);
}
```

使用增强的 for 循环,上面的循环被替换成:

```
for(int element:numArray){
    System.out.print(element);
}
```

从上面的例子可以看到,使用增强的 for 循环语句对数组遍历时,可以简化代码。

3.6.3 多维数组

带有两个以上下标的数组叫做多维数组。在 Java 语言中,多维数组被看做数组的数组。比如二维数组是一个特殊的一维数组,而一维数组中每一个元素又是一个一维数组。下面主要以二维数组为例来说明,高维数组与此类似。

1. 二维数组的声明和创建

二维数组声明的格式为:

```
类型 数组名[ ][ ];
```

例如:

```
int a[ ][ ];                    //声明二维数组,每一个元素类型是 int
```

与一维数组类似，需要用 new 关键字来为二维数组分配内存空间。例如：

```
a= new int[2][3];                //创建 2 行 * 3 列的二维数组对象，每个元素的初值是 0
```

该数组示意图如图 3-23 所示。

	列 0	列 1	列 2
行 0	a[0][0]	a[0][1]	a[0][2]
行 1	a[1][0]	a[1][1]	a[1][2]

图 3-23　2 行 3 列的二维数组

上述二维数组的一条声明语句和一条创建空间语句可组合为一条语句，完成同样的功能：

```
int a[ ][ ]= new int[2][3] ;
```

把二维数组看成是“一维数组的一维数组”，其空间分配还有另外一种形式（动态分配）：即从最高维开始，分别为每一维分配空间。例如：

```
int b[ ][ ] = new int[2][ ];     // 最高维含 2 个元素，每个元素为一个一维整型数组
b[0] = new int[3];               // 最高维第一个元素是一个长度为 3 的整型数组
b[1] = new int[5];               // 最高维第二个元素是一个长度为 5 的整型数组
```

此时该二维数组的元素示意图如图 3-24 所示。

b[0][0]	b[0][1]	b[0][2]		
b[1][0]	b[1][1]	b[1][2]	b[1][3]	b[1][4]

图 3-24　二维数组中的元素示意图

注意：在使用运算符 new 来分配内存时，对于多维数组至少要给出最高维的大小。

二维数组的各行长度（各行包含的元素个数）可不相同。如对上述数组 b 而言，包含 2 行（即 b.length 值为 2），第 0 行包含的元素个数为 3（即 b[0].length 值为 3），第 1 行包含的元素个数为 5（即 b[1].length 值为 5）。

2. 二维数组的初始化

在声明二维数组的同时，可进行数组元素的初始化赋值。系统会根据初始化列表给出的初始值的个数自动算出数组每一维的大小。例如：

```
int b[ ][ ]={{1,2},{2,3},{3,4}};
```

该语句创建包含 3 行 2 列的二维数组，并给每个数组元素赋初值。

又如：

```
int c[ ][ ]={{2,3},{4,5,6}};
```

创建包含 2 行的二维数组 *c*，第 0 行包含的元素个数为 2，第 1 行包含的元素个数为 3。

3. 二维数组元素的引用

二维数组中每个元素的引用方式为：

```
数组名[下标 1][下标 2]
```

其中下标 1、下标 2 为非负的整型常数或表达式。

例如：a[2][3]，b [i+2][j * 3]。

例 3-27　构造杨辉三角形。

将杨辉三角形放在二维数组中。

YangHui.java 代码如下：

```
class YangHui{
public static void main(String[] args){
    int i,j;int yanghui[ ][ ]={{1},{1,1},{1,2,1},{1,3,3,1},{1,4,6,4,1}};
    for(i=0;i<yanghui.length;i++ ){          //外循环控制打印的行数
```

```
        for(j = 0;j<yanghui[i].length;j++ )       //内循环控制打印每1行
          System.out.print("\t" + yanghui[i][j]);
        System.out.println();
        }
    }
}
```

程序运行结果如图 3-25 所示。

```
1
11
121
1331
14641
```

图 3-25　例 3-27 程序运行结果

同样,例 3-27 也可用增强的 for 循环语句实现。其代码如下:

```
class YangHui_Foreach{
    public static void main(String[] args){
        int yanghui[ ][ ] = {{1},{1,1},{1,2,1},{1,3,3,1},{1,4,6,4,1}};
        for(int [] i:yanghui){            //外循环控制打印的行
            for(int j:i)                  //内循环控制打印每一行
                System.out.print("\t" + j);
            System.out.println();
        }
    }
}
```

3.6.4　可变长的方法参数

从 JDK 1.5 版本起新增加了可变长的方法参数,即在方法的声明头中,形式参数个数是可变的。可变长的形参声明格式如下:

```
dataType...parameters
```

其中,省略号"..."表示数据类型为 dataType 的 parameters 参数个数不固定的,可为任意个。在方法调用时,变长形式参数可被替换成 1 个、2 个或多个实参。

例如,下面声明可变长参数的方法:

```
void mymethod(String s , int...numbers)
```

其方法的调用形式可为:mymethod("abc",1)、mymethod("abcd",1,2)或 mymethod("hij",1,2,3)等形式。

在编译时,方法的变长参数将被看成具有相同类型的一维数组。例如,上述声明的方法头将被翻译成如下形式:

```
void mymethod(String s , int numbers [ ])
```

因此在方法体中,可将变长的参数 numbers 当做数组使用。

注意,变长的参数有一些限制:在一个方法中只能定义一个可变长的参数,且必须是方法的最后一个参数。

例 3-28　利用可变长参数,求一组数的平均值。

```
public class VarargsTest
{
    public static float average( float... numbers )
    {
        float sum = 0.0f;
```

```
        for ( float f:numbers )
            sum + = f;
        return sum/numbers.length;
    }
    public static void main( String args[] )
    {     float f1 = 10.0f;
          float f2 = 20.0f;
          float f3 = 30.0f;
          float f4 = 40.0f;
          System.out.printf("f1 = %.1f\nf2 = %.1f\nf3 = %.1f\nf4 = %.1f\n\n",
            f1,f2,f3,f4);
          System.out.printf("Average of f1 and f2 is %.1f\n",
            average( f1, f2 ) );
          System.out.printf("Average of f1, f2 and f3 is %.1f\n",
            average( f1, f2, f3 ) );
          System.out.printf("Average of f1, f2, f3 and f4 is %.1f\n",
            average( f1, f2, f3, f4 ) );
    }
}
```

上述程序的运行结果如图 3-26 所示。

```
f1 = 10.0
f2 = 20.0
f3 = 30.0
f4 = 40.0

Average of f1 and f2 is 15.0
Average of f1, f2 and f3 is 20.0
Average of f1, f2, f3 and f4 is 25.0
```

图 3-26 例 3-28 程序运行结果

3.6.5 Arrays 类

类 Arrays 在包 java.util 中，它封装了一组静态方法用于对数组进行排序、查找、复制等操作。Arrays 类中定义了一组常用的 static 方法，如表 3-11 所示。

表 3-11 Array 类的常用静态方法

方 法	功 能
void sort(type[]a)	对数组 a 以增序的方式进行排序。type 为数据类型，可为基本类型或 Object 类型
int binarySearch(type []a, type key)	对有序的数组 a 使用折半查找算法查找指定的元素值 key 如果找到则返回其下标，否则返回－1
void fill(type[]array, type val)	将值 val 填入到数组 a 的各个元素中
boolean equals(type[] a1, type[] a2)	比较两个数组是否相等：即有相同元素个数，且对应元素相等

数组之间的复制使用方法 System.arraycopy()，其方法头定义如下：

```
void arraycopy(Object src,int srcPos,Object dest, int destPos,int length);
```

方法中的各参数的含义如下：src，源数组；srcPos，源数组中的起始位置；dest，目标数组；destPos，目标数据中的起始位置；length，src 被复制的数组元素的个数。

例 3-29 应用类 Arrays 和 System.arraycopy()方法进行数组操作。

```
import java.util.Arrays;
public class ArrayManipulations
```

```
{
    public static void main(String[] args)
    {
        int[] intArray = {8, 9, 0, 7, 4};
        int intArrayCopy[] = new int[intArray.length];
        System.arraycopy(intArray, 0, intArrayCopy, 0, intArray.length);
        displayArray(intArrayCopy, "intArrayCopy");
        // compare intArray and intArrayCopy for equality
        boolean b = Arrays.equals(intArray, intArrayCopy);
        System.out.printf("intArray = = intArrayCopy ? %b",b );
        // sort intArray into ascending order
        Arrays.sort(intArray);
        displayArray(intArray, "intArray");
        int index = Arrays.binarySearch(intArray,7);
        if (index >= 0)
          System.out.printf( "Found 7 at element %d in intArray", index);
        else
          System.out.println("7 not found in intArray");
        int fillArray[] = new int[5];
        Arrays.fill(fillArray,8);
        displayArray(fillArray, "fillArray");
    }
    // output values in each array
    public static void displayArray(int[] array, String arrayname)
    {
        System.out.printf("\n%s: ", arrayname);
        for (int value : array)
          System.out.printf(" %d ", value);
        System.out.println();
    }
}
```

上面程序运行结果如下：

```
intArrayCopy: 8 9 0 7 4
intArray == intArrayCopy ? true
intArray: 0 4 7 8 9
Found 7 at element 2 in intArray
fillArray: 8 8 8 8 8
```

3.7 小 结

Java 的程序结构是由包声明、导入语句、类定义和接口组成。

Java 中的词法记号包括关键字、标识符、文字、操作符、分隔符、空白符。

Java 中的数据类型包含简单数据类型和复合数据类型，其中简单数据类型由 Java 语言本身提供，可直接在程序中使用；而复合数据类型是由用户定义的类型，包括类、接口和数组，其中数组是类的对象，这和其他程序设计语言是不一样的。

表达式是由运算符和操作数组成。运算符有算术运算符、赋值运算符、关系运算符、逻辑运算符、位运算符、条件运算符等。对一个表达式进行运算时，要注意运算符的优先级与结合性，以及混合运算时数据类型的转换。

条件语句、循环语句和跳转语句是程序控制结构中常用的语句。

方法是完成一个具体的、独立的功能。包括方法的声明、方法的调用和方法的参数传递。

数组是有序相同的数据类型的数据的集合,利用数组数据结构,可以很好地为程序组织起循环。

习　　题

3.1　Java 有哪些基本数据类型?写出 byte 和 int 型所能表达的最大、最小数据。

3.2　根据 Java 对标识符的命名规定,判断下列标识符哪些是合法的,哪些是非法的。

(1) MyGame　　(2)_isHers　　(3)2JavaProgram

(4) Java-Visual-Machine　　(5) $abc　　(6) CONST　　(7)class

3.3　Java 的字符采用何种编码方案?有何特点?写出 5 个常见的转义符。

3.4　写出下面表达式的运算结果,设 a=3,b=-5,f=true。

(1) --a % b++　　(2) (a>=1 && a<=12? a:b)　　(3) f^(a>b)

(4) (--a)<<a　　(5) a+=a　　(6) a*=2+3

(7) (a!=b)&&(3==2+1)||(4<2+5)

3.5　设 double x=2.5; int a=7,b=3 ; float y=4.7f;计算算术表达式:x+a%b*(int)(x+y)%2 的值和 (float)(a+b)/2+(int)x%(int)y 的值。

3.6　设 x 为 float 型变量,y 为 double 型变量,a 为 int 型变量,b 为 long 型变量, c 为 char 型变量,则表达式 x+y*a/x+b/y+c 的值为________类型。

3.7　简述 Java 程序的构成。如何判断主类?下面的程序有几处错误?如何改正?这个程序的源代码应该保存成什么名字的文件?该程序编译后生成什么名字的类文件?

```
public class MyClass
{
      public static void main( );
      {
         byte b = 128;
         char c = "a"
         b = (byte)( - 129);
         System.out.println("b = " + b + ", " + "c = " + c);
      }
      System.out.println("程序结束。");
}
```

3.8　读入一系列整数,统计出正整数个数和负整数个数,读入 0 则结束。

3.9　编写一个 Java 程序,接受用户输入的一个 1~12 之间的整数(如果输入的数据不满足这个条件,则要求用户重新输入),利用 switch 语句输出对应月份的英语单词。

3.10　编写一个字符界面的 Java Application 程序,接受用户输入的一个浮点数,把它的整数部分和小数部分分别输出。

3.11　结构化程序设计有哪 3 种基本流程?分别对应 Java 中的哪些语句?

3.12　在一个循环中使用 break 语句与 continue 语句有什么不同的效果?

3.13　打印以下图案(每行打 5 个星号,每个星号之间空两个空格):

```
*  *  *  *  *
 *  *  *  *  *
  *  *  *  *  *
   *  *  *  *  *
```

3.14 编制程序，打印出100～10 000中所有的“水仙花数”。所谓“水仙花数”是指一个3位数，其各位数字的立方和等于该数本身。例如：153是一个“水仙花数”，因为$153=1^3+5^3+3^3$。

3.15 编制程序完成数制转换：输入一个8位二进制数，将其转换为十进制数输出。

3.16 编写程序求π的值。π的计算公式如下：

$$\pi=16\arctan\left(\frac{1}{5}\right)-4\arctan\left(\frac{1}{239}\right)$$

其中arctan用如下形式的级数计算：

$$\arctan(x)=\frac{x}{1}-\frac{x^3}{3}+\frac{x^5}{5}+\cdots=\sum_{n=0}^{\infty}\frac{(-1)^n x^{2n+1}}{2n+1}$$

直到级数某项绝对值不大于10^{-15}为止。这里π和x均为double型。

3.17 编写一个字符界面的Java Application程序，接受用户输入的字符，以“#”标志输入的结束；比较并输出按字典次序最小的字符。

3.18 编写一个Java Application程序，接受用户输入的10个整数放入数组中，比较并输出其中的最大值、最小值以及整个数组元素。

3.19 编制程序，求出菲波那契数列。菲波那契数列定义如下：

fibonacci(n)=n,(当$n=0,1$)

fibonacci(n)=fibonacci ($n-1$)+fibonacci($n-2$),(当$n\geqslant 2$)

分别用迭代循环法和递归法完成之。

3.20 编程序打印字符‘0’～‘9’、‘a’～‘z’、‘A’～‘Z’的Unicode的码值。

3.21 用位操作把一个128位的有符号数扩大到2倍，并以16进制的形式输出。

3.22 用位运算将一个数扩大100倍。

3.23 把一个二进制数转化为R进制数输出，R可为二进制、八进制或十六进制，从键盘输入。

第4章 类和对象

类和对象是面向对象编程技术中的最基本的概念。本章将介绍类的定义,包括类的作用域的修饰,类中成员变量、成员方法及构造方法的定义;对象的创建、使用及对象的自动回收,并将这些技术应用于一些实例中,通过本章的学习将使读者具备初步的面向对象编程(OOP)的设计能力。

4.1 面向对象程序设计的思想

4.1.1 OOP 思想

面向对象编程技术的关键性观念是它将数据及对数据的操作行为放在一起,作为一个相互依存、不可分割的整体——对象。对相同类型的对象进行分类、抽象后,得出共同的特性而形成了类。面向对象编程就是定义这些类。类是描述相同类型的对象集合。类定义好之后,将作为数据类型用于创建类的对象。程序的执行表现为一组对象之间的交互通信。对象之间通过公共接口进行通信,从而完成系统功能。类中声明的 public 成员组成了对象的对外公共接口。

Java 语言是完全面向对象的语言,程序的结构由一个以上的类组成。所有的过程都被封装起来,并将它们称之为方法。下面将通过一个具体的实例来理解什么是类,如何定义类、创建对象和使用对象。

4.1.2 用类实现抽象数据类型:时钟类

类作为一种抽象的数据类型,封装了对象的数据属性和动态行为,被用来定义类的对象。下面通过一个举例,说明如何用类来实现一个抽象数据类型——时钟类。首先对时钟进行分析。平常看到的大大小小的每一只时钟都具有时、分、秒这 3 个数据用来指示时间的小时数、分钟数、秒数,另外时钟还具有显示时间、调整时间、走时的功能(行为)。对时钟进行分析后,就可以抽象出所有的时钟都具有的特性而形成时钟的类。

时钟类的设计中包含的一组数据属性和行为如下。

(1) 数据属性:时(int)、分 (int)、秒(int)。

(2) 行为:设置时间、显示时间、走时。

用 Java 语言实现类时,数据属性用域(成员变量)实现;行为用方法(或称函数)实现。为了简便起见,在时钟类实现的代码中,时钟的走时行为暂不考虑。

例 4-1 时钟类的实现。

该例子包含两个程序文件——Time1.java 和 TimeTest1.java。Time1.java 用于定义时钟类;TimeTest1 类是一个用于测试时钟类的包含 main 方法的主类,在 main 方法中将创建 Time1 类的一个对象,并调用对象的公共方法。

时钟类 Time1.java 文件的代码如下:

```
//Time1.java
public class Time1 extends Object {
    private int hour;                          // hour 域,取值范围 0～23
    private int minute;                        // minute 域,取值范围 0～59
    private int second;                        // second 域,取值范围 0～59
    public Time1(){                            //构造方法
        setTime( 0, 0, 0 );
    }
    public void setTime( int h, int m, int s ){ // 设置时、分、秒的值
        hour = ((h>= 0 && h<24)? h : 0 );
        minute = ((m>= 0&&m<60)? m : 0 );
        second = ((s>= 0&&s<60)? s : 0 );
    }
    public String toString(){                  //返回时间的字符串
        return hour +":" + minute +":"+ second ;
    }
}
```

测试类 TimeTest1.java 文件的代码:

```
//TimeTest1.java
import javax.swing.JOptionPane;
public class TimeTest1 {
    public static void main( String args[] ){
        //创建 Time1 类的对象 time1
        Time1 time1 = new Time1();
        //显示对象 time1 的值
        String output = "time1 对象建立后\n 时间:" +
            time1.toString() ;
        //设置对象 time1 的值
        time1.setTime( 13, 27, 6 );
        //显示 time1 对象的值
        output += "\n 用 setTime 方法( 13, 27, 6 )修改 time1 对象后\n 时间:" +
            time1.toString() ;
        //创建 Time1 类的对象 time2
        Time1 time2 = new Time1();
        //设置 time2 对象的值
        time2.setTime( 11, 27, 12 );
        //显示 time2 对象的值
        output += "\n\ntime2 对象 \n 时间:" +
            time2.toString() ;
        //设置 time2 对象的值
        time2.setTime( 99, 89, 99 );
        //显示 time2 对象的值
        output += "\n 用 setTime 方法( 99, 89, 99 )修改 time2 对象后\n 时间:" +
            time2.toString() ;
        JOptionPane.showMessageDialog( null, output,
            "时钟类测试:建立了两时钟对象", JOptionPane.INFORMATION_MESSAGE );
```

```
        System.exit( 0 );
    }
}
```

程序运行时,先编译 Time1.java,再编译 TimeTest1.java;选择 TimeTest1.class 运行,得到运行输出结果如图 4-1 所示。

在例 4-1 的时钟类 Time1 中声明了 hour、minute 和 second 3 个整数变量。这 3 个变量称为类的实例变量,变量前的修饰符 private 表明只有该类中才能访问这些变量,而在该类外是不能访问的。另外还声明了一个 public 的构造方法(Time1)和 3 个 public 的成员方法(setTime、toUniversalString、toStandardString),public 表明这些方法可以在类外通过对象调用它们。

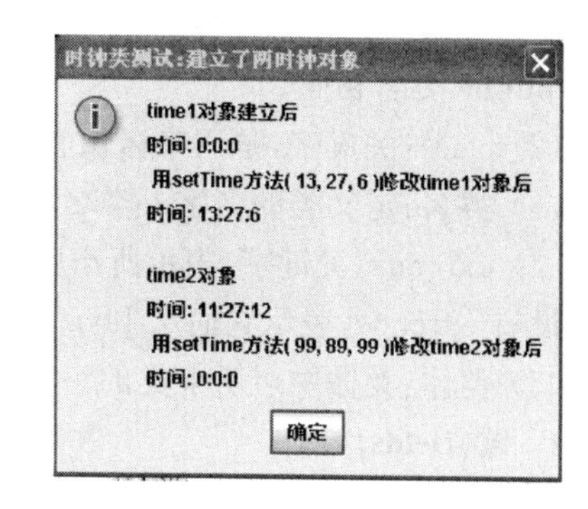

图 4-1　例 4-1 程序运行结果

TimeTest1 是测试主类,创建了两个对象,并调用对象的方法,用于测试 Time1 类的正确性。具体地说,在 TimeTest1 类的 main 方法中,语句

```
Time1 time1 = new Time1();
```

用于创建了 Time1 的第一个对象并赋给了引用变量 time1。创建对象时将自动调用 Time1 类的无参构造方法 Time1()。在构造方法中将调用方法 setTime(0,0,0),以设置 hour、minute 和 second 3 个整数变量的初始值为 0。之后,将通过引用变量 time1 使用第一个对象。

```
String output = "time1 对象建立后\n 时间:" + time1.toString();
```

在字符串的赋值语句中,调用 time1 对象的方法 toString(),返回当前时钟的初始化之后的字符串值,并赋给字符串变量 output。

语句

```
time1.setTime( 13, 27, 6 );
```

是调用 time1 对象的方法 setTime(),重新设置 time1 对象的 hour、minute 和 second 的值。

语句

```
Time1 time2 = new Time1();
```

创建 Time1 类的第 2 个对象,并赋给了引用变量 time2。

```
time2.setTime( 99, 99, 99 );
```

调用 time2 对象的方法 setTime(),重新设置 time2 对象的 hour、minute 和 second 的值都为 99,是一个非法的值。setTime 方法中的条件表达式将会使 hour、minute 和 second 的值设置为 0。

语句

```
JOptionPane.showMessageDialog( null, output,
    "时钟类测试", JOptionPane.INFORMATION_MESSAGE );
```

在对话框中显示字符串变量 output 的值,将得到如图 4-1 所示的程序的运行结果。

4.1.3　类成员:域、方法和构造方法

通过例 4-1 时钟类的实现可以看出,类的定义包含两部分:类的头声明和类体。

定义类的语法格式如下:

```
[类的修饰符] class 类名 [extends 父类名]        //类的头声明
```

```
{
  ...                                        //类体
}
```

上述类的定义中各部分的含义如下。

(1) 修饰符:定义了类的特性。如 Time1 类和 TimeTest1 类中的修饰符为 public,说明这两个类可以被其他的类使用。一个 Java 源文件中最多可能有一个 public 类,而且源文件名必须和 public 类名相同。

(2) class:关键字,是用来标识类的声明。

(3) 类名:定义类的名字。类名必须是合法的 Java 标识符。

(4) extends:关键字,表示所声明的类继承的父类。在 Time1 类和 TimeTest1 类的声明中都没有 extends,表示它们从 Java 的根类 Object 进行继承。

(5) 类体:类体中声明了类的一组成员。类的成员包括域、构造函数和方法。

1. 域(fields)

一个类的数据属性由它的成员变量(域)定义。在类中声明一个域的形式为

```
[域修饰符] 类型 域名;
```

其中,域修饰符常见的有 public、private、protected、static 和 final。这些关键字将在后续章节详细介绍。

类型说明符声明域的类型,它可以是 Java 的基本数据类型(如 int、float、boolean 等),也可以是一个类名,即是一个类的类型。一个具有类的类型的变量在 Java 中称为引用变量。在类中声明的域变量有一个默认的初值。数值型的域变量初值为 0;boolean 型的域变量初值为 false;引用型的域变量初值为 null。

域名必须是合法的 Java 标识符。

2. 方法

方法一般是对类中的数据成员进行操作。如果类中的数据成员是 private 型的,往往定义一个 public 的方法来设置数据成员的值或读取数据成员的值。

3. 构造方法

Time1 类包含一个构造方法 Time1(),构造方法名和类名是一样的。构造方法一般用于初始化某个类的对象。在程序创建 Time1 类的对象时,new 运算符为该对象分配内存,并调用构造方法来初始化该对象,也就是通过调用 Time1 类的构造方法,在构造方法中给对象的各成员变量赋初值。

4.2 类的作用域

在 Java 中,根据类的作用和相关性,将大量的类进行了分组,每组起一个名,称为包名。同一包中的类是可以互相访问的,而不同包中的类能否互相访问要由类的作用域决定。类的作用域定义了类的访问范围。

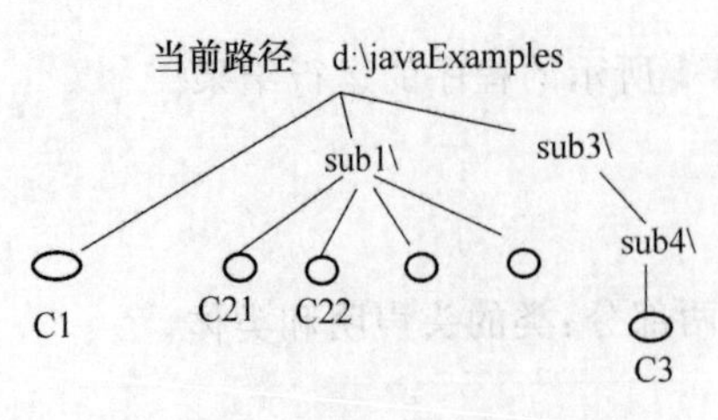

图 4-2 包中的类例子

包物理上对应着操作系统中的目录结构。如图 4-2 中的包的类,假设当前路径 d:\javaExamples 已经放入 classpath 系统环境变量中。类 C1 定义

在无名包中，类 C21、类 C22 定义在相同的包 sub1 中，类 C3 定义在包 sub3、sub4 中。

Java 中用于定义类的作用域的修饰符有两个。

- public(公共作用域)：一个具有公共作用域的类在封装它的包外面是可见的。
- package(包作用域，默认值)：一个具有包作用域的类只是在封装它的包中是可见(用)的，如果一个类声明的前面没有 public 关键字，则此类默认为包作用域。

请看如图 4-2 所示的几个声明类作用域的例子。

①
```
public class C1{
  //…成员
}
```

上述代码声明了类 C1 的作用域为 public 公共作用域，代表 C1 类在包内和包外都可以使用。

②
```
package sub1;
class C21{
  //…成员
}
public class C22 {
//…成员
}
```

上述代码声明类 C21 和类 C22 放在同一个包 sub1 中(关于包的定义和使用将在 4.10 节进行详细介绍)，并声明 C21 作用域是默认的包作用域，代表 C1 只能被和它同在一个包中的其他类所访问(包内使用)；而 C22 类声明为 public 公共作用域。

③
```
package sub3.sub4
public class C3{
  //…成员
}
```

上面的代码声明类 C3 放入包 sub3. sub4 中。C3 类的作用域声明为 public 公共作用域。

上面的例子中，C21 只能被同一包中类(如类 C22)使用，不能被类 C1、C3 访问，因为 C21 是 package 作用域；而类 C1、C21、C3 可以被任何类使用，因为它们是 public 作用域的类。

4.3 成员访问控制

成员访问控制是用来控制类中成员的访问范围，由修饰符声明。表 4-1 列出了常见的成员访问控制修饰符。

表 4-1 成员访问控制修饰符

修饰符	作 用
public	说明公共成员，可以在类外被使用
protected	说明保护成员，在同一包中或子类中被使用
package(缺省值)	说明包作用域成员，该成员只能在同一包中被使用
private	说明私有成员，该成员只能在类中访问

public 的成员组成了类的对外的接口。也就是说，通过对象能访问的成员是类中定义的 public 成员。

通过对象访问的成员格式：

```
对象名.方法名(...);            //调用对象的方法成员
对象名.域;                     //访问对象的数据成员
```

private 成员变量和 private 的成员方法不能在类外被访问。也就是说，private 成员被隐藏在类或对象中。

例 4-2 成员访问控制示例 1。

```
//TimeTest2.java 的代码
public class TimeTest2 {
    public static void main( String args[] ){
        Time1 time = new Time1();
        time.hour = 7;          //错误:hour 是一个 private 的实例变量
        time.minute = 15;       //错误:minute 是一个 private 的实例变量
        time.second = 30;       //错误:second 是一个 private 的实例变量
    }
}
```

在 main 方法中创建了一个 Time1 类的对象 time，并通过 time 对象直接对 private 成员 hour、minute 和 second 设置值，这是不允许的。所以上述程序编译将产生错误信息。

例 4-3 成员访问控制示例 2。

源文件名为 accessModifier.java。定义了 accessModifier 和 accessModifierTest 两个类，在同一包(缺省包)中。在 AccessModifier 类中定义了一个 private 数据成员 a，一个 package 数据成员 b，一个 protected 数据成员 c，一个 public 数据成员 d，一个 public 方法成员 setA 用于给私有成员 a 设置值。在 accessModifierTest 类中，建立 accessModifier 类的一个对象并赋给引用变量 obj，通过引用能访问对象的非 private 成员。accessModifier.java 的代码如下：

```
public class accessModifier {
    private int a;
    int b;
    protected double c;
    public char d;
    public void setA(int a1) {a = a1;}
}
class accessModifierTest {
  public static void main( String args[] ){
    accessModifier obj = new accessModifier();
    obj.setA(1);
    obj.b = 2;
    obj.c = 1.5;
    obj.d = '#';
  }
}
```

4.4 初始化类的对象:构造方法

构造方法用于为类新建的对象分配内存空间和进行变量的初始化，如例 4-1 在 Time1 类中创建了一个不带任何参数的构造方法。在构造方法中对 3 个成员变量进行了初始化。构造方法只能在创建对象时用 new 命令调用。定义构造方法时必须注意两点：

① 构造方法必须与其类名相同；

② 构造方法没有返回类型，但可以有参数，并且可以被重载。

如果一个类中未定义构造方法，则编译时系统提供一个缺省的无参的构造方法，其方法体为空。

例 4-4 构造方法的重载。

该例子包含了两个文件——Time4. java 和 TimeTest4. java。在 Time4. java 中定义了 Time4 类。在 Time4 类中定义了 5 个重载的构造函数。在 TimeTest4. java 中定义了 TimeTest4 应用程序类，在该类中创建了 5 个 Time4 对象，在创建这 5 个 Time4 对象时分别调用 5 个不同的构造方法。

Time4. java 文件代码：

```
//Time4.java
import java.text.DecimalFormat;
public class Time4 {
    private int hour;               // 0 - 23
    private int minute;             // 0 - 59
    private int second;             // 0 - 59
    //5 个重载的构造函数
    public Time4(){
        setTime( 0, 0, 0 );
    }
    public Time4( int h ) {
        setTime( h, 0, 0 );
    }
    public Time4( int h, int m ){
        setTime( h, m, 0 );
    }
    public Time4( int h, int m, int s ) {
        setTime( h, m, s );
    }
    public Time4( Time4 time ){
        setTime( time.hour, time.minute, time.second );
    }
    public void setTime( int h, int m, int s ){
        hour = ( ( h >= 0 && h < 24 ) ? h : 0 );
        minute = ( ( m >= 0 && m < 60 ) ? m : 0 );
        second = ( ( s >= 0 && s < 60 ) ? s : 0 );
    }
    public String toString(){
        DecimalFormat twoDigits = new DecimalFormat( "00" );
        return twoDigits.format( hour ) + ":" +
            twoDigits.format( minute ) + ":" + twoDigits.format( second );
    }
}
```

测试主类 TimeTest4. java 文件代码：

```
//TimeTest4.java
import javax.swing.*;
public class TimeTest4 {
    public static void main( String args[] ){
        Time4 t1 = new Time4();            // 00:00:00
        Time4 t2 = new Time4(2);           // 02:00:00
```

```
            Time4 t3 = new Time4(21,34);        // 21:34:00
            Time4 t4 = new Time4(12,25,42);     // 12:25:42
            Time4 t5 = new Time4(t4);           // 12:25:42
            String output = "对象创建后：" +
                "\n 构造 t1 对象时调用 Time4()，将调用无参构造方法 Time4()，" +
                "\nt1 = " + t1.toString() ;
            output + = "\n 构造 t2 对象时调用 Time4(2)，将调用 Time4(int h)，" +
            " \nt2 = " + t2.toString() ;
            output + = "\n 构造 t3 对象时调用 Time4(21, 34)，将调用 Time4(int h,int m)后，" + "\nt3
= " + t3.toString() ;
            output + = "\n 构造 t4 对象时调用 Time4(12,25,42)，将调用 Time4(int h,int m,int s)后，"
+ " \nt4 = " + t4.toString() ;
            output + = "\n 构造 t5 对象时调用 Time4(t4)，将调用 Time4(Time4 t)后，" +
            "\nt4 = " + t5.toString() ;
            JOptionPane.showMessageDialog( null, output,
                "构造函数的重载", JOptionPane.INFORMATION_MESSAGE );
            System.exit( 0 );
        }
    }
```

程序运行的结果如图 4-3 所示。

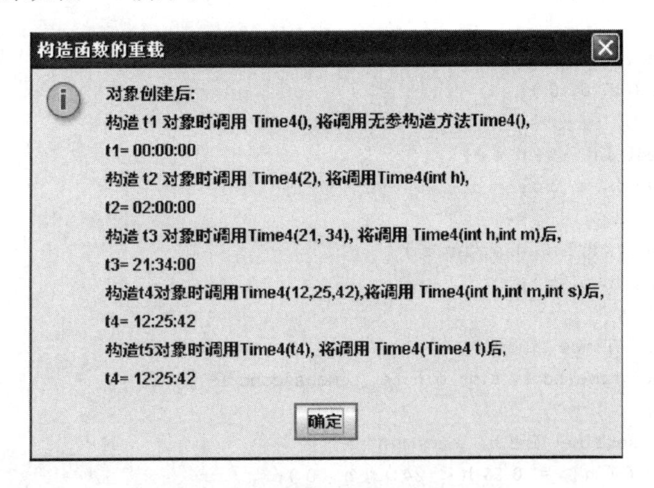

图 4-3　例 4-4 程序运行结果

上述程序中用到类 DecimalFormat，是定义十进制数输出的格式。语句

```
DecimalFormat twoDigits = new DecimalFormat( "00" );
```

创建格式对象定义了输出十进制数的整数部分为两位，不足部分以 0 代替，这使调用方法 twoDigits. format(hour)输出 2 位。

4.5　this

this 定义为被调用的方法的当前对象的引用。this 引用仅能出现在类中的方法体中。

例 4-5　this 的使用。

该例子说明如何隐式地和显式地使用 this 引用。其中 SimpleTime 类的构造方法的 3 个参数名与 SimpleTime 类 3 个成员变量名相同。在这种情况下，局部变量在该方法作用域中屏蔽了同名的域。可以使用 this 来显式地引用屏蔽的域。通过 this 不但可以引用成员变量，也可引用方法，包括构造函数。

```
//ThisTest.java
import javax.swing.*;
import java.text.DecimalFormat;
public class ThisTest {
    public static void main( String args[] ){
        SimpleTime time = new SimpleTime( 12, 30, 19 );
        JOptionPane.showMessageDialog( null, time.toString(),
            "this 的使用",
            JOptionPane.INFORMATION_MESSAGE );
        System.exit( 0 );
    }
}
class SimpleTime {
    private int hour;
    private int minute;
    private int second;
    public SimpleTime( int hour, int minute, int second ){
        this.hour = hour;          //显式地使用 this 访问对象的 hour 变量
        this.minute = minute;      //显式地使用 this 访问对象的 minute 变量
        this.second = second;      //显式地使用 this 访问对象的 second 变量
    }

    public String toString(){
        DecimalFormat twoDigits = new DecimalFormat( "00" );
            //显式地使用 this 访问对象的 hour 变量
        return twoDigits.format( this.hour ) + ":" +
            //显式地使用 this 访问对象的 minute 变量
        twoDigits.format( this.minute ) + ":" +
            //显式地使用 this 访问对象的 second 变量
        twoDigits.format( this.second );
    }
}
```

程序运行的结果如图 4-4 所示。

图 4-4　例 4-5 程序运行结果

4.6　使用 set 和 get 方法

如果类中声明了 private 成员变量，则只能通过该类的方法来操作这些成员变量。类中往往提供 public 型方法以允许通过该类的对象，设置或者读取 private 型的实例变量。设置或读取一个数据成员 varible 的方法名通常命名为 setVarible()和 getVarible()。

例 4-6 set 和 get 方法的使用。

例 4-6 包含文件 Time2. java，定义了 Time2 类，该类包含 hour、minute、second 三个整数类型的 private 成员变量。另外，该类还包含了 setTime、setHour、setMinute、setSecond 四个 set 方法和 getHour、getMinute、getSecond 三个 get 方法。

```
//Time2. java
import java. text. DecimalFormat;
public class Time2 {
    private int hour;
    private int minute;
    private int second;
    public Time2(){
        setTime( 0, 0, 0 );
    }
    public Time2( int h, int m, int s ){
        setTime( h, m, s );
    }
    public void setTime( int h, int m, int s ){           //setTime 方法
        setHour( h );
        setMinute( m );
        setSecond( s );
    }
    public void setHour( int h ){                         //setHour 方法
        hour = ( ( h >= 0 && h < 24 ) ? h : 0 );
    }
    public void setMinute( int m ){                       //setMinute 方法
        minute = ( ( m >= 0 && m < 60 ) ? m : 0 );
    }
    public void setSecond( int s ){                       //setSecond 方法
        second = ( ( s >= 0 && s < 60 ) ? s : 0 );
    }
    public int getHour(){                                 //getHour 方法
        return hour;
    }
    public int getMinute(){                               //getMinute 方法
        return minute;
    }
    public int getSecond(){                               // getSecond 方法
    return second;
    }
    public String toUniversalString(){
        DecimalFormat twoDigits = new DecimalFormat( "00" );
        return twoDigits. format( getHour() ) + ":" +
            twoDigits. format( getMinute() ) + ":" +
            twoDigits. format( getSecond() );
    }
    public String toStandardString(){
        DecimalFormat twoDigits = new DecimalFormat( "00" );
        return ( ( getHour() == 12 || getHour() == 0 ) ? 12 : getHour() % 12 )
  + ":" + twoDigits. format( getMinute() ) + ":" + twoDigits. format( getSecond() ) + ( getHour()
< 12 ? " AM" : " PM" );
    }
}
```

读者可以自己编写使用该类的应用程序测试主类。

初看起来，提供设置和读取方法的能力，在本质上等同于将变量定义为 public 变量。然而读取方法控制着客户访问该变量时的数据正确性。一个公有的 set 方法可以仔细检查对变量值所做的修改，确保新值适合于该数据项。例如，试图将一个月设置为 37 天的做法将遭到拒绝。因此，使用 set 和 get 方法提供对 private 数据的访问，并对访问的数据的正确性进行了检查。

4.7 垃圾收集

Java 平台允许创建任意个对象(当然会受到系统资源的限制)，而且当对象不再使用时会被自动清除，这个过程就是所谓的"垃圾收集"。当一个对象不再被引用的时候，这个对象就会被视为垃圾，由 Java 垃圾收集器(garbage collector)回收该对象占有的空间。指向某一对象的引用的值为 null，将使对象不再被引用。例如：

```
Time1 time = new Time1();
time = null;
```

1. 垃圾收集器

Java 有一个垃圾收集器，它周期性地将不再被引用的对象从内存中清除。这个垃圾收集器是自动执行的，它扫描对象的动态存储区并将引用的对象作标记，将没有引用的对象作为垃圾收集起来。定期地释放那些不再需要的对象所使用的内存空间。

也可以在任何时候在 Java 程序中通过调用 System. gc()方法来显式地运行垃圾收集程序。比如，在创建大量垃圾的代码之后或者在需要大量内存代码之前运行垃圾收集器，垃圾收集器从内存中清除不再被引用的对象。

2. 撤销方法 finalize

在一个对象得到垃圾收集器的收集之前，垃圾收集器给对象一个机会来调用自己的finalize 方法。这个过程就是所说的最终处理(finalize)。finalize 方法的头为：

```
void finalize() {…}
```

例 4-7 将使用 finalize 方法说明垃圾收集程序何时被执行。

4.8 static 方法和域

当声明一个成员变量时，可以指定成员变量是属于一个类的所有对象共享(称为类的变量)，还是属于一个类的各个对象所拥有(称为实例变量)。属于类的变量或方法被称为静态变量或静态方法，以关键字 static 加在变量声明或方法声明的前面。例如：

```
static private int count;
static public int getCount() {…}
```

上面语句分别声明的私有的静态变量 count 和公共的静态方法 getCount()。类的公共静态成员的一般使用格式如下：

```
类名.静态变量名;
类名.静态方法名();
```

例如，Java 的 Math 类中定义了一组静态的 public 型的数学函数。这些数学函数都是通过类名调用的。例如：Math. random()是调用随机数函数。

当在类的静态方法中出现成员变量时，要求成员变量必须是静态的。

例 4-7 静态方法和静态域的应用。

例 4-7 声明了两个类:Employee 雇员类和 EmployeeTest 测试类。Employee 类声明了一个 private 静态变量 count 和 public 静态方法 getCount。类变量 count 保存当前驻留在内存中的 Employee 类的所有对象的数目。这包括已经标记为垃圾收集但尚未回收内存的对象。Employee 类有一个 finalize 方法,该方法说明程序何时执行垃圾收集程序。

Employee.java 文件代码:

```
public class Employee {
    private String firstName;
    private String lastName;
    //定义一个静态变量表示内存中 Employee 类对象的数目
    private static int count = 0;
    public Employee( String first, String last ){     //构造函数
        firstName = first;
        lastName = last;
        ++count;                                      //将静态变量加 1
        System.out.println( "Employee constructor: " +
            firstName + " " + lastName );
    }
    protected void finalize(){                        //定义 finalize 方法
        --count;  //撤销对象时将静态变量 count 的值减 1
        System.out.println( "Employee finalizer: " +
            firstName + " " + lastName + "; count = " + count );
    }
    public String getFirstName(){
        return firstName;
    }
    public String getLastName(){
        return lastName;
    }
    //定义静态方法 getCount,取 count 静态变量的值
    public static int getCount(){
        return count;
    }
}
```

EmployeeTest.java 文件代码:

```
import javax.swing.*;
public class EmployeeTest {
    public static void main( String args[] ){
        String output = "Employees before instantiation: " +
            Employee.getCount();
        Employee e1 = new Employee( "Susan", "Baker" );
        Employee e2 = new Employee( "Bob", "Jones" );
        output += "\n\nEmployees after instantiation: " +
            "\nvia e1.getCount(): " + e1.getCount() +
            "\nvia e2.getCount(): " + e2.getCount() +
            "\nvia Employee.getCount(): " + Employee.getCount();
        output += "\n\nEmployee 1: " + e1.getFirstName() +
        " " + e1.getLastName() + "\nEmployee 2: " +
            e2.getFirstName() + " " + e2.getLastName();
        e1 = null;
```

```
        e2 = null;
        System.gc(); // 显式地运行垃圾收集程序
        output += "\n\nEmployees after System.gc(): " +
            Employee.getCount();
        JOptionPane.showMessageDialog( null, output,
            "静态方法和域", JOptionPane.INFORMATION_MESSAGE );
        System.exit( 0 );
    }
}
```

程序运行的结果如图 4-5 所示。

```
Employee constructor: Susan Baker
Employee constructor: Bob Jones
Employee finalizer: Susan Baker; count = 1
Employee finalizer: Bob Jones; count = 0
```

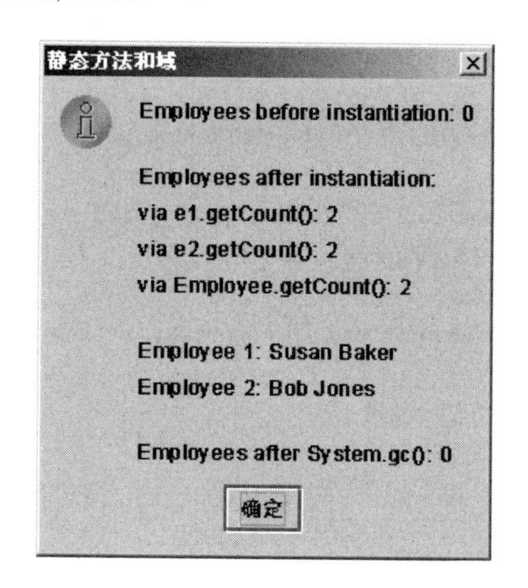

图 4-5　例 4-7 程序运行结果

4.9　类的组合

类的组合是指类之间的一种"has a"的关系，即一个类中的成员可以是其他类（或自身）的引用，把类之间的这种关系叫做类的组合（Composition）。

例 4-8　类的组合。

本例子说明了类的组合，即一个类可以包含其他类的对象的引用。该例子包含了 3 个类：Date、Employee 和 EmployeeTest。Employee 类包含了 4 个成员变量：firstName、lastName、birthDate 和 hireDate。其中两个成员变量 birthDate 和 hireDate 都是 Date 对象的引用。注意，Employee 类并没有导入 Date 类，这是因为这两个类位于同一个包（默认包）中。

Date.java 文件的代码如下：

```
//Date.java
public class Date {
    private int month;
    private int day;
    private int year;
    public Date( int theMonth, int theDay, int theYear ){
        month = checkMonth( theMonth );
```

```java
        year = theYear;
        day = checkDay( theDay );
        System.out.println( "Date object constructor for date " +
            toDateString() );
    }
    private int checkMonth( int testMonth ){                //检测月份的合法性
        if ( testMonth > 0 && testMonth <= 12 )
            return testMonth;
        else {
            System.out.println( "Invalid month (" + testMonth +
                ") set to 1." );
            return 1;
        }
    }
    private int checkDay( int testDay ){                    //检测日期的合法性
        int daysPerMonth[] =
            { 0, 31, 28, 31, 30, 31, 30, 31, 31, 30, 31, 30, 31 };
        if ( testDay > 0 && testDay <= daysPerMonth[ month ] )
            return testDay;
        if ( month == 2 && testDay == 29 && ( year % 400 == 0 ||
            ( year % 4 == 0 && year % 100 != 0 ) ) )
            return testDay;
        System.out.println( "Invalid day (" + testDay + ") set to 1." );
        return 1;
    }
    public String toDateString(){
        return month + "/" + day + "/" + year;
    }
}
```

Employee.java 文件代码：

```java
//Employee.java
public class Employee {
    private String firstName;
    private String lastName;
    private Date birthDate;           //birthDate 的成员变量为 Date 对象的引用
    private Date hireDate;            //hireDate 的成员变量为 Date 对象的引用
    public Employee(String first,String last,Date dateOfBirth,Date dateOfHire){
        firstName = first;
        lastName = last;
        birthDate = dateOfBirth;
        hireDate = dateOfHire;
    }
    public String toEmployeeString(){
        return lastName + ", " + firstName +
            " Hired: " + hireDate.toDateString() +
            " Birthday: " + birthDate.toDateString();
    }
}
```

EmployeeTest.java 文件的代码如下：

```java
//EmployeeTest.java
import javax.swing.JOptionPane;
```

```
public class EmployeeTest {
    public static void main( String args[] ){
        Date birth = new Date( 7, 24, 1949 );
        Date hire = new Date( 3, 12, 1988 );
        Employee employee = new Employee( "Bob", "Jones", birth, hire );
        JOptionPane.showMessageDialog( null, employee.toEmployeeString(),
            "类的组合", JOptionPane.INFORMATION_MESSAGE );
        System.exit( 0 );
    }
}
```

程序运行的结果如图 4-6 所示。

```
Date object constructor for date 7/24/1949
Date object constructor for date 3/12/1988
```

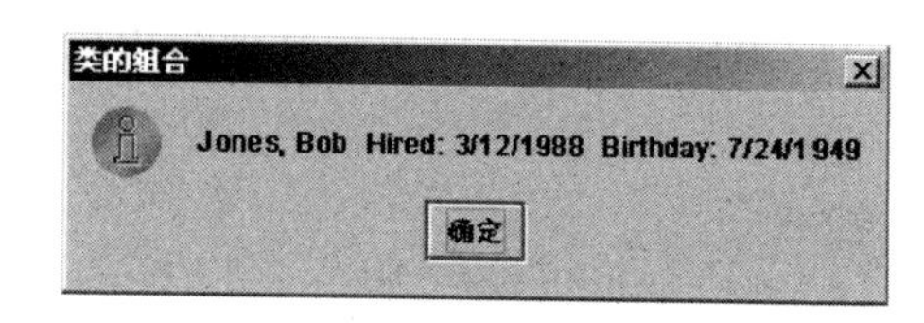

图 4-6　例 4-8 程序运行结果

4.10　Package 包的创建和访问

Java 中的包是相关类和接口的集合，就像文件夹或目录把各种文件组织在一起，使硬盘保存的内容更清晰、更有条理一样。Java 中的包把相关的类组织在一起，并起一个名字，称为包名。

包是组织系统类的单位，使类按功能、来源分为不同的集合，以便组织管理及使用。由于同一包中的类可以互相访问，为了方便编程和管理，通常把需要在一起协同工作的类和接口放在一个包中。

Java 的类库被划分为若干个不同的包，每个包中都有若干个具有特定功能和相互关系的类和接口。下面列出了一些经常使用的 Java 包。

- java.lang：基本语言类，程序运行时自动引入；
- java.io：所有的输入输出类；
- java.util：实用的数据类型类；
- java.awt：构建图形用户界面(GUI)的类；
- java.awt.image：处理和操纵网上图片的工具类；
- java.awt.peer：实现与平台无关的 GUI 的类；
- java.applet：实现 Java Applet 的工具类；
- java.net：实现网络功能的类。

要使用不同包的类，需先从包中导入类，然后再使用。

4.10.1　包的创建

建立一个包时，使用 package 语句。package 语句必须是整个 Java 文件的第一个语句。

创建 package 语句格式为：

```
package 包名；
```

表示在当前目录下创建一个有名字的包，并且该 java 源文件中定义的所有类都隶属于这个有名包中。当一个类所在的源文件无包创建语句时，系统会给源文件创建一个无名包(缺省包)，该 java 文件中定义的所有类都隶属于这个无名包。

例如，在当前目录下创建一个包，包名为 cn. edu. sytu. it. ch04，Time1 放在此包中。

```
package cn.edu.sytu.it.ch04;
public class Time1 {…}
…
```

4.10.2 访问包中的类

import 语句用于导入用户的源代码文件中使用的其他包中的类，这些类和当前类不在同一个包中。例如：

```
import cn.edu.sytu.it.ch04.Time1;
```

表示导入当前目录下的子包 cn. edu. sytu. it. ch04 中一个类 Time1。

```
import cn.edu.sytu.it.ch04.*;
```

* 表示导入整个包，即导入整个子包 cn. edu. sytu. it. ch04 中的所有类，但不包括 cn. edu. sytu. it. ch04 中的子包。

如果没有使用 import 语句导入某个包，但又想使用此包中的某个类，也可以直接在所需要的类名前加上包名前缀。

例如，要使用类 Time1 类创建一个对象 time，语句如下：

```
cn.edu.sytu.it.ch04.Time1 time = new cn.edu.sytu.it.ch04.Time1();
```

Java 的运行系统总是要装入预定义的包，运行系统总是为用户自动装入 java. lang 包。所以 java. lang 包不需要用 import 语句导入源代码文件中。

例 4-9 包的创建及使用。

此例说明了如何创建用户自定义的包，及如何在程序中使用自定义的包中的类。该程序定义两个源文件：Time1. java 和 TimeTest1. java。在 Time1. java 中，用 package 创建了一个包名 cn. edu. sytu. it. ch04，类 Time1 放在包 cn. edu. sytu. it. ch04 中。在 TimeTest1. java 文件中，TimeTest1 类放在当前目录下的无名包中。在当前目录下的 TimeTest1 类中要使用当前目录下的子包 cn. edu. sytu. it. ch04 中的类 Time1 时(因为它们属于不同的包)，则在 TimeTest1. java 文件中必须通过 import 语句导入 Time1 类。

Time1. java 文件代码：

```
package cn.edu.sytu.it.ch04;                //包的声明语句
import java.text.DecimalFormat;
public class Time1 extends Object {
  …//与例 4-1 中代码相同
}
```

TimeTest1. java 文件代码：

```
import javax.swing.JOptionPane;
import cn.edu.sytu.it.ch04.Time1;              //导入包中的类
public class TimeTest1 {
    public static void main( String args[] ){
```

```
        Time1 time = new Time1();
        String output = "The initial universal time is: " +
            time.toUniversalString() + "\nThe initial standard time is: " +
            time.toStandardString();
        time.setTime( 13, 27, 6 );
        output += "\n\nUniversal time after setTime is: " +
            time.toUniversalString() +
            "\nStandard time after setTime is: " + time.toStandardString();
        time.setTime( 99, 99, 99 );
        output += "\n\nAfter attempting invalid settings: " +
            "\nUniversal time: " + time.toUniversalString() +
            "\nStandard time: " + time.toStandardString();
        JOptionPane.showMessageDialog( null, output,
            "包的创建及使用", JOptionPane.INFORMATION_MESSAGE );
        System.exit( 0 );
    }
}
```

对于上面的 Time1.java 文件在编译时,编译器将在当前目录下创建一个与包名相一致的目录结构,即在当前目录(如 d:\javaExamples)创建子目录 cn/edu/sytu/it/ch04,并且把编译后产生的相应的类文件 Time1.class 放在这个子目录中。而在文件 TimeTest1.java 编译时,类 TimeTest1 放在当前目录下中。

程序运行的结果如图 4-7 所示。

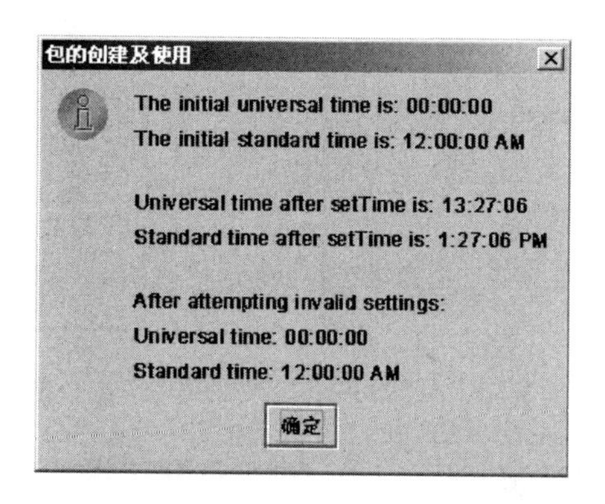

图 4-7　例 4-9 程序运行结果

4.10.3 MyEclipse 环境中包的创建和访问

下面以完成前面的例 4-9 为任务,说明在 MyEclipse 环境下包的创建和访问的过程。

在 MyEclipse 环境下,由于一个包必须放在指定的项目中,所以在包的创建之前先要创建一个 Java Project。

1. 创建一个名字为 ch4 的 Java 项目

进入 MyEclipse 主界面,单击"File—>New—>Java Project",创建一个 Java Project,项目名为 ch4。

2. 创建一个包名为 cn.edu.sytu.it.ch04

单击"File —> New—> Package",进入如图所示 4-8 的输入包名的界面,单击 Source folder 右边的"Browser"按钮,弹出选择 Source folder 的界面,单击文件夹 ch4 下的 src,如

图 4-9所示，单击“ok”，返回先前的界面，如图 4-10 所示，在 Name 中输入包名“cn. edu. sytu. it. ch04”，单击“Finish”即可。这时在 Package View 视图界面，项目 ch4 的 src 下面可看到已经创建好的包名，如图 4-11 所示。

3. 在 cn. edu. sytu. it. ch04 包中创建类 Time1

单击项目 ch4 下的包名“cn. edu. sytu. it. ch04”，单击“File —> New—> Class”，出现输入类名的界面，在 Name 中输入类名“Time1”，单击“Finish”后，这时在 Package View 视图界面可以看到在包“cn. edu. sytu. it. ch04”的下面出现“Time1. java”文件名，在右边的编辑窗口输入 Time1. java 的源文件内容。

4. 在项目的根目录下创建 TimeTest1 类

单击项目 ch4 下的“src”，单击“File —> New—> Class”，出现输入类名的界面，在 Name 中输入类名“TimeTest1”，单击“Finish”后，这时在 Package View 视图界面可以看到在 src 的缺省包名“default package”下出现“TimeTest1. java”文件名，如图 4-11 所示，在右边的编辑窗口输入 TimeTest1. java 的源文件内容。

选中 “TimeTest1. java”，右击将弹出子菜单中，单击“Run As－>Java appication”将执行主类文件 TimeTest1. class。

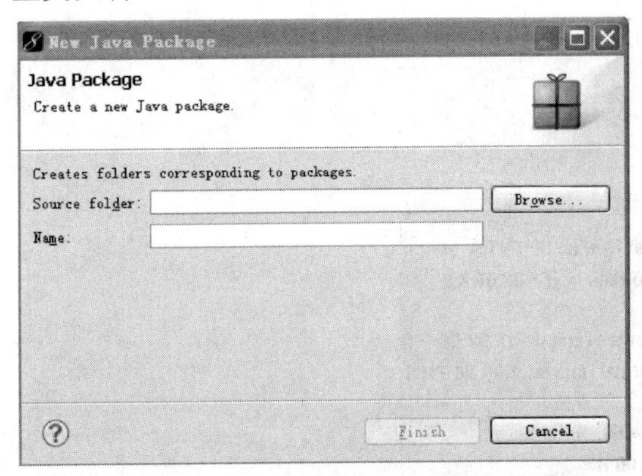

图 4-8　输入包名的界面 1

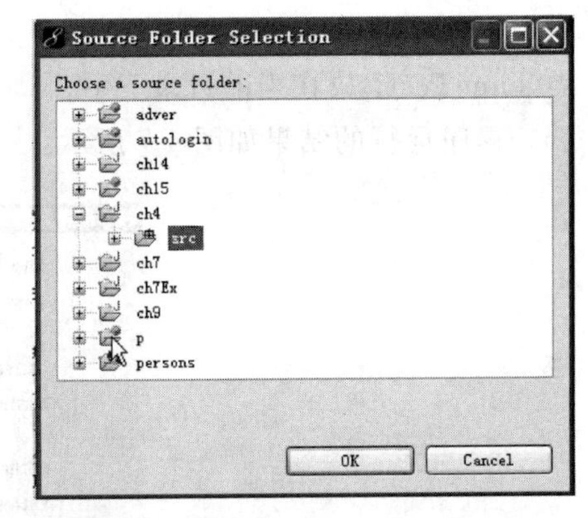

图 4-9　选择父包名的界面

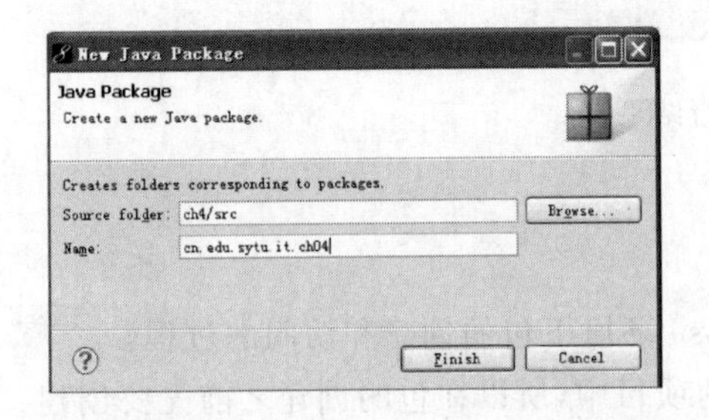

图 4-10　选择父包名的完成界面

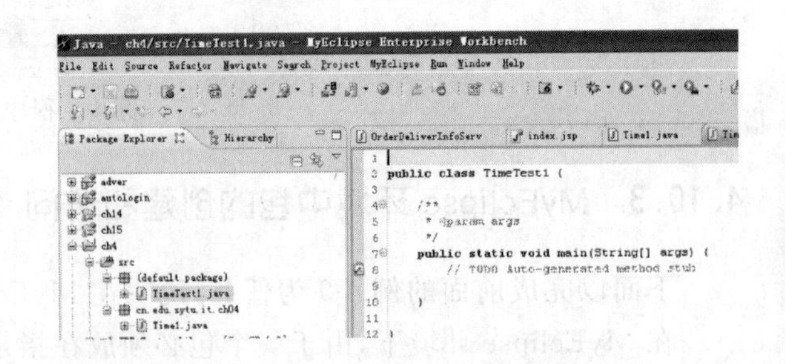

图 4-11　ch4 项目中包创建完成界面

4.10.4　导入 static 成员

JDK1. 5 增加了 import static 语句，可导入某个类的某个静态成员或所有静态成员的功能。import static 语句的格式如下：

import static 包名. 类名. 静态成员名；

上述将导入指定包中的类的静态成员，此成员可为静态数据成员或静态方法。而语句：

```
import static 包名.类名.*;
```

将导入指定包中的类的所有静态成员。

在程序中使用 import static 语言导入的静态成员时，可以直接用类中定义的静态成员，前缀的类名可以不必写了。

例 4-10 应用 import static 的实例。

```
// Using static import to import static methods of class Math.
import static java.lang.Math.*;
public class StaticImportTest
{  public static void main( String args[] )
   {
      System.out.printf( "sqrt( 900.0 ) = %.1f\n", sqrt( 900.0 ) );
      System.out.printf( "PI = %.1f\n", PI);
      System.out.printf( "log( E ) = %.1f\n", log( E ) );
      System.out.printf( "cos( 0.0 ) = %.1f\n", cos( 0.0 ) );
   } // end main
} // end class StaticImportTest
```

程序运行输出结果如下：

```
sqrt( 900.0 ) = 30.0
PI = 3.1
log( E ) = 1.0
cos( 0.0 ) = 1.0
```

图 4-12 例 4-10 程序输出结果

4.11 小结

本章介绍了面向对象程序设计的重要概念——类和对象。类是一种抽象的数据类型，用来表示具有相同性质的一类事物。类封装了对象的属性和方法。类的实例化产生了对象。

一个类的成员包括域、方法和构造函数。方法和构造函数都可以进行重载。为了使一个类的所有对象共享同一个变量的值，可以将成员变量定义成静态(static)的。为了访问静态的 private 成员变量，必须通过 public 静态的成员方法进行访问。

在定义类时可以通过访问控制符来限定其他类对该类及类的成员的访问。类和类之间的访问也可通过包来进行控制。

习题

4.1 修饰类的访问控制符有哪些？修饰类中成员的访问控制符有哪些？各有什么含义作用？请举例说明之。

4.2 构造方法特殊在哪里？构造方法什么时候执行？被谁调用？

4.3 关键字 static 可以修饰类的哪些组成部分？实例变量与类变量的区别是什么？

4.4 静态属性有什么特点？类的对象可以访问或修改静态属性吗？

4.5 构造方法重载的特点是什么？有什么作用？

4.6 什么是包？如何创建包？包物理上对应什么操作系统中的什么实体？

4.7　试写出创建一个名为 MyPackage 包的语句,这个语句应该放在程序的什么位置?

4.8　试写出引用 MyPackage 包中所有类的语句以及引用 MyPackage 包中的一个类 MyClass1 的语句。

4.9　阅读下面的程序,说明它们的输出。

```
class MyClass1{
    int data;
    MyClass1(int d){
        data = d;
    }
    int getData(){
        return data;
    }
    void setData(int d){
        data = d;
    }
}
public class UseRef{
    public static void main(String args[]){
        MyClass1 myobj,myref;
        myobj = new MyClass1( - 1);
        myref = myobj;
        System.out.println("the original data is:" + myobj.getData());
        myref.setData(10);
        System.out.println("now the data is:" + myobj.getData());
    }
}
```

4.10　编写一个类实现秒表的功能。要求至少实现开始计时、停止计时、分针秒针归零、获得分针示数、获得秒针示数、获得总时间的毫秒数等功能。

4.11　编写一个类实现复数的运算。要求至少实现复数相加、复数相减、复数相乘等功能。

4.12　编程创建一个 Box 类。要求:定义 3 个实例变量分别表示立方体的长、宽和高,定义一个构造方法对这 3 个变量进行初始化,然后定义一个方法求立方体的体积。创建一个对象,求给定尺寸的立方体的体积。

4.13　学生类的创建和使用。

(1) 创建一个 student 类,包括的域有学号、班号、姓名、性别、年龄等,且都是 private 类型。

(2) 声明一个构造方法,以初始化对象的所有的域。

(3) 声明分别获得各属性(学号,班号,姓名,性别,年龄)的各个 public 方法。

(4) 声明修改各属性(学号,班号,姓名,性别,年龄)的各个 public 方法。

(5) 声明一个为 public 型的 toString()方法,把该类中的所有域信息组合成一个字符串。

(6) 在类中声明统计班级总人数的私有域 count 得到班级总人数的 public 方法(可在构造方法中进行 Student 对象个数的增加)。

(7) 将类 Student 放在子包 student 中。

(8) 在子包 student 外,创建测试类 Student 的主类。在主类中,使用 Student 类创建两个 Student 对象,输出对象的所有域信息;修改对象的姓名和年龄,修改后显示各对象的姓名和年龄;比较两个 Student 对象的年龄大小,输出年龄较大的 Student 对象。

4.14　定义一个学生类和一个课程类,要求能够表示现实生活中一个学生可以选择多门课程的情况。编写程序进行测试。

4.15　定义一个方程类,以求方程的两个实数根。请将测试主类和方程类定义在不同的包中。

第5章 类的继承和派生

本章讨论类的继承机制。继承(Inheritance)是从已有的类中派生出新的类。通过定义派生类来使用继承技术。定义派生类时,可以通过指定类的成员访问修饰符以控制对类的成员访问;对超类的构造函数和 finalize 方法进行调用,还可对超类的方法进行重新书写。

5.1 继承的概念和软件的重用性

继承是面向对象最显著的一个特性。继承是从已有的类中派生出新的类,新的类能吸收已有类的数据属性和行为,并能扩展新的能力。类和类之间的继承关系可以用 UML 符号表示,如图 5-1 所示,其中父类又叫超类或基类,子类又叫派生类。父类是子类的一般化,子类是父类的特化(具体化)。如表 5-1 所示,列出了几个超类和子类的实际例子。

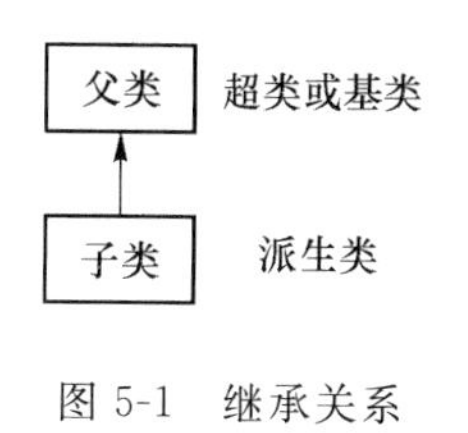

图 5-1 继承关系

表 5-1 继承例子

超类	子类
学生	研究生、本科生、小学生
形状	三角形、圆、矩形
雇员	教师、医生、职员
交通工具	轿车、卡车、公交车

继承是类之间的一种"is-a"的关系。子类可看成是超类的一个特例。子类的对象可当成超类对象。但反过来,不能把超类对象当成子类对象。例如,轿车可看成是交通工具,但不能把交通工具看成就是轿车。而"has-a"关系代表类之间的组合(参见 4.9 节)。在"has-a"关系中一个对象包含一个或多个其他对象的引用成员。例如,轿车由方向盘、轮子、发动机等组成。

继承分为单继承和多继承。单继承是指一个子类最多只能有一个父类。多继承是一个子类可有二个以上的父类。由于多继承会带来二义性,在实际应用中应尽量使用单继承。Java 语言中的类只支持单继承,而接口支持多继承。Java 多继承的功能是通过接口方式来间接实现的。

继承使软件的代码得到重用。在继承关系中,子类通过吸收已有类的数据(属性)和方法(行为),并增加新功能或修改已有功能来创建新类。软件的重用性不仅节省了程序的开发时间,还促进了经过验证和调试的高质量软件的重用,这加快了实现系统的效率。

在 Java 中,Object 类定义和实现了 Java 系统所需要的众多类的共同行为,它是所有类的根类,即所有的类都是由这个类继承、扩充而来的。

5.2 派生类的定义

派生类定义的一般格式为：

```
[类的修饰符] class 子类名 extends 父类名{
    成员变量定义；
    成员方法定义；
}
```

派生类的定义中，用关键字 extends 来明确指出它所继承的超类。

例 5-1 通过继承来定义派生类。

汽车类 Automobile 为父类，卡车类 Truck 为子类。Automobile 类定义了所有汽车具有的共同属性——牌照(Number)——和领取牌照(setNumber)、显示牌照(showNumber)的两个方法。

```
class Automobile{                //超类：汽车类
    int Number;
    void setNumber(int Num){
        Number = Num;
    }
    void showNumber(){
        System.out.println("Automobile Number:" + Number);
    }
}
class Truck extends Automobile{          //派生类：卡车类
    int capacity;
    void setCapacity(int truckCapacity){
        capacity = truckCapacity;
    }
    void showCapacity(){
        System.out.println("Truck Capacity:" + capacity);
    }
}
class AutomobileExtends{          //主类
    public static void main(String args[]){
        Truck tc = new Truck();
        tc.setNumber(8888);     //派生类使用从超类中继承的方法 setNumber
        tc.setCapacity(10);
        tc.showNumber();        //派生类使用从超类中继承的方法 showNumber
        tc.showCapacity();
    }
}
```

该程序运行的结果为：

```
Automobile Number:8888
Truck Capacity:10
```

上述代码中，Truck 类除了具有 Automobile 的成员变量和方法外，还增加了新的成员变量——载重量(capacity)，包括设置载重量(setCapacity)、显示载重量(showCapacity 两个方法。

Truck 子类定义中虽然没有 setNumber、showNumber 两个方法。但通过继承，可将超类

Automobile 中的这两个方法继承过来，所以 Truck 类的对象像使用自己的方法一样使用继承过来的超类的方法。通过这个例子也可以看出继承带来的好处。

5.3 作用域和继承

第 4 章讨论了成员访问控制修饰符 public、private、package 和 protected。

超类的 public 成员既可以在超类中使用，也可以在子类使用，程序可以在任何地方访问 public 超类成员。

超类的 private 成员仅在超类中使用，在子类中不能被访问。

超类的 protected 成员提供了一种介于 public 与 private 之间的访问级别。protected 超类成员可以被子类和同一包内的类所访问。

子类从超类继承类的成员时，超类的所有 public 和 protected 成员在子类中，都保持它们原有的访问修饰符。例如，超类的 public 成员成为子类的 public 成员。超类的 protected 成员也会成为子类的 protected 成员。子类只能通过超类所提供的非 private 方法来访问超类的 private 成员。

5.4 方法的重新定义

如果在子类中定义的某个方法与父类的某个方法有相同方法署名（方法头），则称子类重新定义或重写（overriding）了父类的该方法。子类的对象调用这个方法时，将使用子类中定义的方法，对它而言，父类中定义的方法就“看不见”了。如果在子类的方法中要调用超类的这个被重写的方法，可以用关键字 super 实现，具体的格式如下：

```
super. 超类同名方法(…);
```

即在超类方法名之前加上关键字 super 和点运算符(.)，就可以在子类中访问超类同名方法。

例 5-2 方法的重写。

该例子说明了继承和方法的重新定义。点可以用 x 和 y 一对坐标来表示。而圆可以用表示圆心的一个点和半径来表示。所以，将圆类（Circle）声明为点类（Point）的子类。该例子的继承层次如图 5-2 所示。程序还包括一个 CircleTest 类，是应用程序主类。

Point 类的设计如下。

成员变量：

```
int x,y
```

成员方法：

```
setX(int),getX(),setY(int),getY(),toString();
Point(),Point( int xValue, int yValue )
```

Circle 类的设计如下。

成员变量：

```
int x,y                                //继承自 Point 类
double radius
```

成员方法：

```
setX(int),getX(),setY (int),getY()     //继承自 Point 类
setRadius(double),getRadius(),getDiameter(),
```

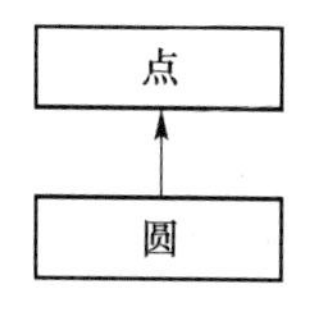

图 5-2 点-圆继承层次

```
getCircumference(),getArea()
toString()                          //重写父类同名方法
Circle(),Circle( int xValue, int yValue, double radiusValue )
```

父类 Point 类中有 toString 方法，子类 Circle 类中也有同名 toString 方法。在子类 Circle 中对父类中的 toString 方法进行了重写。这样就可以使 Point 类和 Circle 类的同名方法 toString 显示输出的内容不一样。

Point. java 的代码如下：

```
public class Point {
  protected int x;
  protected int y;
  public Point(){
  }
  public Point( int xValue, int yValue ){
    x = xValue;
    y = yValue;
  }
  public void setX( int xValue ){
    x = xValue;
  }
  public int getX(){
    return x;
  }
  public void setY( int yValue ){
    y = yValue;
  }
  public int getY(){
    return y;
  }
  public String toString(){
    return "[" + x + "," + y + "]";
  }
}
```

Circle. java 的代码如下：

```
public class Circle extends Point {
    private double radius;
    public Circle(){
    }
    public Circle( int xValue, int yValue, double radiusValue ){
      x = xValue;
      y = yValue;
      setRadius( radiusValue );
    }
    public void setRadius( double radiusValue ){
      radius = ( radiusValue < 0.0 ? 0.0 : radiusValue );
    }
    public double getRadius(){
      return radius;
    }
    public double getDiameter(){
      return 2 * radius;
```

```
    }
    public double getCircumference(){
      return Math.PI * getDiameter();
    }
    public double getArea(){
      return Math.PI * radius * radius;
    }
    public String toString(){//重写了超类 Point 类中的 toString 方法
      return "Center = "
        //通过 super 调用超类中的被重写的 toString 方法
        + super.toString() + " Radius = " + radius;
    }
}
```

CircleTest.java 的代码如下：

```
import javax.swing.JOptionPane;
public class CircleTest {
    public static void main( String[] args ){
        Point point = new Point(20,30);
        Circle circle = new Circle( 37, 43, 2.5 );
        String output = "The location of point are: " + point.toString();
        output += "\nThe location and radius of circle are\n" + circle.toString();
        JOptionPane.showMessageDialog( null, output );
        System.exit( 0 );
    }
}
```

程序运行的结果如图 5-3 所示。

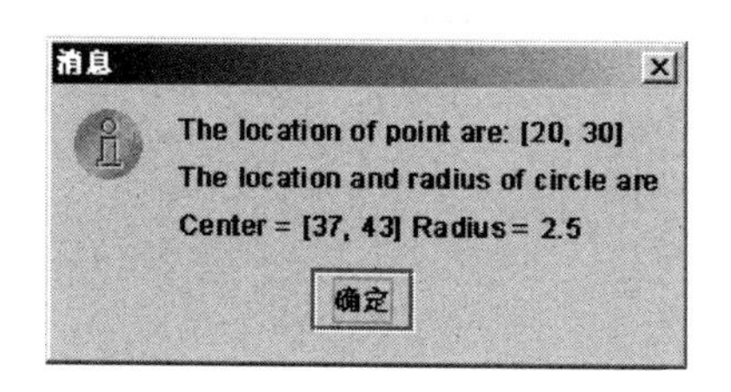

图 5-3　例 5-2 程序运行结果

5.5　继承下的构造方法和 finalize 方法

一组类之间通过多重继承，就形成类似一棵树的继承层次结构。继承层次下的构造方法是如何调用的呢？子类构造方法在执行自己的任务之前，子类对象的实例化过程开始于一系列的构造方法的调用。

1. 继承层次结构下的构造方法的调用次序

在创建子类对象时，必须先调用直接父类的构造方法，然后才调用子类本身的构造方法。调用直接父类的构造方法可显式地〔子类方法体中的第一条语句是 super(…)〕或隐式地〔子类方法体中无语句 super(…)时，将调用直接父类的默认构造函数或无参数构造方法〕进行。类似地，如果父类派生于另一个类，则要求父类的构造方法调用层次结构中上一级超类的构造方法，依此类推。在调用请求中，最先调用的构造方法总是根类 Object 的构造方法，最后才会执行类层次结构中最底层的子类构造方法。超类的构造函数不能够被继承。

2. 继承层次结构下的 finalize 方法的调用次序

与构造方法的调用次序恰恰相反，在类层次结构中子类 finalize 方法调用应先于超类的 finalize 方法，直至最后调用 Object 超类的 finalize 方法。

如果类层次结构中的父类声明了自己的 finalize 方法，则子类中的方法 finalize 的最后一个操作应调用超类的 finalize 方法，以确保在垃圾收集器回收对象内存时，能够正确地结束对象的所有部分。

例 5-3 继承下的构造函数和 finalize 方法。

利用点-圆层次结构来声明一个新的 Point1 类和一个新的 Circle1 类。每个类都声明了构造函数和 finalize 方法，在这些方法中，将打印出提示消息，说明继承下的构造函数和 finalize 方法的调用次序。

Point1.java 的代码如下：

```
public class Point1 {
    private int x;
    private int y;
    public Point1(){                        //Point1 类的无参构造函数
        System.out.println( "Point1 no-argument constructor: " + this );
    }
    public Point1( int xValue, int yValue ){     //Point1 类的带有参数的构造函数
        x = xValue; y = yValue;
        System.out.println( "Point1 constructor: " + this );
    }
    protected void finalize(){                  //Point1 类的 finalize 方法
        System.out.println( "Point1 finalizer: " + this );
    }
    public void setX( int xValue ){
        x = xValue;
    }
    public int getX(){
        return x;
    }
    public void setY( int yValue ){
        y = yValue;
    }
    public int getY(){
        return y;
    }
    public String toString(){
        return "[" + getX() + ", " + getY() + "]";
    }
}
```

Circle1.java 的代码如下：

```
public class Circle1 extends Point1 {
    private double radius;
    public Circle1(){
        //隐式地调用超类 Point1 的无参构造函数
        System.out.println( "Circle1 no-argument constructor: " + this );
    }
```

```
    public Circle1( int xValue, int yValue, double radiusValue ){
        super( xValue, yValue );           //显式地调用超类的带两个参数的构造函数
        setRadius( radiusValue );
        System.out.println( "Circle1 constructor: " + this );
    }
    protected void finalize(){
        System.out.println( "Circle1 finalizer: " + this );
        super.finalize();                  //调用超类的 finalize 方法
    }
    public void setRadius( double radiusValue ){
        radius = ( radiusValue < 0.0 ? 0.0 : radiusValue );
    }
    public double getRadius(){
        return radius;
    }
    public double getDiameter(){
        return 2 * getRadius();
    }
    public double getCircumference(){
        return Math.PI * getDiameter();
    }
    public double getArea(){
        return Math.PI * getRadius() * getRadius();
    }
    public String toString(){
        return "Center = " + super.toString() + "; Radius = " + getRadius();
    }
}
```

ConstructorFinalizerTest.java 的代码如下：

```
public class ConstructorFinalizerTest {
    public static void main( String args[] ){
        Point1 point;
        Circle1 circle1, circle2;
        //创建 Point1 类的对象将调用该类的构造函数
        point = new Point1( 11, 22 );
        System.out.println();
        //创建 Circle1 类的对象将调用该类的构造函数
        //在调用 Circle1 类的构造函数前,首先要调用超类的构造函数
        circle1 = new Circle1( 72, 29, 4.5 );
        System.out.println();
        //创建 Circle1 类的对象将调用该类的构造函数
        //在调用 Circle1 类的构造函数前,首先要调用超类的构造函数
        circle2 = new Circle1( 5, 7, 10.67 );
        point = null;           //将 point 的引用设置为 null
        circle1 = null;         //将 circle1 的引用设置为 null
        circle2 = null;         //将 circle2 的引用设置为 null
        System.out.println();
        //显式调用垃圾收集程序将调用设为 null 的对象的 finalize 方法
        System.gc();
    }
}
```

程序运行的结果如图 5-4 所示。

```
Point1 constructor: [11, 22]

Point1 constructor: Center = [72, 29]; Radius = 0.0
Circle1 constructor: Center = [72, 29]; Radius = 4.5

Point1 constructor: Center = [5, 7]; Radius = 0.0
Circle1 constructor: Center = [5, 7]; Radius = 10.67

Point1 finalizer: [11, 22]
Circle1 finalizer: Center = [72, 29]; Radius = 4.5
Point1 finalizer: Center = [72, 29]; Radius = 4.5
Circle1 finalizer: Center = [5, 7]; Radius = 10.67
Point1 finalizer: Center = [5, 7]; Radius = 10.67
```

图 5-4　例 5-3 程序运行结果

注意,在类的语句表达式中出现关键字 this,例如:

```
System.out.println( "Point1 constructor: " + this );
```

相当于 this.toSring(),其值与对象的类型有关:若 this 指向 Point1 对象,则调用 Point1 类中定义的方法 toSring();若指向 Circle1 对象,则调用 Circle1 类中的 toSring()方法,从程序的运行输出结果中可进一步得到验证。

5.6　超类和子类的关系

下面再次使用点-圆继承层次来讨论超类与子类的关系。为了使圆类继承点类并能访问点类中的成员变量,可将点类中的 x 和 y 定义成 protected 的成员,这样就可以在点类的子类(圆类)中访问点类中的 x 和 y 变量了。下面的例 5-4 中,Circle2 类通过继承 Point2 类,就可以在 Circle2 类中访问它的超类(Point2 类)的 protected 和 public 成员了。

例 5-4　使用 protected 数据的点-圆层次。

Point2.java 的代码如下:

```
public class Point2 {
    protected int x;        //protected 的数据成员 x 将被子类继承
    protected int y;        //protected 的数据成员 y 将被子类继承
    public Point2(){
    }
    public Point2( int xValue, int yValue ){
        x = xValue;
        y = yValue;
    }
    public void setX( int xValue ){
        x = xValue;
    }
    public int getX(){
        return x;
    }
    public void setY( int yValue ){
        y = yValue;
    }
    public int getY(){
```

```
        return y;
    }
    public String toString(){
        return "[" + x + "," + y + "]";
    }
}
```

Circle2.java 的代码如下：

```
public class Circle2 extends Point2 {
    private double radius;
    public Circle2( int xValue, int yValue, double radiusValue ){
        //隐式地调用父类无参构造方法
        x = xValue;           //使用继承自超类 Point2 的变量 x
        y = yValue;           //使用继承自超类 Point2 的变量 y
        setRadius( radiusValue );
    }
    public void setRadius( double radiusValue ){
        radius = ( radiusValue < 0.0 ? 0.0 : radiusValue );
    }
    public double getRadius(){
        return radius;
    }
    public double getDiameter(){
        return 2 * radius;
    }
    public double getCircumference(){
        return Math.PI * getDiameter();
    }
    public double getArea(){
        return Math.PI * radius * radius;
    }
    public String toString(){
        //使用继承自超类 Point2 的变量 x 和 y
        return "Center = [" + x + "," + y + "]; Radius = " + radius;
    }
}
```

CircleTest2.java 的代码如下：

```
import java.text.DecimalFormat;
import javax.swing.JOptionPane;
public class CircleTest2 {
    public static void main( String[] args ) {
        Circle2 circle = new Circle2( 37, 43, 2.5 );
        //使用继承自 Point2 的方法 getX 和 getY
        String output = "X coordinate is " + circle.getX() +
            "\nY coordinate is " + circle.getY() +
            "\nRadius is " + circle.getRadius();
    circle.setX( 35 );          //使用继承自 Point2 的方法 setX
    circle.setY( 20 );          //使用继承自 Point2 的方法 setY
    circle.setRadius( 4.25 );
        output += "\n\nThe new location and radius of circle are\n" +
            circle.toString();
        DecimalFormat twoDigits = new DecimalFormat( "0.00" );
```

```
        output += "\nDiameter is " +
            twoDigits.format( circle.getDiameter() );
        output += "\nCircumference is " +
            twoDigits.format( circle.getCircumference() );
        output + = "\nArea is " + twoDigits.format( circle.getArea() );
        JOptionPane.showMessageDialog( null, output );
        System.exit( 0 );
    }
}
```

程序运行的结果如图 5-5 所示。

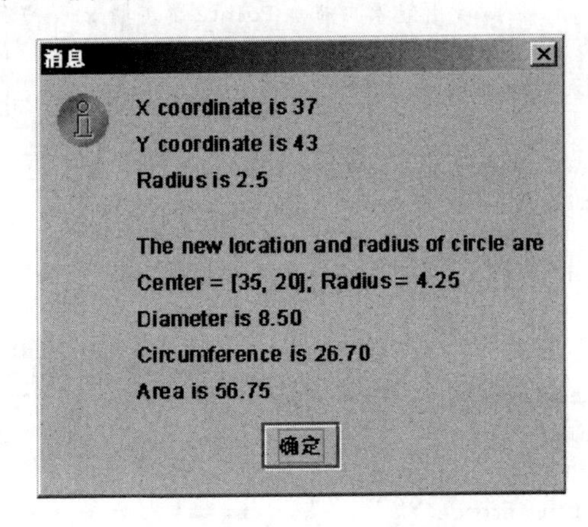

图 5-5　例 5-4 程序运行结果

例 5-4 中,将超类(Point2 类)的成员变量 x、y 声明为 protected,以便子类能够继承。但在使用 protected 成员变量时,会产生以下两个问题。

(1) 子类可不必使用一个方法来设置超类变量的值,而是直接将非法值赋给超类变量,导致该超类变量处于非法状态。例如,如果将 Circle2 的成员变量 radius 声明为 protected,则它的子类也许会将负值赋给 radius 变量。

(2) 编写的子类方法依赖于超类实现,而不是依赖于超类的服务接口(即非 private 方法)。对于超类的 protected 成员变量,如果超类实现发生改变,则需要修改超类的所有子类。例如,如果由于某种原因将成员变量 x 和 y 的名称改为 xCoordinate 和 yCoordinate,则子类直接引用这些超类成员变量的所有地方都必须进行相应的修改。在这种情况下,容易破坏软件,因为超类中的一处小修改可以"破坏"子类实现。程序员应能够随意地修改超类实现,并仍为子类提供同样的服务。

为了避免上述在子类中因为使用超类 protected 成员变量而带来的问题,应尽量使子类依赖于超类的服务,而不是依赖于超类的实现。例 5-5 把超类中的成员变量声明为 private,并在超类中定义访问这些 private 成员变量的 public 型的方法,这样在子类中不直接访问超类中声明为 private 的成员变量,而是通过使用从超类中继承过来的 public 方法对这些成员变量进行访问。

例 5-5　使用 private 数据的点-圆层次。

Point3.java 的代码如下:

```
public class Point3 {
```

```
        private int x;        //将超类中的成员变量 x 声明为 private
        private int y;        //将超类中的成员变量 x 声明为 private
        …
}
```

Circle3.java 的代码如下：

```
public class Circle3 extends Point3 {
        private double radius;
        public Circle3( int xValue, int yValue, double radiusValue ){
                super( xValue, yValue );          //显式调用超类的构造方法
                setRadius( radiusValue );
        }
        …
        public String toString(){
                //调用超类的 toString 方法
                return "Center = " + super.toString() + "; Radius = " + getRadius();
        }
}
```

CircleTest3.java 的代码与 CircleTest2.java 相同(完整程序见光盘)。

例 5-5 和例 5-4 完成的功能是一样的。但例 5-5 将超类(Point3)中的成员变量声明为 private,避免了超类中的 protected 成员带来的问题。

5.7 继承的程序设计举例

下面来看一个具有 3 级继承层次的例子。这 3 级为点-圆-圆柱体。它们之间的继承关系如图 5-6 所示。

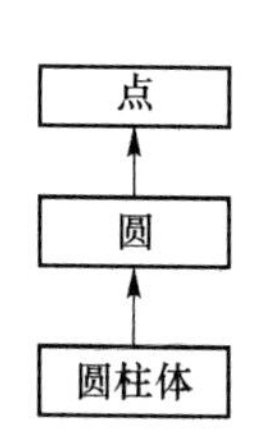

图 5-6 3 级继承层次

Point3 类的设计如下。

成员变量：

```
int x,y
```

成员方法：

```
setX,setY,getX,getY,toString
```

圆类 Circle3 类继承了 Point3 类。

Circle3 类的设计如下。

成员变量：

```
double radius
```

成员方法：

```
setX,setY,getX,getY          //继承自 Point3 类
setRadius,getRadius,getDiameter,getCircumference,getArea
toString                     //重写 Point3 类中的 toString 方法
```

其中：setRadius 方法用来设置圆的半径；getRadius 方法用来读取圆的半径；getDiameter 方法用来求圆的直径；getCircumference 方法用来求圆的周长；getArea 方法用来求圆的面积。

圆柱体类 Cylinder 派生于 Circle3 类。

Cylinder 类的设计如下。

成员变量：

```
double height
```

成员方法：

```
setX,setY,getX,getY                              //间接继承自 Point3 类
setRadius,getRadius,getDiameter,getCircumference  //继承自 Circle 3 类
getArea                                           //重写 Circle3 类中的 getArea 方法
setHeight,getHeight,getVolume
toString                                          //重写 Circle3 类中的 toString 方法
```

其中：setHeight 方法用来设置圆柱体的高；getHeight 方法用来读取圆柱体的高；getVolume 方法用来求圆柱体的体积。因为圆的面积和圆柱体的表面面积的求解算法是不一样的，所以在 Cylinder 类中对从 Circle3 类继承来的 getArea 进行了重写。

CylinderTest 类为应用程序主类。例 5-6 仅列出了 Cylinder 类和 CylinderTest 类的代码。Point3 类的代码和 Circle3 类的代码可以见 5.6 节例 5-5。

例 5-6 继承的程序设计举例。

Cylinder.java 的代码如下：

```
public class Cylinder extends Circle3 {
    private double height;                          //圆柱体的高
    public Cylinder(){
    }
    public Cylinder(int xValue,int yValue,double radiusValue,double heightValue){
        super( xValue, yValue, radiusValue );   //显式调用超类的构造函数
    setHeight( heightValue );
    }
    public void setHeight( double heightValue ){
        height = ( heightValue < 0.0 ? 0.0 : heightValue );
    }
    public double getHeight(){
        return height;
    }
    public double getArea(){                        //重写超类 Circle3 的 getArea 方法
        return 2 * super.getArea() + getCircumference() * getHeight();
    }
    public double getVolume(){
        return super.getArea() * getHeight();
    }
    public String toString(){                       //重写超类 Circle3 的 toString 方法
        return super.toString() + "; Height = " + getHeight();
    }
}
```

CylinderTest.java 的代码如下：

```
import java.text.DecimalFormat;
import javax.swing.JOptionPane;
public class CylinderTest {
    public static void main( String[] args ){
        Cylinder cylinder = new Cylinder( 12, 23, 2.5, 5.7 );
        String output = "X coordinate is " + cylinder.getX() +
            "\nY coordinate is " + cylinder.getY() + "\nRadius is " +
            cylinder.getRadius() + "\nHeight is " + cylinder.getHeight();
        cylinder.setX( 35 );                    //调用继承自 Point3 中的 setX 方法
        cylinder.setY( 20 );                    //调用继承自 Point3 中的 setY 方法
        cylinder.setRadius( 4.25 );             //调用继承自 Circle3 中的 setRadius 方法
```

```
        cylinder.setHeight( 10.75 );
        output + =
            "\n\nThe new location, radius and height of cylinder are\n" +
            cylinder.toString();
        DecimalFormat twoDigits = new DecimalFormat( "0.00" );
         output + = "\n\nDiameter is " +
            twoDigits.format( cylinder.getDiameter() );
        output + = "\nCircumference is " +
            twoDigits.format( cylinder.getCircumference() );
        output + = "\nArea is " + twoDigits.format( cylinder.getArea() );
        output + = "\nVolume is " + twoDigits.format( cylinder.getVolume() );
        JOptionPane.showMessageDialog( null, output ); // display output
        System.exit( 0 );
    }
}
```

程序运行的结果如图 5-7 所示。

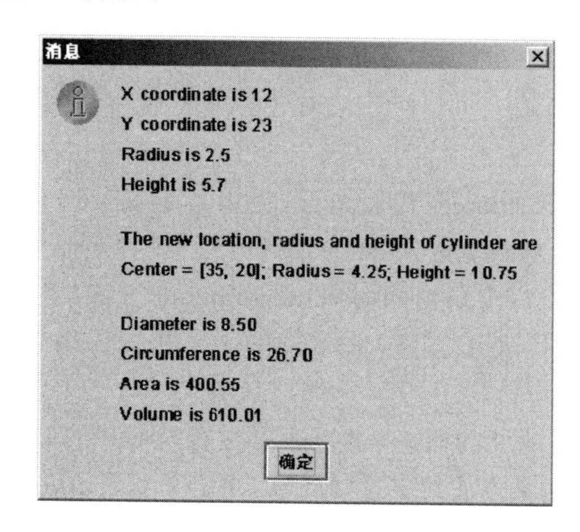

图 5-7　例 5-6 程序运行结果

5.8 小　结

本章介绍了面向对象程序设计的一个重要特性——继承。通过继承在原有类的基础上派生出高效的子类。在子类可以增加新的成员,使用超类中被声明为 protected 和 public 的成员,也可以在子类中重写超类中的方法,使该方法的执行更符合子类的情况。

继承层次结构下的构造方法的调用次序:在创建子类对象时,必须先调用直接父类的构造方法,然后才调用子类本身的构造方法。继承层次结构下的 finalize 方法的调用次序恰恰与此相反。

习　题

5.1　试描述继承下的父类(超类)和子类的概念。父类、子类间有何关系?

5.2　什么是单重继承?什么是多重继承?Java 采用什么继承?

5.3　Java 中如何定义继承关系?写出定义类库中的类 java.awt.Frame 的子类 My-

Frame 的类头的语句。

5.4　关键字 protected 的作用是什么？什么情况下使用比较合适？

5.5　什么叫方法的重新定义？

5.6　试解释构造函数重载的作用。一个构造函数如何调用同类的其他构造函数？如何调用父类的构造函数？

5.7　阅读下面的程序并写出程序的执行结果，并说明为什么？

```
class S1{
    public static void main(String[] args){
        new S2 ( );
    }
    S1 ( ){
        System.out.println("S1");
    }
}
class S2 extends S1{
    S2( ){
        System.out.println("S2");
    }
}
```

5.8　定义一个类 MyRectangle 代表矩形，为矩形定义 getLength 方法（获得矩形的长度）、getWidth 方法（获得矩形的宽度）、setLength 方法（设置矩形的长度）、setWidth 方法（设置矩形的宽度）、getArea 方法（求矩形的面积）和 toString 方法（显示矩形的格式），为矩形派生出一个子类 MySquare 代表正方形，并对 getArea 和 toString 进行重写。编写程序进行测试。

5.9　编写一个类实现地址的概念，包括的属性有“国家”、“省份”、“市县”、“街道”、“门牌”、“单位”、“邮编”，自行定义方法封装这些属性，并定义一个方法按照标准格式打印出寄给该地址的信封。从该地址类派生出国内、国际两种地址，两种地址的格式不同。重载打印信封的方法，新方法不再在方法内部直接执行打印操作，而是返回一个按格式组合好的地址字符串。

第6章　多 态 性

多态性是面向对象编程中的重要特性。本章讨论的内容包括：抽象方法与抽象类的声明、接口的声明和实现；嵌套类和内部类的定义和使用；基本类型包装类的定义和使用。

6.1　多态性概念

多态性(Polymorphism)是指：在继承层次结构中，超类中定义的属性或行为被子类继承之后，可以具有不同的数据类型或表现出不同的行为。这使得同一个属性或行为在超类及其各个子类中具有不同的语义。

在继承层次结构中，超类可以定义为抽象类或接口，通过在子类中实现超类中的抽象方法，从而实现对象的多态性。

6.2　继承层次结构中对象间的关系

继承层次结构中的子类对象可以视为超类的对象，这样就可以将子类对象赋给超类变量。尽管子类对象的类型不同，但这么做是允许的，因为每个子类对象就是超类的对象。然而，超类对象并不是其任何子类的对象，即不能将超类对象赋给子类引用。

```
Point3 point  =  new Point3( 30, 50 );
Circle4 circle  =  new Circle4( 120, 89, 2.7 );
Point3 pointRef  =  circle;      (允许)
Point3 pointRef = new Circle4(120,12,0)     (允许)
pointRef.toString(); //call Circle4.toString();
pointRef = point;
pointRef.toString() //call Point.toString();
Circle circle =  point;        // 不允许,编译出错
```

下列语句是将 point 对象强制转换成 Circle 对象，circle. getX() 语句是正确的，而 circle. getRadius()语句在运行时会产生错误。因为 circle 引用指向的 point 对象根本无 getRadius 方法。

```
Circle circle = (Circle) point;
circle. getX() ;
circle. getRadius() ; //run-time error
```

例 6-1　继承层次结构中对象间的关系。

该程序使用了 5.6 节例 5-5 中的 Point3 类和 Circle3 类。

```
//HierarchyRelationshipTest1.java
import javax.swing.JOptionPane;
```

```
public class HierarchyRelationshipTest1 {
    public static void main( String[] args )
    {   Point3 point = new Point3( 30, 50 );
        Circle3 circle = new Circle3( 120, 89, 2.7 );
        String output = "Call Point3's toString with superclass" +
            " reference to superclass object: \n" + point.toString();
        output + = "\n\nCall Circle3's toString with subclass" +
            " reference to subclass object: \n" + circle.toString();
        Point3 pointRef = circle;
        output + = "\n\nCall Circle3's toString with superclass" +
            " reference to subclass object: \n" + pointRef.toString();
        JOptionPane.showMessageDialog( null, output );
        System.exit( 0 );
    }
}
```

程序运行结果如图 6-1 所示。

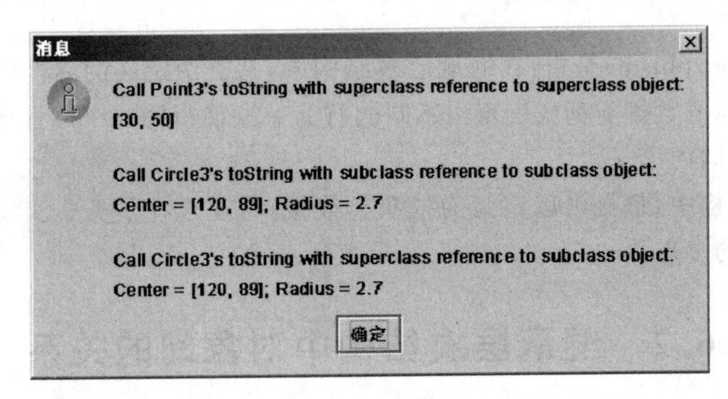

图 6-1　例 6-1 程序运行结果

上述代码首先创建了一个 Point3 对象，并将该对象赋给了 Point3 型的引用变量 point，然后又创建了一个 Circle3 对象，并将该对象赋给了 Circle3 型的引用变量 circle，接着将circle 引用指向的子类对象赋给了 Point3 型的变量 pointRef。因为可以将一个 Circle3 对象看成是一个 Point3 对象，所以子类对象能赋给超类的引用变量。而当通过 pointRef 引用变量调用 Circle3 型对象的 toString 方法时，即 pointRef.toString()实际上调用 Circle3 类的 toString 方法。

在例 6-1 中，是将子类对象 circle 的引用赋给了超类 Point3 的变量 pointRef，在 Java 中是允许的，因为一个 Circle3 是一个 Point3。但如果将超类对象的引用赋给一个子类类型的变量，则是不允许的。例 6-2 使用了 Point3 类和 Circle3 类。

例 6-2　超类对象赋给子类的引用变量，引起编译错误。

```
//HierarchyRelationshipTest2.java
public class HierarchyRelationshipTest2 {
    public static void main( String[] args )
    {
        Point3 point = new Point3( 30, 50 );
        Circle3 circle;
        circle = point; //超类对象赋给子类类型的引用变量，是不允许的
    }
}
```

该程序编译时将出现错误。

6.3 抽象类和抽象方法

6.3.1 抽象类和具体类的概念

类是一组相同类型对象的描述，即通过类可抽象出一组对象的公共特性。而一组类的公共特性可以用超类来描述。同样，一组超类的公共特性或公共接口的进一步抽象，可以用抽象类描述。类的公共接口用抽象方法描述。每个抽象类中至少包含一个抽象方法。

抽象类和具体类的区别为：抽象类只能作为继承层次结构中的超类，不能创建抽象类的对象；而具体类（非抽象类）可以用来创建对象。

抽象超类是一组类的一般化描述，它们仅仅指定子类的共同点。

6.3.2 抽象方法的声明

使用关键字 abstract 声明抽象方法，形如

```
public abstract 返回类型 draw();
```

抽象方法仅仅声明了功能的接口（声明了方法的头），并不提供功能的实现（即没有定义方法体）。包含抽象方法的类必须声明为抽象类。抽象类的所有具体子类都必须为超类的抽象方法提供具体实现。

6.3.3 抽象类的声明

使用关键字 abstract 声明抽象类，形如

```
public abstract class Shape
```

抽象类通常包含一个或多个抽象方法（静态方法不能为抽象方法）。

抽象超类不能实例化。但可以使用抽象超类来声明引用变量，用以保存抽象类所派生的任何具体类的对象。应用程序通常使用这种变量来多态地操作子类对象。

6.3.4 抽象类程序设计的举例

下面举例说明抽象类的使用。该例子所使用到的类的层次结构如图 6-2 所示。类的层次以抽象超类 Shape 为开始，派生出 Point 类，然后由 Point 类派生出 Circle 类，再由 Circle 类派生出 Cylinder 类。其中 Shape 以斜体字出现表示它是抽象类。

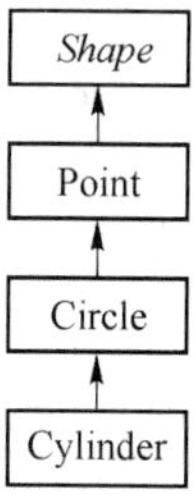

Shape
Point
Circle
Cylinder

图 6-2 Shape 类的层次结构

例 6-3 抽象类的程序设计示例。

Shape.java 文件代码：

```
public abstract class Shape extends Object {       //定义 Shape 抽象类
    public double getArea(){
        return 0.0;
    }
    public double getVolume(){
        return 0.0;
    }
    public abstract String getName();              //声明 getName 抽象方法
}
```

Point.java 类文件代码：

```
public class Point extends Shape {    //Point 类继承抽象类 Shape
    private int x;
    private int y;
    public Point(){
    }
    public Point( int xValue, int yValue ){
        x = xValue;
        y = yValue;
    }
    public void setX( int xValue ){
        x = xValue;
    }
    public int getX(){
        return x;
    }
    public void setY( int yValue ){
        y = yValue;
    }
    public int getY(){
        return y;
    }
    public String getName(){             //具体实现 Shape 类中的 getName 方法
        return "Point";
    }
    public String toString(){            //重写超类 Object 中的 toString 方法
        return "[" + getX() + ", " + getY() + "]";
    }
}
```

文件 Circle.java 代码：

```
public class Circle extends Point {
    private double radius;
    public Circle(){
    }
    public Circle( int x, int y, double radiusValue ){
        super( x, y );
        setRadius( radiusValue );
    }
    public void setRadius( double radiusValue ){
        radius = ( radiusValue < 0.0 ? 0.0 : radiusValue );
    }
    public double getRadius(){
        return radius;
    }
    public double getDiameter(){
        return 2 * getRadius();
    }
    public double getCircumference(){
        return Math.PI * getDiameter();
    }
    public double getArea(){          //重写超类中的 getArea 方法
```

```
        return Math.PI * getRadius() * getRadius();
    }
    public String getName(){          //重写超类中的 getName 方法
        return "Circle";
    }
    public String toString(){         //重写超类中的 toString 方法
        return "Center = " + super.toString() + "; Radius = " + getRadius();
    }
}
```

文件 Cylinder.java 的代码：

```
public class Cylinder extends Circle {
    private double height;
    public Cylinder(){
    }
    public Cylinder( int x, int y, double radius, double heightValue ){
        super( x, y, radius );
        setHeight( heightValue );
    }
    public void setHeight( double heightValue ){
        height = ( heightValue < 0.0 ? 0.0 : heightValue );
    }
    public double getHeight(){
        return height;
    }
    public double getArea(){          //重写超类中的 getArea 方法
        return 2 * super.getArea() + getCircumference() * getHeight();
    }
    public double getVolume(){        //重写超类中的 getVolume 方法
        return super.getArea() * getHeight();
    }
    public String getName(){          //重写超类中的 getName 方法
        return "Cylinder";
    }
    public String toString(){         //重写超类中的 toString 方法
        return super.toString() + "; Height = " + getHeight();
    }
}
```

文件 AbstractInheritanceTest.java 的代码：

```
import java.text.DecimalFormat;
import javax.swing.JOptionPane;
public class AbstractInheritanceTest {
    public static void main( String args[] ){
        DecimalFormat twoDigits = new DecimalFormat( "0.00" );
        Point point = new Point( 7, 11 );          // Point 对象的引用 point
        Circle circle = new Circle( 22, 8, 3.5 ); // Circle 对象的引用 circle
        //cylinder 为 Cylinder 对象的引用
        Cylinder cylinder = new Cylinder( 20, 30, 3.3, 10.75 );
        String output = point.getName() + ": " + point + "\n" +
            circle.getName() + ": " + circle + "\n" +
            cylinder.getName() + ": " + cylinder + "\n";
        Shape arrayOfShapes[] = new Shape[3]; //声明 Shape 类型的数组
```

```
        arrayOfShapes[0] = point;       //将 point 赋给 Shape 类型的数组元素 1
        arrayOfShapes[1] = circle;      //将 circle 赋给 Shape 类型的数组元素 2
        arrayOfShapes[2] = cylinder;    //将 cylinder 赋给 Shape 类型的数组元素 3
        for ( int i = 0; i < arrayOfShapes.length; i++ ) {
            output += "\n\n" + arrayOfShapes[ i ].getName() + ": " +
                arrayOfShapes[ i ].toString() + "\nArea = " +
                twoDigits.format( arrayOfShapes[ i ].getArea() ) +
                "\nVolume = " +
                twoDigits.format( arrayOfShapes[ i ].getVolume() );
        }
        JOptionPane.showMessageDialog( null, output );
        System.exit( 0 );
    }
}
```

程序运行的结果如图 6-3 所示。

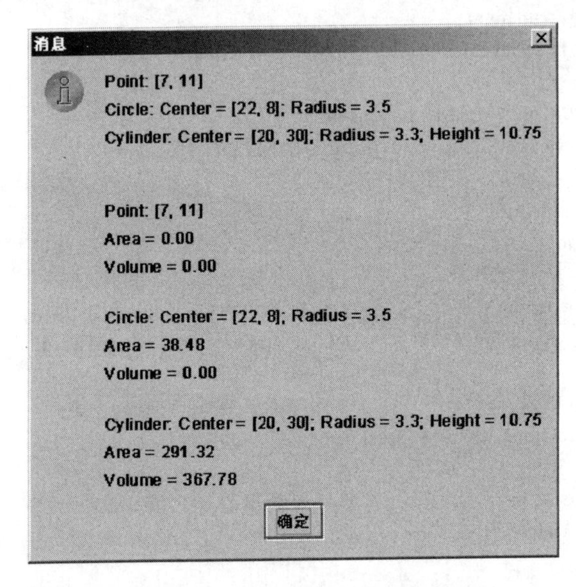

图 6-3　例 6-3 程序运行结果

6.4　接口的声明和实现

6.4.1　接口的概念

在 Java 中接口主要用于声明一组类的公共操作(功能)的接口。

接口由一组常量和一组抽象方法组成。接口中不包括变量和有具体实现的方法。

在 Java 中不直接支持类间的多重继承,但接口支持多重继承,即一个接口可以有一个以上的父接口。在解决实际问题的过程中,往往需要这种机制。

需要特别说明的是,接口仅仅规定了一组功能的对外协议和规范,而并没有实现这组功能,这个功能的真正实现是在实现这个接口的各个子类中完成的,即由这些子类来具体定义接口中各抽象方法的方法体,以适合某些特定的行为。在 Java 中,用关键字 implements 实现接口功能。

接口与抽象类的区别如下：

（1）接口中不能实现任何方法，而抽象类可以；

（2）一个类可以实现许多接口，但只继承一个父类。

6.4.2 接口的声明

接口由一组常量和抽象方法组成。声明一个接口与声明一个类相似。用关键字 interface 声明一个接口，其格式如下：

```
[public] interface 接口名 [extends 父接口名列表]
{  [public][final][static] 类型 变量名 = 常量值；      //一组常量声明
      …
   [public ][abstract] 返回类型 方法名(参数列表)；      //一组抽象方法声明
   }
```

public 修饰的接口是公共接口，可以被所有的类和接口使用；没有 public 修饰符的接口是包作用域，即只能被同一个包中的其他类和接口利用。因为在接口中声明的数据成员只能是 public、static 和 final 型的，所以这些修饰符在接口中可缺省不写。同样，所有声明在接口中的方法默认为 public 和 abstract，所以不需用修饰符限定它们。

接口之间支持继承性，"父接口名列表"中以逗号分隔所有的父接口名，这些接口可以被新的接口所继承。

接口的定义举例如下：

```
public interface Shape{
    public final double PI = 3.1415926;
    public abstract void draw(Graphics g);     }
```

该接口声明了一个常量，一个抽象方法 draw()。

6.4.3 接口的实现

为了使用接口，要编写实现接口的类。如果一个类实现一个接口，那么这个类应提供在接口中声明的所有方法的实现。那么这个类才是具体的类，否则它还是一个抽象的类。具体的类才能用来定义对象，抽象的类是不能实例化的。

声明一个类来实现一个接口或多个接口时，在类的声明中用关键字 implements。格式如下：

```
[public] class 类名 [extends 父类名] [implements 接口 1,接口 2,…,接口 n]
```

一个类可以实现多个接口时，这些接口以逗号分隔。

6.4.4 接口的程序设计举例

下面举例说明接口的使用。该例子中各类之间的层次关系类似图 6-2，区别仅仅在于用接口 Shape 替代了抽象超类 Shape。

例 6-4 区别接口程序设计示例。

文件 Shape.java 的代码：

```
public interface Shape {                //声明 Shape 接口
    public double getArea();            //声明 getArea 抽象方法
    public double getVolume();          //声明 getVolume 抽象方法
```

```
    public String getName();            //声明 getName 抽象方法
}

//Point.java, Point 类实现 Shape 接口
public class Point extends Object implements Shape {
    private int x;
    private int y;
    public Point(){
    }
    public Point( int xValue, int yValue ){
        x = xValue;
        y = yValue;
    }
    public void setX( int xValue ){
        x = xValue;
    }
    public int getX(){
        return x;
    }
    public void setY( int yValue ){
        y = yValue;
    }
    public int getY(){
        return y;
    }
    public double getArea(){            //实现 Shape 接口中的 getArea 方法
        return 0.0;
    }
    public double getVolume(){          //实现 Shape 接口中的 getVolume 方法
        return 0.0;
    }
    public String getName(){            //实现 Shape 接口中的 getName 方法
        return "Point";
    }
    public String toString(){           //覆盖 Object 类中的 toString 方法
        return "[" + getX() + "," + getY() + "]";
    }
}
```

文件 Circle.java 的代码：

```
public class Circle extends Point {
    private double radius;
    public Circle(){
    }
    public Circle( int x, int y, double radiusValue ){
        super( x, y );
        setRadius( radiusValue );
    }
    public void setRadius( double radiusValue ){
        radius = ( radiusValue < 0.0 ? 0.0 : radiusValue );
    }
    public double getRadius(){
```

```
        return radius;
    }
    public double getDiameter(){
        return 2 * getRadius();
    }
    public double getCircumference(){
        return Math.PI * getDiameter();
    }
    public double getArea(){          //覆盖超类中的 getArea 方法
        return Math.PI * getRadius() * getRadius();
    }
    public String getName(){          //覆盖超类中的 getName 方法
        return "Circle";
    }
    public String toString(){         //覆盖超类中的 toString 方法
        return "Center = " + super.toString() + "; Radius = " + getRadius();
    }
}
```

文件 Cylinder.java 的代码：

```
public class Cylinder extends Circle {
    private double height;
    public Cylinder(){
    }
    public Cylinder( int x, int y, double radius, double heightValue ){
        super( x, y, radius );
        setHeight( heightValue );
    }
    public void setHeight( double heightValue ){
        height = ( heightValue < 0.0 ? 0.0 : heightValue );
    }
    public double getHeight(){
        return height;
    }
    public double getArea(){          //覆盖超类中的 getArea 方法
        return 2 * super.getArea() + getCircumference() * getHeight();
    }
    public double getVolume(){        //覆盖超类中的 getVolume 方法
        return super.getArea() * getHeight();
    }
    public String getName(){          //覆盖超类中的 getName 方法
        return "Cylinder";
    }
    public String toString(){         //覆盖超类中的 toString 方法
        return super.toString() + "; Height = " + getHeight();
    }
}
```

文件 InterfaceTest.java 的代码：

```
import java.text.DecimalFormat;
import javax.swing.JOptionPane;
public class InterfaceTest {
    public static void main( String args[] ){
```

```
        DecimalFormat twoDigits = new DecimalFormat( "0.00" );
        Point point = new Point( 7, 11 );
        Circle circle = new Circle( 22, 8, 3.5 );
        Cylinder cylinder = new Cylinder( 20, 30, 3.3, 10.75 );
        String output = point.getName() + ": " + point + "\n" +
            circle.getName() + ": " + circle + "\n" +
            cylinder.getName() + ": " + cylinder + "\n";
        Shape arrayOfShapes[] = new Shape[ 3 ]; //声明 Shape 接口类型的数组
        arrayOfShapes[ 0 ] = point;
        arrayOfShapes[ 1 ] = circle;
        arrayOfShapes[ 2 ] = cylinder;
        for ( int i = 0; i < arrayOfShapes.length; i++ ) {
            output += "\n\n" + arrayOfShapes[ i ].getName() + ": " +
                arrayOfShapes[ i ].toString() + "\nArea = " +
                twoDigits.format( arrayOfShapes[ i ].getArea() ) +
                "\nVolume = " +
                twoDigits.format( arrayOfShapes[ i ].getVolume() );
        }
        JOptionPane.showMessageDialog( null, output );
        System.exit( 0 );
    }
}
```

程序运行的结果如图 6-4 所示。

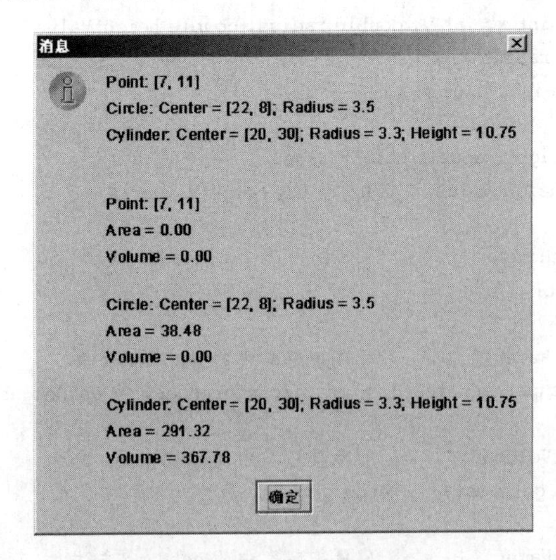

图 6-4　例 6-4 程序运行结果

6.5　final 方法和 final 类

变量可以声明为 final,说明该变量是常量,常量的值不能在声明后进行修改,并且在声明时必须对它进行初始化。

final 修饰符也能用于声明方法和类。

子类不能重载超类中声明为 final 的方法。因为子类不可能重载 private 方法,所以声明为 private 的方法隐式地为 final 方法。声明为 static 的方法也隐式地为 final 方法,因为只能重载非静态方法。

声明为 final 的类不能为超类。final 类中的所有方法都隐式地为 final 方法。

将类声明为 final 可以防止程序员创建绕过安全限制的子类。

例如,String 类是 final 类,不能扩展该类,使用字符串的程序可以调用 String 对象在 Java API 中指定的函数。

6.6 嵌套类和应用实例

到目前为止,所接触到的类都是顶级(top)类,即没有定义在其他类的内部或方法的内部。在 Java 中允许在一个类的内部能定义另一个类,从而创建所谓的嵌套类。

6.6.1 内部类的概念

嵌套在其他类里面的类称为内部类(inner class)。内部类主要用于事件处理。内部类因嵌套在外部类中,而作为外部类的一个成员,可以在内部类中访问外部类内定义的所有成员。

6.6.2 内部类的声明

内部类的声明格式如下:

```
[修饰符] class outerClass{
    …
    [修饰符] class innerClass{
        …
    }
    …
}
```

在外部类 outerClass 的类体中包含了一个内部类 innerClass 的定义。

例 6-5 内部类应用的例子。

从键盘输入一个数,计算此数的平方并输出结果。

程序的显示界面使用 javax. swing 包中的 JFrame 类生成。定义扩展 JFrame 的子类 TestWindow,程序说明如下。

(1) 在其构造方法中生成 GUI 的 4 个组件:一个 JLabel 对象,两个 JTextField 对象和两个 JButton 按钮对象并赋值给 5 个引用变量 aLabel、aField、displayField、computeButton、exitButton。

(2) 将 JFrame 的内容面板容器的布局方式设置为 Flowlayout (即将加入的组件集合按从左到右、从上到下的方式布局)。

```
Container container = getContentPane();
container.setLayout( new FlowLayout() );
```

(3) 将某个组件加入到容器中的方法如下:

```
container.add(组件);
```

(4) 给两个 JButton 对象 computeButton 和 exitButton 注册事件处理监听对象 handler,此监听对象是实现了 Java 的事件监听接口 ActionListener 的内部类 ActionEventHandlcr 的一个对象。

```
ActionEventHandler handler = new ActionEventHandler();
computeButton.addActionListener( handler );
exitButton.addActionListener( handler );
```

(5) 内部类用于处理事件。内部类 ActionEventHandler 实现了 Java 中事件监听接口 ActionListener 中声明的抽象方法 actionPerformed(),在此方法中定义用户按 JButton 按钮时的事件处理的操作代码。

当用户与某一 GUI 组件对象交互时,Java 运行系统将产生类型 ActionEvent 的事件对象 event。封装在事件对象中的方法 event.getSource()返回用户当前正在交互的 GUI 组件名。在方法 actionPerformed 的实现体中,当用户通过鼠标选中按钮 computeButton〔即 event.getSource()==computeButton 为 true〕时,取用户输入的数,计算其平方并在组件 displayField 中显示;当用户通过鼠标选中按钮 exitButton〔即 event.getSource()==exitButton 为 true〕时,程序中断运行。

程序 TestWindow.java 的代码如下:

```
import java.awt.*;
import java.awt.event.*;
import javax.swing.*;
public class TestWindow extends JFrame {
    private JLabel aLabel;
    private JTextField aField, displayField;
    private JButton computeButton,exitButton;
    public TestWindow() // set up GUI
    {       // call JFrame constructor to set title bar string
        super( "内部类的使用:计算一个数的平方" );
            // use inherited method getContentPane to get window's content pane
        Container container = getContentPane();
        container.setLayout( new FlowLayout() ); // change layout
        aLabel = new JLabel( "输入一个整数:" );
        aField = new JTextField( 10 );
        container.add( aLabel );
        container.add( aField );
            // set up displayField
        displayField = new JTextField( 30 );
        displayField.setEditable( false );
        container.add( displayField );
            // set up computeButton
        computeButton = new JButton( "计算平方" );
         container.add( computeButton );
            // set up exitButton
        exitButton = new JButton( "退出" );
        container.add( exitButton );
            // create an instance of inner class ActionEventHandler
        ActionEventHandler handler = new ActionEventHandler();
            //register computeButton event handler
        computeButton.addActionListener( handler );
            //register exitButton event handler
        exitButton.addActionListener( handler );
        setSize( 400, 140 );
        setVisible( true );
    }
```

```
    public static void main( String args[] )
    {          TestWindow window = new TestWindow();
    }
        // inner class declaration for handling JButton events
    private class ActionEventHandler implements ActionListener {
        // method to handle action events
        public void actionPerformed( ActionEvent event )
        {    if ( event.getSource() == exitButton )
                System.exit( 0 ); // terminate the application
            else if ( event.getSource() == computeButton ) {
                String a = aField.getText( );
                int ai = Integer.parseInt(a);
                ai = ai * ai;
                String b = String.valueOf(ai);
                displayField.setText( a +" 的平方是：" + b );
            }
        } // end method actionPerformed
    } // end inner class ActionEventHandler
} // end class TestWindow
```

程序运行的结果如图 6-5 所示。

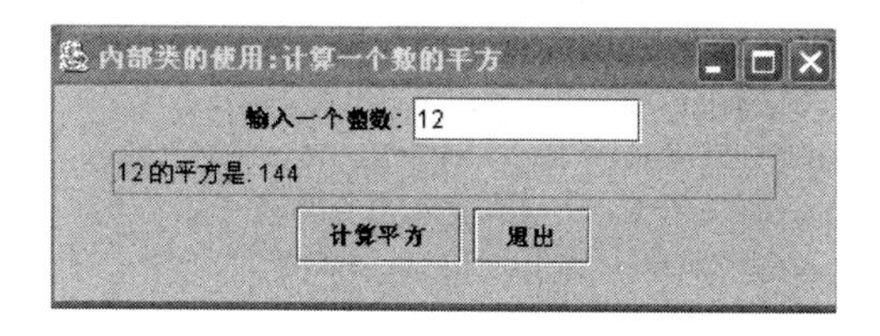

图 6-5　例 6-5 程序运行结果

6.6.3　匿名内部类声明

所谓匿名内部类是指定义的内部类没有类名。当程序中使用匿名内部类时，在定义匿名内部类的地方往往直接创建该类的一个对象。匿名内部类的声明格式如下：

```
new ParentName() {
    ...          //内部类的定义
}
```

其中，ParentName 是匿名内部类要扩展的父类名，或是匿名内部类要实现的接口名。

通过以上格式直接创建 ParentName 类型的子类的一个对象。ParentName 后面的小括号表示调用这个匿名内部类的默认构造函数，而小括号后面的一对大括号中的内容为该匿名内部类的定义。匿名内部类往往用于注册 Java 组件的事件处理对象。

例 6-6　匿名内部类应用的例子。

本例是用匿名内部类完成例 6-5 的功能，程序的输出结果也一样。

TestWindow1. java 代码如下：

```
import java.awt.*;
import java.awt.event.*;
import javax.swing.*;
public class TestWindow1 extends JFrame {
    private JLabel aLabel;
```

```
    private JTextField aField, displayField;
    private JButton computeButton,exitButton;
    public TestWindow1() // set up GUI
    {   // call JFrame constructor to set title bar string
        super("匿名内部类的使用:计算一个数的平方");
        // use inherited method getContentPane to get windows content pane
        Container container = getContentPane();
        container.setLayout( new FlowLayout() ); // change layout
        aLabel = new JLabel( "输入一个整数:" );
        aField = new JTextField( 10 );
        container.add( aLabel );
        container.add( aField );
        // set up displayField
        displayField = new JTextField( 30 );
        displayField.setEditable( false );
        container.add( displayField );
        // set up computeButton
        computeButton = new JButton( "计算平方" );
        container.add( computeButton );
        // set up exitButton
        exitButton = new JButton( "退出" );
        container.add( exitButton );
        //register computeButton event handler
        computeButton.addActionListener( new ActionListener() {
                                        //声明匿名内部类
        public void actionPerformed( ActionEvent event )
        {   String a = aField.getText( );
            int ai = Integer.parseInt(a);
            ai = ai * ai;
            String b = String.valueOf(ai);
            displayField.setText( a + " 的平方是: " + b );
        } // end method actionPerformed
    }
    );
    //register exitButton event handler
    exitButton.addActionListener( new ActionListener() { //声明匿名内部类
        public void actionPerformed( ActionEvent event )
        {   System.exit( 0 ); // terminate the application
        }
    }
    );
    setSize( 400, 140 ); setVisible( true );
  } // end constructor
  public static void main( String args[] )
  {
    TestWindow1 window = new TestWindow1();
  } // end main
} // end class TestWindow1
```

在声明和使用嵌套类时,应该注意以下事项。

(1) 包含嵌套类的类在编译时,将为每个类产生单独的.class 文件。嵌套类文件名为 OuterClassName$InnerClassName.class。匿名内部类的文件名为 OuterClassName$#.

class,# 从 1 开始,编译时,每遇到一个匿名内部类,# 递增 1。

(2) 带有名称的内部类可以声明为 public、protected、包访问或 private,并且它们的使用限制与其他类成员相同。

(3) 内部类能以 OuterClassName.this 的形式访问其外部类的 this 引用。

(4) 外部类负责创建内部类的对象。为创建另一个类的内部类对象,首先创建该外部类的一个对象,并将该对象的引用赋给外部类类型的变量(假设它是 outref)。然后使用如下形式创建内部类的对象:

```
OuterClassName.InnerClassName innerRef = outref.new InnerClassName();
```

(5) 静态嵌套类不需创建外部类的对象,静态嵌套类不能访问外部类的非静态成员。

6.7 基本数据类型的包装类

在 Java 中的数据类型分成两类:基本类型和引用类型。

基本类型包括 boolean、char、byte、short、int、long、float 和 double。每个基本类型都有一个相应的位于 java.lang 包中的包装类。每个类型包装类使用户能够像操作对象一样操作基本类型的数据。

基本类型 boolean、char、byte、short、int、long、float、double 对应的包装类为 Boolean、Char、acter、Byte、Short、Integer、Long、Float 和 Double。每个类型包装类均声明为 final,因此它们的方法隐式地为 final 方法,不能重新定义这些方法。

类型包装类中的方法一般声明为静态方法。如果程序需要操作基本类型的值,可以通过类名来调用这些静态方法。

例如,将一个字符串转变为一个 double 值:

```
Double.parseDouble("28.8")
```

将参数中的字符串转变为 Double 型的数据。其中 parseDouble 为 Double 类型包装类的一个静态方法,直接通过类名.静态方法名()进行调用。

用基本类型包装类能创建封装基本类型的数据的对象,使操作基本类型的数据就像操作一个对象一样。如

```
Integer i1 = new Integer(1234);
```

6.8 小　结

本章介绍了多态性程序设计。Java 中用抽象类、接口与抽象方法的覆盖实现多态性。

final 用于修饰常量、方法和类时,final 类不能被继承;final 方法不能在子类被重写。

Java 中除定义顶层类外,还可以定义嵌套类。嵌套在一个类中的内部类主要用于事件处理和访问外部类的所有成员。

用基本类型包装类能创建封装基本类型的数据的对象,使操作基本类型的数据就像操作一个对象一样。

习　题

6.1　抽象方法有什么特点?抽象方法的方法体在何处定义?定义抽象方法有什么好处?

6.2　final 修饰符可以用来修饰什么？被 final 修饰后有何特点？

6.3　什么是多重继承？Java 是否支持多重继承？

6.4　什么是接口？它起到什么作用？试比较接口与抽象类的异同。

6.5　书写语句创建一个名为 MyInterface 的接口，它是继承了 MySuperInterface1 和 MySuperInterface2 两个接口的子接口。

6.6　一个类实现接口的抽象方法时有什么条件？实现接口的类是否必须覆盖该接口的所有抽象方法？

6.7　内部类的作用是什么？在内部类如何访问外部类的成员？如何访问外部类的当前对象？

6.8　设计一个人员类(Person)，其中包含一个方法 pay，代表人员的工资支出。再从 Person 类派生出教师类(Teacher)和大学生类(CollegeStudent)，其中

教师的工资支出为：基本工资＋授课时数×30

大学生的工资支出为：奖学金支出

(1) 将人员类定义为抽象类，pay 为抽象方法，设计程序实现多态性。

(2) pay 定义在接口中，设计程序实现多态性。

6.9　通过嵌套类的程序设计，实现如下图所示的功能：输入姓名按回车键后在下面的文本框中显示出来。

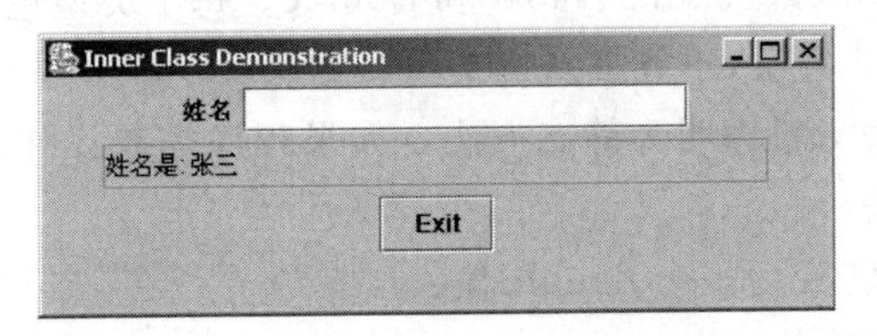

6.10　通过嵌套类的程序设计实现如下图所示的功能：用户在输入时、分、秒后，分别按回车键后，都能在下面的文本框中显示出用户设置的时间值。

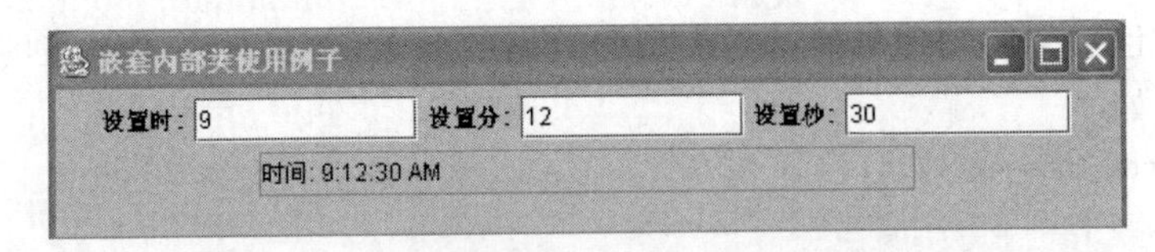

6.11　对第 4 章习题 4.15 进行修改，设计如下的 GUI 界面，通过嵌套匿名内部类求方程的两个实数根。

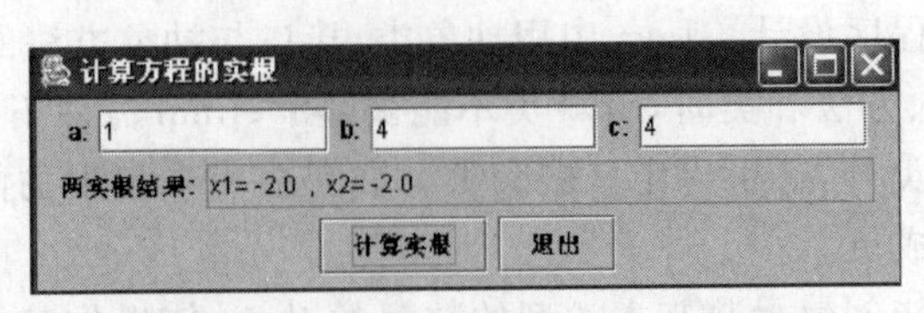

第7章　Java实用包

本章主要介绍 Java 中所提供的实用包，包括 java. lang 包和 java. util 包中的类，这些类在平时的编程中会经常使用到。

7.1　Math 类

java. lang. Math 类是标准的数学类，封装了一些常用的数学函数和常量，比如三角函数、指数函数、随机数函数等。Math 类的所有方法和变量都是静态的。Math 类是 final 型的，不能被子类化或实例化。

例 7-1　Math 类中方法的应用。

```
public class MathDemo{
  public static void main(String args[]){
    System.out.println("Math.E = " + Math.E);        //输出常量 e
    //四舍五入取整
    System.out.println("Math.round(Math.E) = " + Math.round(Math.E));
    System.out.println("Math.PI = " + Math.PI); //输出常量 p
    //输出大于等于 e 的最小双精度数
    System.out.println("ceil(E) = " + Math.ceil(Math.E));
    //输出小于等于 e 的最大双精度数
    System.out.println("floor(E) = " + Math.floor(Math.E));
    //将双精度值 p 转化为一个整数值，输出 double 型
    System.out.println("rint(PI) = " + Math.rint(Math.PI));
    System.out.println("lnE = " + Math.log(Math.E)); // 计算 e 的自然对数值
    //计算 PI/4 的正弦值
    System.out.println("sin(pi/4) = " + Math.sin(Math.PI/4));
    System.out.println(Math.random()); //产生 0 到 1 之间的 double 型数
    //产生 1 到 100 之间的整数
    System.out.println((int)(100 * Math.random() + 1));
  }
}
```

程序运行结果如图 7-1 所示。

```
Math.E=2.718281828459045
Math.round(Math.E)=3
Math.PI=3.141592653589793
ceil(E)=3.0
floor(E)=2.0
rint(PI)=3.0
lnE=1.0
sin(pi/4)=0.7071067811865475
0.24979398154230137
43
```

图 7-1　例 7-1 程序运行结果

7.2 字符串类 String

字符串处理是许多程序的重要内容，Java 中字符串是 String 类的对象。String 类位于 java.lang 包中。

7.2.1 String 构造函数

String 类支持几种构造函数，以不同的方式来初始化 String 对象。表 7-1 列出了 String 类常见的 6 个构造函数。

表 7-1 String 类的常见构造函数

构造函数	用　途
String()	创建一个 String 实例，该实例中不包含字符
String(byte[] bytes)	创建一个被 byte 数组初始化的字符串
String(byte[] bytes, int offset, int length)	创建一个被 byte 数组的子区域所初始化的字符串。参数 offset 指定子区域开始的下标（注意：第一个字符的下标为 0），参数 length 指定所使用 byte 的长度
String(char[] value)	创建一个被字符数组初始化的字符串
String(char[] value, int offset, int count)	创建一个被字符数组的子区域所初始化的字符串。参数 offset 指定子区域开始的下标（注意：第一个字符的下标为 0），参数 count 指定所用字符的个数
String(String original)	构造一个 String 对象，该对象包括了与 String 对象 original 相同的字符序列

例 7-2 的程序演示了 String 类常见的 6 个构造函数的使用。

例 7-2 String 类构造函数的使用。

```
//StringConstructors.java
import javax.swing.*;
public class StringConstructors {
    public static void main( String args[] ){
    char charArray[] = { 'b', 'i', 'r', 't', 'h', ' ', 'd', 'a', 'y' };
    //调用构造函数来创建 String 对象
    String s1 = new String();
    String s2 = new String( charArray );
    String s3 = new String( charArray
    , 6, 3 );
    String output = "s1 = " + s1 + "\ns2 = " + s2 + "\ns3 = " + s3 +"\n";
    JOptionPane.showMessageDialog( null, output,
        " String 类构造函数的使用", JOptionPane.INFORMATION_MESSAGE );
    System.exit( 0 );
  }
}
```

程序运行的结果如图 7-2 所示。

图 7-2 例 7-2 程序运行结果

7.2.2 String 类的方法和应用实例

String 类中提供了大量的方法，通过这些方法的使用，能更快、更方便地完成对字符串的处理。表 7-2 列出 String 类常用的一些方法。

表 7-2 String 类的常用方法

方 法	用 途
char charAt(int index)	获取当前字符串的某一指定位置的字符，index 从 0 开始。调用方法 charAt 的字符串成为当前字符串
void getChars(int srcBegin, int srcEnd, char[] dst,int dstBegin)	以参数 srcBegin 作为起始下标，以参数 srcEnd-1 作为结束下标，从调用方法的字符串中取子串复制到以参数 dstBegin 为起始下标的字符数组中
int length()	获取当前字符串包含的长度(字符个数)
int compareTo(String anotherString)	当前字符串与 anotherString 字符串进行逐个字符比较(按字符的 unicode 值比较大小)。若相等，则返回 0；如小于，则返回一个负数；若大于，则返回一个正数
boolean equals(Object anObject)	与 anObject 进行比较，若相等返回 true，否则 flase
boolean equalsIgnoreCase(String anotherString)	判断当前字符串与另一字符串 anotherString 是否相等，判断时不区分字符的大小写
boolean regionMatches(int toffset, String other, int ooffset, int len)	判断当前字符串的子串(子串的开始位置为 toffset，长度为 len)与另一字符串的子串 other(子串的开始位置为 ooffset，长度为 len)是否相等
String concat(String str)	连接两个 String 对象，并返回一个新的 String 对象，该对象包含两个源字符串的拼接
boolean endsWith(String suffix)	判断是否以 suffix 字符串为结尾
int indexOf(String str)	定位 str 在字符串中第一次出现的位置
int lastIndexOf(String str)	定位 str 在字符串中最后一次出现的位置
int lastIndexOf(String str, int fromIndex)	以 fromIndex 为起始索引，进行反向查找第一次出现 str 的位置
String replace(char oldChar, char newChar)	返回一个新的 String 对象：在该对象中，字符串中的 oldChar 字符均被替代成新字符 newChar
String substring(int beginIndex, int length)	从当前字符串中截取从参数 beginIndex 所指定的下标开始的长度为 length 的一个子字符串
char[] toCharArray()	将 String 对象转化成字符型数组
String toLowerCase()	将字符串中的所有大写字母改为小写，并返回该字符串
String toUpperCase()	将字符串中的所有小写字母改为大写，并返回该字符串
String trim()	将字符串中出现在开头和结尾的空格删除，并返回一个新的字符串
static String valueOf(type b)	String 提供静态方法 valueOf，用于将基本类型、字符数组或 Object 转化成字符串对象。type 是基本类型、字符数组或 Object 类型
String[] split(String regex)	以 regex 为统一的定界符，将字符串用分离成一个个单词，返回存放单词的 String 型数组

下面举例说明 String 类的常见方法的应用。

例 7-3 String 类的 length、charAt 和 getChars 方法的应用。

```
//StringMiscellaneous.java
import javax.swing.*;
public class StringMiscellaneous {
    public static void main( String args[] ){
        String s1 = "hello there";
        char charArray[] = new char[ 5];
        String output = "s1: " + s1;
        output += "\nLength of s1: " + s1.length();
        output += "\nThe string reversed is: ";
        for ( int count = s1.length() - 1; count >= 0; count-- )
            output += s1.charAt(count) + "";      //调用 String 类的 charAt 方法
        s1.getChars(0,5,charArray,0);             //调用 String 类的 getChars 方法
        output += "\nThe character array is: ";
        for ( int count = 0; count < charArray.length; count++ )
            output += charArray[ count];
        JOptionPane.showMessageDialog( null, output,
            " String 类的 length、charAt 和 getChars 方法的使用",
            JOptionPane.INFORMATION_MESSAGE );
        System.exit( 0 );
}
    }
```

程序运行的结果如图 7-3 所示。

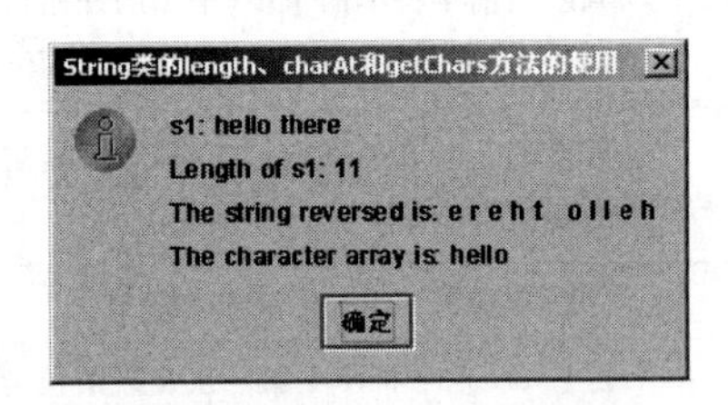

图 7-3 例 7-3 程序运行结果

例 7-4 字符串的比较。

通过使用 String 类的 equals、equalsIgnoreCase、compareTo 方法来进行字符串的比较。

```
//StringCompare.java
import javax.swing.JOptionPane;
public class StringCompare {
    public static void main( String args[] ){
        String s1 = new String( "hello" );
        String s2 = "goodbye";
        String s3 = "Happy Birthday";
        String s4 = "happy birthday";
        String output = "s1 = " + s1 + "\ns2 = " + s2 + "\ns3 = " + s3 +
            "\ns4 = " + s4 + "\n";
        if ( s1.equals( "hello" ) ) //判断字符串是否相等
            output += "s1 equals \"hello\"\n";
        else
            output += "s1 does not equal \"hello\"\n";
                //在不区分大小写的情况下判断两字符串是否相等
        if ( s3.equalsIgnoreCase( s4 ) )
            output += "s3 equals s4\n";
        else
```

```
            output + = "s3 does not equal s4\n";
        //进行两字符串的大小比较
        output + = "s1.compareTo( s2 ) is " + s1.compareTo( s2 ) +
            "\ns2.compareTo( s1 ) is " + s2.compareTo( s1 ) +
            "\ns1.compareTo( s1 ) is " + s1.compareTo( s1 ) +
            "\ns3.compareTo( s4 ) is " + s3.compareTo( s4 ) +
            "\ns4.compareTo( s3 ) is " + s4.compareTo( s3 ) + "\n";
        JOptionPane.showMessageDialog( null, output,
            "字符串的比较", JOptionPane.INFORMATION_MESSAGE );
        System.exit( 0 );
    }
}
```

程序运行的结果如图 7-4 所示。

下面的例子通过调用 String 类的 indexOf 和 lastIndexOf 方法来完成定位字符串中的字符和子字符串。

例 7-5 字符串检索。通过调用 String 类的 indexOf 和 lastIndexOf 方法完成定位字符串中的字符和子字符串。

```
//StringIndexMethods.java
import javax.swing.*;
public class StringIndexMethods {
    public static void main( String args[] ){
        String letters = "abcdefghabcdefgh";
        //调用 String 类的 indexOf 方法定位某个字符在字符串中第一次出现的位置
        String output = "'c' is located at index " + letters.indexOf( 'c' );
        output += "\n'a' is located at index " + letters.indexOf( 'a', 1 );
        output += "\n'$' is located at index " + letters.indexOf( '$' );
        //调用 lastIndexOf 方法定位某个字符在字符串中最后一次出现的位置
        output += "\nLast 'c' is located at index " +
            letters.lastIndexOf( 'c' );
        output += "\nLast 'a' is located at index " +
            letters.lastIndexOf( 'a', 10 );
        output += "\nLast '$' is located at index " +
            letters.lastIndexOf( '$' );
        //调用 indexOf 方法定位某个子字符串在字符串中第一次出现的位置
        output += "\n\"def\" is located at index " +
            letters.indexOf( "def" );
        output += "\n\"def\" is located at index " +
            letters.indexOf( "def", 7 );
        output += "\n\"hello\" is located at index " +
            letters.indexOf( "hello" );
        //调用 lastIndexOf 方法定位某个子字符在字符串中最后一次出现的位置
        output += "\nLast \"def\" is located at index " +
            letters.lastIndexOf( "def" );
        output += "\nLast \"def\" is located at index " +
            letters.lastIndexOf( "def", 25 );
        output += "\nLast \"hello\" is located at index " +
            letters.lastIndexOf( "hello" );
        JOptionPane.showMessageDialog( null, output,
            "字符串检索", JOptionPane.INFORMATION_MESSAGE );
        System.exit( 0 );
    }
}
```

程序运行的结果如图 7-5 所示。

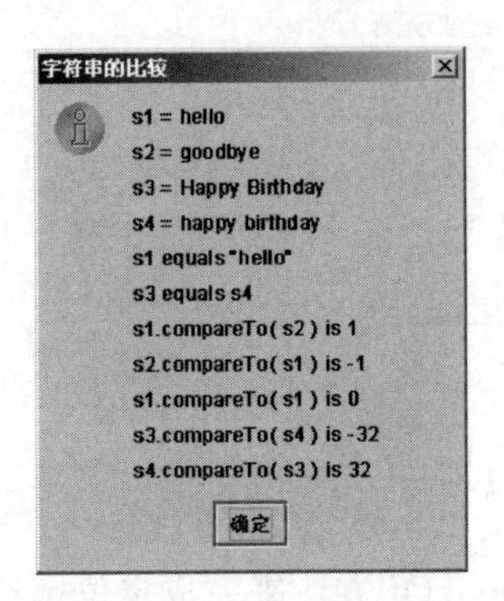

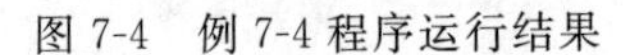
图 7-4 例 7-4 程序运行结果

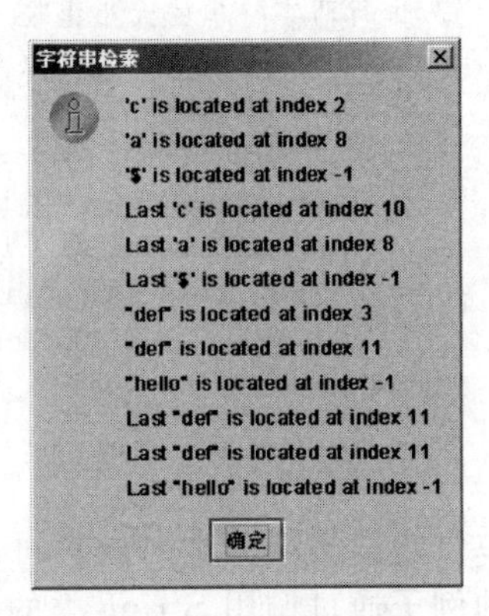

图 7-5 例 7-5 程序运行结果

下面的例子通过调用 String 类的 substring 方法来实现在字符串中提取一个子字符串。

例 7-6 字符串截取。

```
//SubString.java
import javax.swing.*;
public class SubString {
    public static void main( String args[] ){
        String letters = "abcdefghabcdefgh";
        //调用 substring 方法在字符串中提取一个子字符串
        String output = "Substring from index 10 to end is " +
        "\"" + letters.substring( 10 ) + "\"\n";
    output += "Substring from index 3 up to 6 is " +
        "\"" + letters.substring( 3, 6 ) + "\"";
    JOptionPane.showMessageDialog( null, output,
        "字符串截取", JOptionPane.INFORMATION_MESSAGE );
    System.exit( 0 );
  }
}
```

程序运行的结果如图 7-6 所示。

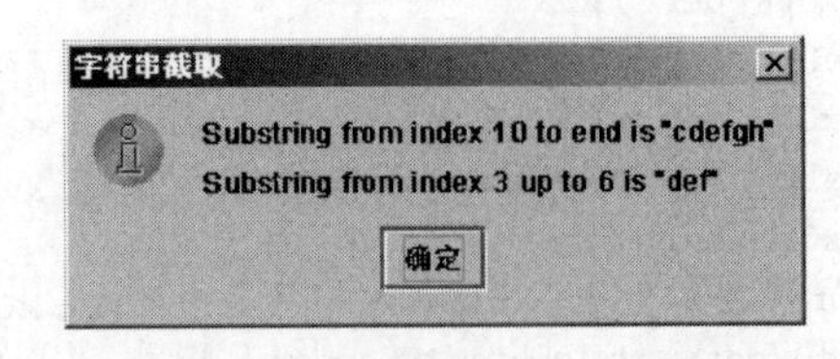

图 7-6 例 7-6 程序运行结果

例 7-7 调用 String 类的 spilt()方法，从文本中分离出单词。

程序完成在窗口的一个 JTextField 中，输入一行字符串，按回车后，将分离出的一组单词结果显示在窗口的 JTextArea 中。假设用一个空格字符作为单词之间的分隔符。

解题分析：String 类的单词解析方法为：String[] split(String regex)，其中 regx 的内容可以是单词之间的分解符，可以是单个分解符，也可以是用"|"分隔的多个分解符。如：

```
String a [];
String s = "I am a student"
```

```
a = s.split(" ");
```

则返回含有4个单词的字符串数组。而：

```
String s = "Hello,I am a studentent"
a = s.split(" |,");
```

表示单词之间的分隔符是空格或逗号,所以返回含有5个单词的字符串数组。

程序源代码如下：

```
//TokenTest.java
import java.util.*;
import java.awt.*;
import java.awt.event.*;
import javax.swing.*;
public class TokenTest extends JFrame {
   private JLabel promptLabel;
   private JTextField inputField;
   private JTextArea outputArea;
   public TokenTest(){
       super( "测试 String 的方法 split" );
       Container container = getContentPane();
       container.setLayout( new FlowLayout() );
       promptLabel = new JLabel( "输入一个句子,然后按回车" );
       container.add( promptLabel );
       inputField = new JTextField( 20 );
       inputField.addActionListener(
          new ActionListener() {
              public void actionPerformed( ActionEvent event ){
                     //得到 JTextField 中的输入字符串
                     String input = event.getActionCommand() ;
                     //调用 split()方法获得空格分隔的一组单词
                     String[]tokens = input.split(" ");
                     outputArea.setText( "Number of elements: " +
                     tokens.length + "\nThe tokens are:\n" );
                     for ( String token:tokens)
                       outputArea.append( token + "\n" );
              }
          }
       );
     container.add( inputField );
     outputArea = new JTextArea( 10, 20 );
     outputArea.setEditable( false );
     container.add( new JScrollPane( outputArea ) );
     setSize( 275, 240 );
     setVisible( true );
   }
   public static void main( String args[] ){
      TokenTest application = new TokenTest();
      application.setDefaultCloseOperation( JFrame.EXIT_ON_CLOSE );
   }
}
```

程序运行结果为：

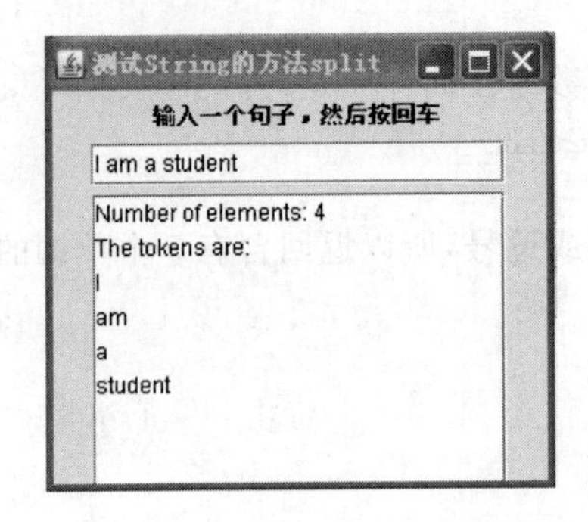

图 7-7　例 7-7 程序运行结果

7.3　StringBuilder 类

String 类的对象一旦被创建后，则它的内容就永远不允许改变，所以把 String 类的对象称为是常量字符串。而位于 java. lang 包中的 StringBuilder 类，用来表示内容可以变化的字符串。在对字符串处理方法上，StringBuilder 也提供了与 String 类似的方法。此外，Java 语言中还提供了类似于 StringBuilder 的 StringBufffer，但 StringBuilder 对字符串操作的速度比 StringBufffer 要快，只是 StringBufffer 是线程安全的，下面将重点介绍 StringBuilder 类的使用。

7.3.1　StringBuilder 构造函数

表 7-3 所示列出了 StringBuilder 常用的三个构造函数。每个 StringBuilder 对象能存储由其容量指定的字符串。如果超出了 StringBuilder 对象的容量，则存储容量就会被自动地扩大，以容纳多出来的字符。

表 7-3　StringBuilder 类的构造函数

构造函数	用　途
StringBuilder()	默认构造函数，创建一个不包含字符且初始容量为 16 个字符的 StringBuilder 对象
StringBuilder(int length)	创建一个不包含字符、初始容量由参数 length 指定的 StringBuilder 对象
StringBuilder(String str)	创建一个 StringBuilder 对象，该对象包含参数 str 所指定的字符串，且初始容量等于参数 str 所指定的字符串的长度再加上 16

7.3.2　StringBuffer 的方法

类似于 String 类，StringBuilder 类也提供了大量处理字符串的方法，如表 7-4 所示。

表 7-4　StringBuilder 类的常用方法

方　法	用　途
StringBuilder append(基本类型 b)	将基本类型的数据 b 转换为一个字符串，然后将该字符串添加到 StringBuilder 对象的末尾
StringBuilder append(char[] str, int offset, int len)	将参数 str 所指定的字符数组的内容从 offset 开始，长度为 len 的子字符串添加到 StringBuilder 对象的末尾

续 表

方 法	用 途
StringBuilder append(String str)	将参数 str 所对应的字符串添加到 StringBuilder 对象的末尾
StringBuilder append(StringBuilder sb)	将参数 sb 所对应的字符串添加到 StringBuilder 对象的末尾
StringBuilder delete(int start, int end)	删除从 start 为开始索引到 end-1 索引之间的字符串
StringBuilder insert(int offset, 基本类型 c)	将基本类型的数据 c 转换为字符串,插入到 offset 所对应的位置
StrnggBuilder insert(int offset, char[] str)	将 str 转换为字符串,插入到 offset 所对应的位置
StringBuilder insert(int offset, String str)	将字符串 str 插入到 offset 所对应的位置
StringBuilder replace(int start, int end, String str)	从 start 位置开始到 end-1 处的子串用 str 替换
StringBuffer reverse()	颠倒 StringBuffer 对象的内容
int capacity()	在不需另外分配内存的情况下,StringBuilder 对象可以存储的字符数目
int length()	返回 StringBuilder 对象的当前字符数目
void setLength(int newLength)	重新设置字符串的长度为 newLength
char charAt(int index)	返回 StringBuilder 对象中以参数 index 为索引的字符
void getChars(int srcBegin, int srcEnd, char[] dst, int dstBegin)	将 StringBuilder 对象中由参数 srcBegin 作为开始索引,以参数 srcEnd-1 作为结束索引的字符串复制到以 dstBegin 为起始位置的字符数组 dst 中
void setCharAt(int index, char ch)	将 StringBuilder 对象中由 index 所指定位置的字符设置为参数 ch 中的字符

StringBuilder 类提供 11 个重载的 append 方法,以允许将各种类型的值添加到 StringBuilder 对象的末尾。这些方法可以处理各种基本类型数据、字符数组、String 对象、Object 对象和 StringBuilder 对象。每个 append 都接收一个参数,并将该参数转换为一个字符串,然后将该字符串添加到 StringBuilder 对象的末尾。

同样地,StringBuilder 类提供 10 个重载的 insert 方法,以允许在 StringBuilder 对象的任何位置插入各种类型的值。每个 insert 方法都能将它的第二个参数转换为一个字符串,并将该字符串插入到第一个参数所指定的索引的前面。

StringBuilder 类还提供 delete 和 deleteCharAt 方法,用于删除 StringBuilder 对象的任何位置上的字符。

例 7-8 StringBuilder 类的应用

```
//StringBuilderOperation.java
public class StringBuilderOperation {
   public static void main( String args[] ){
      char charArray[] = { 'a', 'r', 'e', ' ', 'e', 'f' };
      String student = "student.";
      //调用默认构造函数创建 StringBuilder 对象 buffer1
      StringBuilder buffer1 = new StringBuilder();
      //调用指定初始字符串的构造函数创建 StringBuilder 对象 buffer3
      StringBuilder buffer2 = new StringBuilder("You");
      buffer1.append(buffer2);
```

```
        buffer1.append(' ');
        buffer1.append(charArray,0,3);
        buffer1.append(" a ");
        buffer1.append(student);
        System.out.printf("%s.\n",buffer1.toString());
        buffer1.insert(10,"good ");//在位置10处插入子串"good "
        System.out.printf("After insert:\n%s\n",buffer1.toString());
        buffer1.replace(10,14,"best"); //用best替换从位置10到位置13的子串
        System.out.printf("After replace:\n%s\n",buffer1.toString());
        buffer1.setLength(3);//重新设置字符串的长度为3
        System.out.printf("After setLength:\n%s\n",buffer1.toString());
    }
}
```

程序运行的结果为：

```
You are a student..
After insert:
You are a good student.
After replace:
You are a best student.
After setLength:
You
```

7.4 StringTokenizer 类

位于 java.util 包中的 StringTokenizer 类可以将字符串分解为组成它的语言符号(单词)。下面首先介绍 StringTokenizer 类的构造函数，如表 7-5 所示。在构造函数中分隔单词的定界符，可以是多个不同的字符。

表 7-5　StringTokenizer 类的构造函数

构造函数	用　途
StringTokenizer(String str)	以字符串 str 为参数，以默认定界符(空格、换行符、制表符、回车符)建立一个 StringTokenizer 对象
StringTokenizer(String str, String delim)	以字符串 str 为参数，以 delim 为定界符字符串建立一个 StringTokenizer 对象
StringTokenizer(String str, String delim, boolean returnDelims)	以字符串 str 为参数，以 delim 为定界符字符串，以 returnDelims 参数确定该定界符是否也作为语言符号返回建立一个 StringTokenizer 对象

在 StringTokenizer 类中提供了操作 StringTokenizer 对象的方法，表 7-6 列出了 StringTokenizer 类中常用的方法。

表 7-6　StringTokenizer 类的常用方法

方　法	用　途
int countTokens()	返回要进行语言符号化的字符串中语言符号的数目
boolean hasMoreTokens()	检测正在进行语言符号化的字符串中是否还有更多的语言符号
String nextToken()	获取 StringTokenizer 类中的下一个语言符号

例 7-9　StringTokenizer 类的使用。

```
//TokenExample.java
import java.util.*;
class TokenExample {
   public static void main( String[ ] args ) {
       String s = "This%is! the$way#the@world%ends";
       String delimiters = "!%$#@"; //定义分隔单词的多个定界符
       StringTokenizer st = new StringTokenizer( s, delimiters );
       while ( st.hasMoreTokens() )
         System.out.println( st.nextToken() );
   }
}
```

程序运行结果为:

```
This
is
the
way
the
world
ends
```

7.5　泛型集合

Java 的集合类,主要用于存放一组对象的引用变量集合,又称为容器类。Java 集合类支持泛型(generic)程序设计的思想,表现在两方面:一是容器中存放的元素类型是通用的,使用时能够定制类型;二是对容器的常用操作算法(如插入、修改、删除、查询和排序等)已经封装在集合类中,并且也为通用的(即算法可以作用于不同的数据类型上),所以被称为泛型集合类。程序设计时通过使用这些泛型集合类,能达到快速、方便地开发程序,提高软件的生产率。表 7-7 列出了一些 Java 泛型集合的特性描述。Collection 和 Map 是 Java 集合框架的根接口。Java 所有集合类都位于包 java.util 中。

表 7-7　Java 泛型集合

接　口	描　述
Collection	是 List、Set 和 Queue 接口的父接口,声明了一组公共操作的方法
Set	Set 集合中存放的元素是无序的,且元素值不能重复
List	List 中存储的元素是有序的,且元素值能重复
Map	Map 映像中存放的是键值对,且键要唯一性、不允许出现重复键
Queue	Queue 队列存放的元素具有先进先出(first-in first-out)的特点
Stack	Stack 堆栈存放的元素具有后进先出(last-in first-out)的特点

7.5.1　Collection 接口与 Iterator 迭代器

1. Collection 接口

Collection 接口定义了 List、Set 和 Queue 的一组公共操作方法,如表 7-8 所示。

表 7-8 Collection 接口的常用方法

方 法	描 述
boolean add(Object o)	添加元素到列表中。添加成功时返回 ture
boolean addAll(Collection c)	将集合 c 中的所有元素添加到指定的列表中
boolean remove(Object o)	删除列表中指定元素 o,添加成功时返回 ture
boolean removeAll(Collection c)	从列表中删除集合 c 里包含的元素
bolean contains(Object o)	返回列表里是否包含指定元素
boolean containsAll(Collection c)	返回列表里是否含集合 c 里的所有元素
Iterator iterator()	返回一个 Iterator 对象,用于遍历列表里的元素
int size()	返回列表中的元素个数
Object[] toArray()	把列表转换成一个数组
boolean isEmpty()	返回列表是否为空

2. 迭代器 Iterator

Iterator 接口用于遍历集合中的元素。Java 每一个集合类,都拥有一个方法 iterator(),其返回一个迭代器对象。例如,下面语句创建一个迭代器:

```
Iterator it = test.iterator();          //test 是一个集合对象。
```

使用迭代器对象,能够取出集合中的一个个元素。Iterator 具有以下 3 个方法。

```
boolean hasNext():如果被迭代的集合元素,还有没被遍历,则返回 true。
Object next():返回集合里下一个元素。
void remove():删除集合里上一次 next 方法返回的元素。
```

7.5.2 List(ArrayList、LinkedList、Vector)和 Enumeration 类的应用举例

1. List 接口概述

List 集合中存放的是有序的一组元素,其元素之间的顺序关系由插入的时间先后决定。List 使用类似于数组的索引(从 0 开始)来表示元素在 List 中的位置。List 接口中定义的一组常用方法如表 7-9 所示,其中 T 是泛型类型,即参数化类型,也就是所操作的数据类型被指定为一个参数。泛型类型可以用在类、接口和方法的创建时被指定。

表 7-9 List 接口中的常用方法

方 法	描 述
boolean add(T element) void add(int index,T element)	向列表尾添加一个元素 向列表指定的位置 index 添加一个元素
boolean remove(int index)	删除指定位置的一个元素,删除成功时返回 true
T get(int index)	获取位置 index 上的元素
T set(int index, T element)	将指定位置的元素替换成指定的值,返回前一位置的元素
Object set(int index,object element)	将 index 位置处的元素修改为 element
boolean contains(Object o)	判断列表是否包含指定的元素
List subList(int from,int to)	截取子 List,从 from 开始到 to 结束,包含头但不含尾
int size()	返回列表中的元素个数
int indexOf(Object ele)	返回给定元素在列表中第一次出现处的索引。如果没有找到元素 elem,则返回-1

ArrayList、LinkedList 和 Vector 是实现 List 接口的类。

2. ArrayList 类

ArrayList 类实现了一个可变长的动态数组，即数组元素个数能够动态地被改变。每个 ArrayList 对象都有一个容量(Capacity)，容量可随着不断添加新元素而自动增加。ArrayList 类的查询速度(set、get)快，但增删元素操作稍慢。

ArrayList 以初始长度创建，当长度超过时，列表空间自动变大；当删除对象时，列表空间自动变小。ArrayList 的随机访问(set、get)速度快，但是向表中插入和删除比较慢。当需要插入大量元素时，在插入前可以调用列表对象的 ensureCapacity 方法来增加存储容量以提高插入效率。

例 7-10　ArrayList 类和 Iterator 类的应用：建立 ArryList 对象，然后对列表添加、修改和遍历。

```
import java.util.ArrayList;
import java.util.Iterator;
import java.util.List;
public class ArrayListTest {
public static void main(String args[]){
    List<String> list = new ArrayList<String>();
                                          //声明一个元素类型为 String 的 ArrayLis 对象
      list.add("Red"); //添加三个元素
      list.add("Yellow");
      list.add("Blue");
      for(int i = 0;i < list.size(); i + + ){   //遍历方法 1
        System.out.printf(" % s ",list.get(i));
      }
      System.out.println();
      list.set(0,"Black"); //修改索引号为 0 的元素
      Iterator it1 = list.iterator();
      while(it1.hasNext()){ //遍历方法 2
          System.out.printf(" % s " ,it1.next());
      }
      list.remove(1);       //删除索引号为 1 的元素
      System.out.println();
      for(String element:list){     //遍历方法 3
         System.out.printf(" % s ",element);
      }
   }
}
```

程序运行结果如下：

```
Red Yellow Blue
Black Yellow Blue
Black Blue
```

3. Vector 类

Vector 向量是类似于 ArrayList 的一种顺序存储的数据结构。表 7-10 和表 7-11 列出了 Vector 类的构造方法和的常用方法。

表 7-10　Vector 类的构造方法

构造函数	用　途
Vector()	创建一个 Vector 原始容量为 10 个元素，容量增量等于 0。在需要变大时，该 Vector 将大小增大一倍
Vector(Collection c)	创建集合元素 c 的副本，并将它们存储到 Vector 中
Vector(int initialCapacity)	创建一个 Vector 其原始容量由 initialCapacity 指定
Vector(int initialCapacity, int capacityIncrement)	创建一个 Vector 其原始容量由 initialCapacity 指定，并且它的增量由 capacityIncrement 指定

表 7-11　Vector 类中的常用方法

方　法	用　途
boolean add(Object o)	将对象 o 加到 Vector 的末尾
int capacity()	返回不再分配内存时 Vector 中所能存储的元素个数
boolean contains(Object elem)	判断 Vector 中是否包含对象 elem
Enumeration elements()	返回一个包含 Vector 中所有元素的 Enumeration
Object firstElement()	返回 Vector 中第一个元素的引用
int indexOf(Object elem)	确定元素 elem 在 Vector 中第一次出现处的索引。如果在 Vector 中没有找到元素 elem，则返回－1
boolean isEmpty()	判断 Vector 是否为空
Object lastElement()	返回 Vector 中最后一个元素的引用
boolean remove(Object o)	从 Vector 中删除第一次出现对象 o 处的元素，删除元素后，Vector 中位于该元素之后的所有元素都向前移动一个位置
int size()	确定当前 Vector 中的元素个数

要注意的是：能存放到 Vector 中的数据元素必须是引用类型，但不能是基本的数据类型。

4. Enumeration 类

枚举类 Enumeration 定义在 java. util 包中，具有包含一组枚举值的能力。Enumeration 类中封装的方法，用于逐个地提取枚举对象中包含的一组枚举值。

用于操作 Enumeration 对象的方法有以下两个。

(1) boolean hasMoreElements()：测试 Enumeration 对象中是否含有元素。

(2) object nextElement()：返回 Enumeration 对象中的下一元素。

下面的代码用于从 Vector 的对象 v 中创建一个 Enumeration 对象，通过 Enumeration 对象挨个地读取向量中的元素。

```
Enumerate enum = v.elements();
while (enum.hasMoreElements()) {
    System.out.println(enum.nextElement());
}
```

或

```
for (Enumeration e = v.elements() ; e.hasMoreElements() ;)
{ System.out.println(e.nextElement());
  }
```

例 7-11 Vector 类和 Enumeration 类的使用。

```
//VectorTest.java
import java.util.*;
public class VectorTest {
   private static final String colors[] = { "red", "white", "blue" };
   public VectorTest(){                        //建立 VectorTest 类的构造函数
       Vector vector = new Vector();           //调用 Vector 类的默认构造函数
       printVector( vector );                  //调用 printVector 方法
       vector.add( "magenta" );                //调用 Vector 类的 add 方法
       for ( int count = 0; count < colors.length; count++ )
          vector.add( colors[ count ] );
       vector.add( "cyan" );
       printVector( vector );
       try {
          //调用 Vector 类的 firstElement 方法
          System.out.println( "First element: " + vector.firstElement() );
          //调用 Vector 类的 lastElement 方法
          System.out.println( "Last element: " + vector.lastElement() );
       }
       catch ( NoSuchElementException exception ) {
          exception.printStackTrace();
       }
       if ( vector.contains( "red" ) )  //调用 Vector 类的 contains 方法
            System.out.println( "\"red\" found at index " +
               vector.indexOf( "red" ) );//调用 Vector 类的 indexOf 方法
       else
            System.out.println( "\"red\" not found\n" );
       vector.remove( "red" );                //调用 Vector 类的 remove 方法
       System.out.println( "\"red\" has been removed" );
       printVector( vector );
       if ( vector.contains( "red" ) )
            System.out.println( "\"red\" found at index " +
               vector.indexOf( "red" ) );
       else
          System.out.println( "\"red\" not found" );
       //调用 Vector 类的 size 和 capacity 方法
       System.out.println( "Size: " + vector.size() +
          "\nCapacity: " + vector.capacity() );
   }
   private void printVector( Vector vectorToOutput ){
       if ( vectorToOutput.isEmpty() ) //调用 Vector 类的 isEmpty 方法
            System.out.print( "vector is empty" );
       else {
            System.out.print( "vector contains: " );
            //调用 Vector 类的 elements 方法
            Enumeration items = vectorToOutput.elements();
            //调用 Enumeration 的 hasMoreElements 方法
            while ( items.hasMoreElements() )
           //调用 Enumeration 的 nextElement 方法
           System.out.print( items.nextElement() + " " );
       }
       System.out.print( "\n" );
```

```
    }
    public static void main( String args[] ){
        new VectorTest();                    //创建 VectorTest 对象
    }
}
```

上述程序定义的方法 printVector(),功能是输出 Vector 向量对象中的所有元素。

该程序的输出如下:

```
vector is empty
vector contains: magenta red white blue cyan
First element: magenta
Last element: cyan
"red" found at index 1
"red" has been removed
vector contains: magenta white blue cyan
"red" not found
Size: 4
Capacity: 10
```

与 ArrayList 的主要区别:Vector 类是线程安全的,但访问速度较慢;而 ArrayList 是线程不安全的,但访问速度较快。所以在单线程的情形下,使用 ArrayList 的程序效率高。

5. LinkedList 类

ArrayList 和 Vector 底层的数据结构使用的是数组结构。而 LinkedList 底层数据结构使用的是双向链表结构。LinkedList 特点:增删速度很快,查询很慢。LinkedList 类中增加了访问链表头结点和尾结点的方法。

例 7-12 应用 LinkedList 类对链表对象进行创建、添加和遍历。

```
import java.util.LinkedList;
import java.util.Arrays;
public class LinkedListTest {
    public static void main(String[] args)
    {  String[] colors = {"Black", "Blue", "Yellow"};
       LinkedList<String> links = new LinkedList<String>(Arrays.asList(colors));
                                                              //从数组内容创建链表
       links.addLast("Red");    // 在链表尾加入一项
       links.add("Pink");       // 在链表尾加入一项
       links.add(3, "Green");   // 在索引号为 3 位置上插入以元素
       links.addFirst("White");// 在链表头插入一项
       colors = links.toArray(new String[links.size()]); //将链表转换成数组
       for (String color : colors)      //打印数组内容
          System.out.printf("%s  ",color);
    }
}
```

程序运行结果如下:

```
White  Black  Blue  Yellow  Green  Red  Pink
```

7.5.3 Collections 的算法应用实例

Collections 类提供了一组常用算法,主要是针对集合对象的操作。Collections 类封装的都是静态方法,如表 7-12 所示。

表 7-12　Collections 中的常用静态方法

方　法	描　述
void sort(List<T> list)	对 list 按增序排序
int binarySearch(list,Object obj)	对 list 进行二分搜索法，返回索引；若找不到指定的元素，则返回负数
T max(Collection<? extends T> list)	求集合的最大值
T min(Collection<? extends T> list)	求集合的最小值
void reverse(List<? > list)	倒排列表的全部元素

例 7-13　应用 Collections 类的静态方法对列表进行排序，折半查找，求最大值、最小值。

```
import java.util.List;
import java.util.Arrays;
import java.util.Collections;
public class Algorithms
{    public static void main(String[] args)
   {
       Integer[] ints = {4,2,1,5};
       List<Integer> list = Arrays.asList(ints); //以数组内容创建列表
       System.out.println("Before sorted: ");
       output(list);
       System.out.println("\nAfter sorted: ");
       Collections.sort(list); // 以增序排序列表
       output(list);
       Collections.sort(list,Collections.reverseOrder()); // 以降序排列列表
       System.out.println("\nAfter reversed sorted: ");
       output(list);
        System.out.printf("\nMax: %d   Min: %d", Collections.max(list),Collections.min
              (list));//求列表的最大值、最小值
       int key = 4;
     //对有序列表，折半查找一元素
       System.out.printf("\nSearching for %d, found at index %d。",key,Collections.binarySearch
             (list, key));
   }
   // output List information
   private static void output(List<Integer> listRef){
   for (Integer element : listRef)
           System.out.printf(" %d ", element);
   }
}
```

程序运行结果为：

```
Before sorted:
4 2 1 5
After sorted:
1 2 4 5
After reversed sorted:
5 4 2 1
Max: 5   Min: 1
Searching for 4, found at index 1.
```

7.5.4 Set (HashSet 与 TreeSet)应用实例

Set 集合中的元素是无序的(存入与取出的顺序不一定一致),且元素不可以重复,没有索引。实现 set 接口的类有 HashSet 和 TreeSet。HashSet 底层的存储数据结构是哈希表,而 TreeSet 底层的存储数据结构是二叉树。Set 集合的功能与 Collection 是一致的。

例 7-14 利用 HashSet 集合的元素不重复特性,对 List 列表去掉重复的元素。

```
import java.util.List;
import java.util.Arrays;
import java.util.HashSet;
import java.util.Set;
import java.util.Collection;
public class SetTest
{
   public static void main(String[] args) {
     String[] schools = {"江南大学","北京大学","清华大学","江南大学","北京大学","南京大学"};
      List<String> list = Arrays.asList(schools);
      System.out.printf("List: %s\n", list);
      printNonDuplicates(list);
   }
   // 从 list 集合对象为内容,创建 Set 对象,以达到去除集合中的重复元素
   private static void printNonDuplicates(Collection<String> values)
   {
      Set<String> set = new HashSet<String>(values);
      System.out.printf("Nonduplicate set : ");
      for (String value : set)
         System.out.printf("%s  ", value);
      System.out.println();
   }
}
```

程序运行输出结果如下:

```
List:[江南大学,北京大学,清华大学,江南大学,北京大学,南京大学]
Nonduplicate set :江南大学  南京大学  北京大学  清华大学
```

7.5.5 Map 应用实例

Map 提供了一组键值的映射。其中存储的每一项 entry 都由键值(key-value)组成,又叫 mapping。关键字决定了 entry 在 map 中的存储位置。关键字应该是唯一的,不允许出现重复。每个 key 只能映射一个 value。

类 HashMap、HashTable 和 TreeMap 实现了 Map 接口。HashMap 和 HashTable 将元素存储在哈希表结构中,而 TreeMap 将元素存放在树形结构中。Map 接口的一组常用方法如表 7-13 所示。

表 7-13 Map 中的常用方法

方 法	描 述
V put(K key, V value)	将一对键值添加到集合中。如果集合中已存在重复键,则替换旧值,并返回旧值,否则返回 null

续表

方　法	描　述
V get(Object key)	返回键映射的值。如果该键不存在，则返回 null
boolean containsKey(Object key)	判断该集合中是否有这个键的映射
int size()	返回 map 中键值映射对数

例 7-15　从键盘输入一行句子，统计各个单词出现的频数。设单词之间的分隔符为一个空格。

解题分析：将单词和单词的计数，作为 key-value 存放到 map 对象中。map 对象的创建：

```
Map<String, Integer> map = new HashMap<String, Integer>();
```

算法具体过程如下：

(1) 从键盘输入一行句子放入字符串 line 变量中。

(2) 从 line 分离出单词放入数组变量中。

```
String[] tokens = line.split(" ");
```

(3) 扫描 tokens 中的每一单词 word，判断 map 中是否含有关键字 word，如果不含有，则将映射(word,1)放入 map 中，即：

```
map.put(word, 1);
```

如果含有，则将已存放的映射(word,count)修改为(word,count+1)，即：

```
int count = map.get(word);  // 读取 word 的计数器 count
map.put(word, count + 1);   // 将 count + 1
```

(4) 输出单词的统计结果。即读出 map 中的全部键-值对：

```
Set<String> keys = map.keySet();                              // 取 Map 中所有 keys
TreeSet<String> sortedKeys = new TreeSet<String>(keys);  // 排序 keys
for (String key : sortedKeys)              //通过遍历 key,读出 key 对应的 value
  System.out.printf("%s%s\n", key, map.get(key));
```

源程序完整代码如下：

```
import java.util.Map;
import java.util.HashMap;
import java.util.Set;
import java.util.TreeSet;
import java.util.Scanner;
public class WordTypeCount
{
   public static void main(String[] args)   {
       // 创建 HashMap 对象,存放映射:String keys (单词)and Integer values(个数)
       Map<String, Integer> map = new HashMap<String, Integer>();
       buildMap(map);   // 生成 map 内容
       displayMap(map);  // 显示 map 内容
   }
   private static void buildMap(Map<String, Integer> map) // 生成 map 内容
   {
       Scanner scanner = new Scanner(System.in);
       System.out.println("Enter a line:");
       String input = scanner.nextLine();
       String[] tokens = input.split(" ");
       for (String token : tokens)
```

```
        {
            String word = token.toLowerCase();
            if (map.containsKey(word)) {
                int count = map.get(word);
                map.put(word, count + 1);
            }
            else
                map.put(word, 1);
        }
    }
    // 显示 map 内容
    private static void displayMap(Map<String, Integer> map)
    {
        Set<String> keys = map.keySet(); // 读取 map 中 keys
        TreeSet<String> sortedKeys = new TreeSet<String>(keys);  // 排序 keys
        System.out.printf("%nmap contains:\n%-10s%-10s\n","Key","Value");
        for (String key : sortedKeys)
            System.out.printf("%-10s%-10s\n", key, map.get(key));
        System.out.printf("Map size: %d\n", map.size());
    }
}
```

程序运行结果：

```
Enter a line:
To use arrays to store data in and retrieve data
Map contains:
Key       Value
and         1
arrays      1
data        2
in          1
retrieve    1
store       1
to          2
use         1
Map size : 8
```

7.6 小　结

本章节介绍了一些常用的类：Math 类、字符串类 String 、String Builder 类、String Tokenizer 类和各种泛型类、泛型接口，详细介绍了它们的应用。通过这些接口和类的应用，可以方便我们更好地进行程序设计。

习　题

7.1　Java 中最基础的类库是什么？谁是所有类的根类？作算术运算应该使用什么类？

7.2　Math.random()方法用来完成什么功能？如何利用这个方法生成 1～100 之间的随机数(含 1 和 100)？如何利用这个方法生成 1～100 之间的随机偶数？

7.3　语句 Math.floor(x)将舍去浮点数 x 的小数部分，而返回其整数部分(返回值也是一个浮点数),如何利用这个方法实现浮点数第一位小数的四舍五入?如何利用这个方法实现浮点数第二位小数的四舍五入?

7.4 StringTokenizer 类的作用是什么?

7.5 编写一个程序,通过使用 String concat(String str)方法,连接两个串得到一个新串,并输出这个新串。

7.6 编写程序使用 char charAt(int index)方法,得到一个字符串中的第一个和最后一个字符。

7.7 假设一个学生所选课程为语文、数学、英语、政治、物理、化学,给出此学生所选课程的 Vector 列表,并访问物理存放在 Vector 中的位置。

7.8 对学生类的对象,建立列表管理,使能够对学生对象进行插入、修改、删除、浏览、统计人数等功能。

提示:利用 java 的实用包的类 ArrayList1。

7.9 从键盘输入一行字符串。

① 将字符按 Unicode 字典顺序排序后输出。

② 统计每个字符出现的次数和出现的位置。

③ 统计含有的单词个数。

④ 正向和反向输出单词。

7.10 (1) 利用字符串类 String 的 split 方法,设计一个功能似于类 StringTokenizer 的类。

(2) 键盘输入一行句子,统计含有的单词个数。请利用(1)的功能。

提示:类 String 的 split 方法的使用:

```
String[] words;   String inputText,s;
s = new String(inputText);
words = s.split("\\n|\\t|\\r| "); //用回车或 tab 键或空格分隔单词
```

第8章　图形和Java 2D

本章讲述了在Java中绘制二维图形、控制颜色和设置文本字体。然后，介绍了功能更强大的Java 2D API的特性，利用Java 2D API，使程序员能描绘出任意的二维图形，并能设置图形的线条样式和填色效果，能对图形作旋转、缩放、扭曲等。

8.1　Java图形环境与图形对象

Java的图形环境使程序员能在屏幕上绘图。Graphics类（在java.awt包中）对象用于管理图形环境如何绘图。Graphics类提供了一组方法，以在屏幕上绘制文本、图形（如线条、椭圆、矩形和其他多边形等）、图像以及设置字体和控制颜色等。

Graphics类是一个抽象类，这是由Java的可移植性决定的。在各个平台实现Java时，系统将创建Graphics的一个子类，以实现绘图功能。Graphics对用户隐藏了这种实现，同时又提供了接口，使程序员能够以独立于平台的方式来使用图形。

Component类是Java中许多GUI组件（在java.awt包中）的超类。Component类中有一个paint方法，以Graphics对象为参数，其格式如下：

```
public void paint(Graphics g)
```

当要在某个Component组件上进行paint操作时，系统会将一个定制的Graphics对象传递给paint方法中的Graphics类型的引用g。在paint的方法体中，程序员使用g绘制各种图形。

当由于图形的数据发生了变化，需要在Component组件上重新绘制图形时，程序员很少直接地调用paint方法，因为绘制图形是一个事件驱动过程。为此，Component类中提供了一个无形参的repaint()方法，该方法首先调用update()进行背景清除，然后由update()调用paint()进行绘图。

8.2　颜色控制

Color类（在java.awt包中）用于设置绘图或显示GUI组件的颜色。Color类中共定义了13种静态颜色常量，如表8-1所示。

表8-1　Color类中常用的颜色常量

颜色名称和颜色常量	RGB值
白色 Color.white	255,255,255
黑色 Color.black	0,0,0
浅灰 Color.lightGray	192,192,192

续 表

颜色名称和颜色常量	RGB 值
灰色 Color. gray	128,128,128
暗灰 Color. darkGray	64,64,64
红色 Color. red	255,0,0
绿色 Color. green	0,255,0
蓝色 Color. blue	0,0,255
黄色 Color. yellow	255,255,0
紫红 Color. magenta	255,0,255
青蓝 Color. cyan	0,255,255
粉红 Color. pink	255,175,175
橘色 Color. orange	255,200,0

表 8-1 中,RGB 分别代表红、绿、蓝 3 种原色,每种原色的色值从 0～255,色值越大,浓度越大,将不同色值的 3 种原色组合起来,就可以获得各种不同的颜色。

从 Color 类创建颜色对象时,可通过如下 3 种方法。

① 通过构造方法 Color(int red,int green,int blue) :3 个参数分别指定 3 种原色的色值,范围为 0～255。

② 通过构造方法 Color(float red,float green,float blue):3 个浮点型参数分别是红、绿、蓝 3 种原色的相对值,范围为 0.0～1.0 。

③ 直接使用 Color 中的常量。

例如,要创建红色,可使用 Color. red 或 new Color(255,0,0)。

要设置图形环境的颜色,使用 Graphics 对象提供的方法如下。

① void setColor(Colorc):设置图形环境用于绘图的当前颜色;

② Color getColor():返回图形环境当前颜色的值。

例如,设 g 是 graphics 对象,设置绘图的当前颜色为绿色的语句为

```
g.setColor(Color.green); 或 g.setColor(new Color(0,255,0));
```

要设置一个 GUI 组件的显示颜色,可调用 GUI 组件对象的方法如下。

① 组件对象. setBackground(Color c):设置背景色;

② 组件对象. setForeground(Color c):设置前景色。

例如,设置一个 JLabel 组件的背景色是黄色,前景色为红色的语句为

```
JLabel Jl1 = new Jlabel("姓名");
Jl1.setBackground(Color.YELLOW);
Jl1.setForeground(Color.RED);
```

例 8-1 颜色控制示例。

该应用程序绘制不同颜色的填充矩形和字符串。要注意的是:Java 的图形系统的坐标系中,像素原点(0,0)在 JFrame 的左上角。

```
// ShowColors.java
import java.awt.*;
import javax.swing.*;
public class ShowColors extends JFrame {
        public ShowColors() {
```

```
        super( "颜色的使用" );
        setSize( 400, 130 ); setVisible( true );
    }
    // draw rectangles and Strings in different colors
    public void paint( Graphics g )
    {   // call superclass's paint method
        super.paint( g );
        // set new drawing color using integers
        g.setColor( new Color( 255, 0, 0 ) );
        g.fillRect( 25, 25, 100, 20 );
        g.drawString( "Current RGB: " + g.getColor(), 130, 40 );
        // set new drawing color using floats
        g.setColor( new Color( 0.0f, 1.0f, 0.0f ) );
        g.fillRect( 25, 50, 100, 20 );
        g.drawString( "Current RGB: " + g.getColor(), 130, 65 );
        // set new drawing color using static Color objects
        g.setColor( Color.blue);
        g.fillRect( 25, 75, 100, 20 );
        g.drawString( "Current RGB: " + g.getColor(), 130, 90 );
    } // end method paint
    public static void main( String args[] )
    { ShowColors application = new ShowColors();
        application.setDefaultCloseOperation( JFrame.EXIT_ON_CLOSE );
    }
}
```

程序运行的结果如图 8-1 所示。

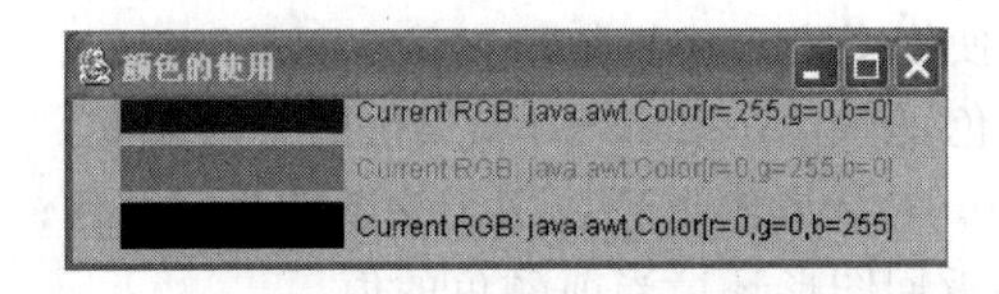

图 8-1 例 8-1 程序运行结果

8.3 字体控制

Font 类(在 Java. awt 包中)用于创建字体显示效果。字体的设置包括 3 个方面的内容:字体名称(name)、样式(style)和大小(size)。所以,创建字体对象的 Font 类的构造方法格式如下:

```
Font(String name,int style,int size);
```

其中 3 个参数的含义如下。

(1) name:字体名称,可以是运行程序的当前系统所支持的任何字体。使用 getFontList()函数获取系统可以使用的所有字体名称。

(2) style:字体风格,可以是 Font. BOLD(粗体)、Font. PLANE(正常)、Font. ITALIC(斜体),它们都是 Font 类的静态域。字体风格还可以组合使用,例如:Font. BOLD+ Font. ITALIC(粗斜体)。

(3) size:字体大小,它以磅(point)进行度量。1 磅等于 1/72 英寸。例如:

```
Font MyFont = new Font ("TimesRoman", Font.BOLD, 12 );
```

MyFont 对应的是 12 磅 TimesRoman 类型的黑体字。

Graphics 对象的 setFont 方法将当前绘制字体(将用于文本显示的字体)设置为该方法的 Font 参数。setFont 方法的格式为：

```
void setFont(Font f);
```

一旦调用了 Graphics 类的 setFont 方法，所有在调用后显示的文本均以新字体的形式进行显示，直至重新调用 setFont 方法改变该字体为止。

一个 Component 组件也是用 Component 对象的 setFont 方法设置当前绘制字体的显示效果。

例 8-2 以不同的字体效果，显示文本。

```
// Using fonts.
import java.awt. * ;
import javax.swing. * ;
public class Fonts extends JFrame {
    public Fonts()
    {   super( "设置字体" );
        setSize( 420, 125 );
        setVisible( true );
    }
    // display Strings in different fonts and colors
    public void paint( Graphics g ) {
        super.paint( g );
        g.setFont( new Font( "楷体", Font.BOLD, 12 ) );
        g.drawString( "楷体 12 bold .", 20, 50 );
        g.setFont( new Font( "楷体", Font.ITALIC, 24 ) );
        g.drawString( "楷体 24 italic.", 20, 80 );
        g.setFont( new Font( "宋体", Font.PLAIN, 14 ) );
        g.drawString( "宋体 14 plain .", 20, 100 );
        g.setColor( Color.RED );
        g.setFont( new Font( "宋体", Font.BOLD + Font.ITALIC, 18 ) );
        g.drawString( g.getFont().getName() + " " + g.getFont().getSize() +
            " bold italic.", 20, 120 );
    }
    public static void main( String args[] )
    { Fonts application = new Fonts();
        application.setDefaultCloseOperation( JFrame.EXIT_ON_CLOSE );
    }
}
```

程序运行的结果如图 8-2 所示。

图 8-2 以不同字体来显示文本

一旦调用了 Graphics 类的 setFont 方法，所有在调用后显示的文本均以新字体的形式进行显示，直至重新调用 setFont 方法改变该字体为止。

8.4 使用 Graphics 绘制图形

使用 Graphics 对象的画图方法绘制图形，包括绘制直线、矩形、圆、椭圆、弧、多边形和折线等。

1. 画直线

void drawLine(int x1,int y1,int x2,int y2)：(x1,y1)表示线段的一个坐标点,(x2,y2)表示线段的另一个坐标点。

2. 画多种类型矩形

void drawRect(int x, int y, int width, int height)：画空心矩形。(x,y)表示矩形左上角的坐标；width、height 表示矩形的宽度和高度。

void fillRect(int x, int y, int width, int height)：画填充型矩形。

void drawRoundRect(int x, int y, int width, int height, int arcWidth, int arcHeight)：画圆角矩形。即矩形的四个顶角呈圆弧状。arcWidth 代表了圆角弧的横向直径；arcHeight 代表了圆角弧的纵向直径。

void fillRoundRect(int x, int y, int width, int height, int arcWidth, int arcHeight)：画填充型圆角矩形。

void draw3DRect(int x, int y, int width, int height, boolean raised)：画 3D 矩形。raised 是定义该立体矩形是具有凸出(值为 true)还是凹下(值为 false)的效果。

fill3DRect(int x, int y, int width, int height, boolean raised)：画填充型 3D 矩形。

3. 画椭圆

void drawOval(int x, int y, int width, int height)：画空心椭圆。(x,y)表示椭圆所在的矩形左上角的坐标；width、height 表示矩形的宽度和高度。

void fillOval(int x, int y, int width, int height)：画填充型椭圆。

4. 绘制弧

弧为椭圆的一部分。用度数来衡量弧的角度,从起始角扫过弧张角(arc angle)所指定的度数,就绘制了一条弧。起始角是弧形开始时的度数。弧张角指弧形扫过的度数,弧张角为正数时,表示逆时针方向扫过;为负数时,表示顺时针方向扫过。

void drawArc(int x,int y,int width,int height, int startAngle, int arcAngle)：绘制一条弧形,弧段从 startAngle 开始绘制,并扫过 arcAngle 度数。(x,y)表示相对于边界矩形的左上角坐标,width、height 表示矩形的宽度和高度。

void fillArc(int x,int y,int width,int height, int startAngle, int arcAngle)：绘制填充型弧。

5. 绘制多边形和折线

Polygon 表示封闭式多边形(即起点和终点用直线连接)。Polyline 表示折线,即绘制一系列相连的线段(起点和终点不用直线连接)。绘制多边形或折线时,用一组点表示要绘制图形的顶点坐标。其中,这组点的 X 族的坐标用一维数组 int xPoints[]表示,Y 族的坐标用一维数组 int yPoints[]表示,这组点的点数用整型变量 point 表示。

void drawPolygon(int xPoints[], int yPoints[], int points)：画空心的封闭式多边形。

void fillPolygon(int xPoints[], int yPoints[], int points)：画填充型的封闭式多边形。

void drawPolygon(Polygon p)：画空心的封闭式多边形。

void fillPolygon(Polygon p)：画填充型的封闭式多边形。

Polygon(int xValues[], int yValues[], int numberOfPoints)：构造一个多边形对象。

void drawPolyline(int xPoints[], int yPoints[], int points)：画折线。

例 8-3 绘制线条、矩形、椭圆、弧、多边形和折线。

```
// Drawing graphics.
import java.awt.*;
import javax.swing.*;
public class DrawingGraphics extends JFrame {
        public DrawingGraphics() {
        //set window's title bar String
      super( "Drawing lines, rectangles and ovals" );
        setSize( 850, 170 );
        setVisible( true );
    }
    public void paint( Graphics g )
    {   super.paint( g ); // call superclass's paint method
        g.setColor( Color.RED );
        g.drawLine( 5, 33, 350, 33 );
        g.setColor( Color.BLUE );
        g.drawRect( 5, 40, 90, 55 );
        g.fillRect( 100, 40, 90, 55 );
        g.setColor( Color.CYAN );
        g.fillRoundRect( 195, 40, 90, 55, 50, 50 );
        g.drawRoundRect( 290, 40, 90, 55, 20, 20 );
        g.setColor( Color.YELLOW );
        g.draw3DRect( 5, 100, 90, 55, true );
        g.fill3DRect( 100, 100, 90, 55, false );
        g.setColor( Color.MAGENTA );
        g.drawOval( 195, 100, 90, 55 );
        g.fillOval( 290, 100, 90, 55 );
        int i = 400;
        g.setColor( Color.YELLOW );
        g.drawRect( 15 + i, 35, 80, 80 );
        g.setColor( Color.BLACK );
        g.drawArc( 15 + i, 35, 80, 80, 0, 360 );
        // start at 0 and sweep 110 degrees
        g.setColor( Color.YELLOW );
        g.drawRect( 100 + i, 35, 80, 80 );
        g.setColor( Color.BLACK );
        g.drawArc( 100 + i, 35, 80, 80, 0, 110 );
        // start at 0 and sweep -270 degrees
        g.setColor( Color.YELLOW );
        g.drawRect( 185 + i, 35, 80, 80 );
        g.setColor( Color.BLACK );
        g.drawArc( 185 + i, 35, 80, 80, 0, -270 );
        // start at 0 and sweep 360 degrees
        g.fillArc( 15 + i, 120, 80, 40, 0, 360 );
        // start at 270 and sweep -90 degrees
        g.fillArc( 100 + i, 120, 80, 40, 270, -90 );
        // start at 0 and sweep -270 degrees
        g.fillArc( 185 + i, 120, 80, 40, 0, -270 );
        int j = 700;
        int xValues[] = { 20 + j, 40 + j, 50 + j, 30 + j, 20 + j, 15 + j };
        int yValues[] = { 50, 50, 60, 80, 80, 60 };
        Polygon polygon1 = new Polygon( xValues, yValues, 6 );
        g.drawPolygon( polygon1 );
        int xValues2[] = { 70 + j, 90 + j, 100 + j, 80 + j, 70 + j, 65 + j, 60 + j };
        int yValues2[] = { 100, 100, 110, 110, 130, 110, 90 };
```

```
        g.drawPolyline( xValues2, yValues2, 7 );
    } // end method paint
    public static void main( String args[] )
    { DrawingGraphics application = new DrawingGraphics();
        application.setDefaultCloseOperation( JFrame.EXIT_ON_CLOSE );
    }
}
```

程序运行结果如图 8-3 所示。

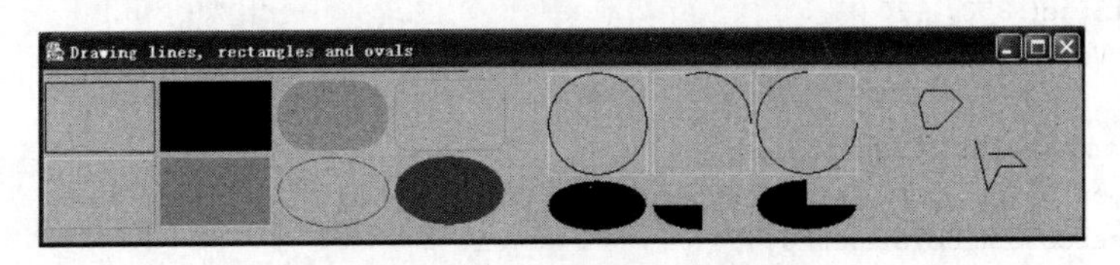

图 8-3　绘制线条、矩形、椭圆、弧、多边形和折线

8.5　Java 2D API

利用 Java 2D API,使程序员能描绘出任意的二维图形,并能设置图形的线条样式和填色效果,能对图形做旋转、缩放、扭曲等。与 Java 2D API 相关的包如下。

(1) java. awt:包含了一些新增的 2D API 类和接口。其中 Graphics2D 类扩展了 Graphics,以便对形状、文本和图像的展示提供更加完善的控制。Graphics2D 类新增了许多状态属性,如 Stroke (线条样式)、Paint(填充图案)、Clip(剪切路径)、Transform(坐标变换)等。与 Graphics2D 有关的状态属性集被称为 Graphics2D 上下文。

Java 2D 绘制进程是通过 Graphics2D 对象及其状态属性来控制的。要绘制文本、形状或图像,需设置 Graphics2D 上下文,然后调用一种 Graphics2D 绘制方法(例如 draw 或 fill)。常用的一组绘制方法的格式如下。

① void draw(Shape s):利用 Graphics2D 上下文中的 Stroke 和 Paint 对象描绘 Shape 的路径。

② void fill(Shape s):利用 Graphics2D 上下文中的 Paint 填充 Shape。

(2) java. awt. geom:包含能被绘制的图形类。所有的 2D 图形类都实现了 java. awt 的 Shape 接口,如 Line2D、Rectangle2D、Arc2D 、Eclipse2D、CubicCurve2D 等。这些类在指定图形的尺寸时,有 Float 精度和 Double 精度两种形式的坐标。例如,Line2D. Float(表示 Line2D 外部类中的嵌套静态类 Float)和 Line2D. Double(表示 Line2D 外部类中的嵌套静态类 Double)。要使用静态嵌套类,通过外部类名限定。

GeneralPath(通用路径)类能勾画出由直线和复杂曲线构成的任意几何图形。通用路径表示了生成一个几何图形的轨迹。要创建一个 GenaralPath 的对象,可通过调用构造方法

```
new GenaralPath();
```

GeneralPath 类提供一组方法用于定义通用路径。常用的方法如下。

① void moveTo(double x, double y):通过移动到指定坐标,实现增加一个点到路径中。

② void lineTo(double x, double y):通过从当前坐标到新指定坐标绘制一条直线到路径中。

(3) java. awt. font:包含用于文本布局和字体定义的类和接口。如碰撞侦测与打光(highlighting)等,以及决定 text 的适当位置与顺序。

(4) java. awt. color:用于颜色空间定义和颜色监视的类和接口。

(5) java. awt. image 和 java. awt. image. renderable:用于图像定义和绘制的类和接口。

(6) java. awt. print:用于打印所有基于 Java 2D 的文本、图形和图像的类和接口。

8.5.1 设置 Graphics2D 上下文

1. 设置画笔样式

画笔样式由实现 Stroke 接口的对象定义。画笔样式的引入可以为直线和曲线指定不同宽度和虚线图案。

BasicStroke 对象用于定义 Graphics2D 上下文的 Stroke 属性。BasicStroke 的构造方法用于定义的图形轮廓特性,包括线条宽度(width)、线条端点样式(end caps)、线段连接样式(line joins)和短画线图案(dash)等。它的构造方法一般形式为

```
BasicStroke(float width,int cap, int join, float miterlimit, float[] dash, float dash_phase)
```

从第二个参数开始的后面的各参数可逐个缺省,缺省时将采用默认值。参数的含义如下。

① width:画笔的线条宽度,是垂直于画笔轨迹的测量值。

② caps:是线段末端的装饰。有 3 个不同的 BasicStroke 类中的常量值:CAP_BUTT(端点无修饰)、CAP_ROUND(圆形端点) 和 CAP_SQUARE(方形端点)。

③ join:定义两个线段汇合处的连接样式。有 3 个不同的 BasicStroke 类中的常量值:JOIN_BEVEL(扁平角)、JOIN_MITER(尖角)和 JOIN_ROUND(圆角)。

④ miterlimit :对具有 JOIN_MITER 装饰的线段连接点的剪裁限制。miterlimit 必须大于或等于 1.0 f。

⑤ dash:数组中的元素交替代表短画线的尺寸和短画线之间的间距。元素 0 代表第一个短画线,元素 1 代表第一个间距。

⑥ dash_phase:定义短画线图案开始位置的偏移量。

要设置或改变 Graphics2D 上下文中的 Stroke 属性,请调用方法 setStroke(Stroke s)。

2. 指定绘图颜色和填充图案

用 Graphics2D 的 setPaint 方法设置 Paint 对象,以确定绘图颜色和填充图案。

GradientPaint 和 TexturePaint 是实现 Paint 接口的类。GradientPaint 定义在两种颜色间渐变的填充图案。而 TexturePaint 则利用重复图像片段的纹理方式定义一种填充图案。

GradientPaint 类提供了使用一种颜色到另一种颜色的渐变填充形状的方法。创建 GradientPaint 对象的构造方法如下。

(1) `GradientPaint(float x1, float y1, Color color1, float x2, float y2, Color color2)`

功能:构造一个非周期性的 GradientPaint 对象。指定开始位置(x1,y1)和颜色(color1)到结束位置(x2,y2)和颜色(color2)。填充颜色将沿着两个位置之间的直线成比例地从一种颜色变为另一种颜色。

(2) `GradientPaint(float x1, float y1, Color color1, float x2, float y2, Color color2,boolean cyclic)`

功能:cyclic 值为 true 时支持从开始位置(x1,y1)到结束位置(x2,y2)的颜色渐变,即从 color1 变化到 color2 再变化到 color1 的渐变;在填充区域内可重复应用渐变,从而形成花纹效果。

要使用颜色渐变填充一个 Shape 图形对象,其过程如下:

① 创建 GradientPaint 对象;

② 调用 Graphics2D. setPaint;

③ 创建 Shape；

④ 调用 Graphics2D. fill(shape)。

在下面的例 8-4 中，将使用蓝—黄—蓝色渐变来填充椭圆形。

TexturePaint 类利用纹理填充图形。要使用纹理填充一个 Shape 图形对象，其过程如下：

① 创建 TexturePaint 对象；

② 调用 Graphics2D 对象的 setPaint 方法；

③ 创建 shape；

④ 调用 Graphics2D. fill(shape)。

在下面的例 8-4 中，将使用由缓冲区图像创建的简单纹理来填充矩形。

8.5.2 使用 Graphics2D 绘制图形

利用 Graphics2D 对象绘制 2D 图形的基本步骤如下。

① 通过如下语句产生 Graphics2D 对象。

```
Graphics2D g2d = (Graphics2D)g;  //g 是 Graphics 对象
```

② 创建要绘制的 Shape 图形对象。

③ 使用 Graphics2D 对象的 set 方法，设定 Graphics2D 上下文的状态属性。

④ 调用 Graphics2D 对象所提供的方法 draw 或 fill，完成绘制图形的操作。

例 8-4 应用画笔样式、颜色渐变和纹理填充，绘制 Java 2D 图形。

```
import java.awt. * ;
import java.awt.geom. * ;
import java.awt.image. * ;
import javax.swing. * ;
public class Shapes extends JFrame {
    public Shapes()
    {  super( "画 2D 图形" );
       setSize( 600, 190 );
       setVisible( true );
    }
    //画 2D 图形
    public void paint( Graphics g )
    {   super.paint( g ); // call superclass's paint method
        Graphics2D g2d = ( Graphics2D ) g; // cast g to Graphics2D
        // 使用蓝-黄-蓝色渐变来填充 2D 椭圆。
        g2d.setPaint( new GradientPaint( 5, 40, Color.BLUE, 15, 50,
            Color.YELLOW, true ) );
        g2d.fill( new Ellipse2D.Double( 5, 40, 65, 110 ) );
        // 使用颜色 red 和线宽 10.0f 画 2D rectangle
        g2d.setPaint( Color.RED);
        g2d.setStroke( new BasicStroke( 10.0f ) );
        g2d.draw( new Rectangle2D.Double( 80, 40, 65, 110 ) );
          //利用纹理填充图形
          // 创建大小为 5×5 的缓冲区纹理图像片
        BufferedImage bi = new BufferedImage(5, 5, BufferedImage.TYPE_INT_RGB);
        Graphics2D big = bi.createGraphics();
        // 绘制 BufferedImage 图形以创建纹理
        big.setColor(Color.blue);
        big.fillRect(0,0,5,5);          // draw a filled rectangle
        big.setColor(Color.yellow);
```

```
        big.fillOval(0,0,5,5);          // draw a filled Oval
        // 由缓冲区图像创建纹理画图
        Rectangle r = new Rectangle(0,0,5,5);
        TexturePaint tp = new TexturePaint(bi,r);
        // 将纹理画图添加到图形上下文中
        g2d.setPaint(new TexturePaint(bi,new Rectangle(0,0,5,5)));
        // 创建并绘制使用该纹理填充的矩形
        g2d.fill( new RoundRectangle2D.Double( 155, 40, 75, 110, 50, 50 ) );
        // 使用颜色 green 和线宽 6.0f 填充 2D 弧
        g2d.setPaint( Color.green );
        g2d.setStroke( new BasicStroke( 6.0f ) );
        g2d.draw( new Arc2D.Double( 240, 40, 75, 110, 0, 270, Arc2D.PIE ) );
        // 使用颜色 yellow 和线宽 6.0f 画 2D 线
        g2d.setPaint( Color.yellow );
        g2d.draw( new Line2D.Double( 395, 30, 320, 150 ) );
        // 使用颜色 Red 和线宽 6.0f 画 2D 虚线
        float dashes[] = { 10,6 };
        g2d.setPaint( Color.red);
        g2d.setStroke( new BasicStroke( 4, BasicStroke.CAP_ROUND,
            BasicStroke.JOIN_ROUND, 10, dashes, 0 ) );
        g2d.draw( new Line2D.Double( 320, 30, 395, 150 ) );
        // 使用通用路径画四面体
        float x[] = new    float[4];
        float y[] = new    float[4];
        x[0] = 500;         y[0] = 30;
        x[1] = 500;         y[1] = 130;
        x[2] = 450;         y[2] = 180;
        x[3] = 550;         y[3] = 180;
        g2d.setPaint(Color.yellow);
        g2d.setStroke( new BasicStroke( 4));
        GeneralPath    tour = new    GeneralPath();
        tour.moveTo(x[0],y[0]);
        for(int i = 1;i<x.length;i ++ )
            tour.lineTo(x[i],y[i]);
        tour.lineTo(x[1],y[1]); tour.lineTo(x[2],y[2]);
        tour.lineTo(x[3],y[3]); tour.lineTo(x[0],y[0]);
        tour.lineTo(x[2],y[2]); tour.closePath();
        g2d.draw(tour);
    }
    public static void main( String args[] ) {
        Shapes application = new Shapes();
        application.setDefaultCloseOperation( JFrame.EXIT_ON_CLOSE );
    }
}
```

程序运行结果如图 8-4 所示。

图 8-4　绘制 Java 2D 图形

8.6 小 结

Graphics 对象提供一组方法，以绘制各种图形、文本和控制颜色等。

Component 类的 paint 方法，以 Graphics 对象为参数，在响应某个事件时调用，以达到在 Component 组件上绘图。repaint()方法用于间接地调用 paint()方法，以达到在 Component 组件上重新绘图。

Graphics 对象绘制直线、矩形、椭圆、弧的方法分别是 drawLine、drawRect、drawOval、drawArc，绘制封闭式多边形和非封闭式多边形折线的方法是 drawPolygon 和 drawPolyline，绘制填充型的图形则是将对应的绘图方法中的 draw 改成 fill。

Color 类声明了用于操作颜色的方法和常量。Graphics 对象的 setColor 和 getColor 方法用于设置和得到当前的绘图颜色。Component 类的 setBackground 和 setForground 方法用于设置组件的背景色和前景色。

Font 类用于设置字体。Graphics 对象与 Component 对象调用 setFont 方法将设置当前绘制字体。

Java 2D 绘制进程是通过 Graphics2D 对象及其状态属性来控制的。要绘制文本、形状或图像，需设置 Graphics2D 上下文，然后调用一种 Graphics2D 绘制方法(draw 或 fill)。

习 题

8.1 概念题。

① Graphics 类的作用？如何设置 Graphics 对象的绘图颜色？如何设置 Graphics 对象的绘制文本字体效果？绘制一个绘制封闭式的填充型多边形的方法是什么？

② Paint 方法和 repaint 方法的作用分别是什么？

③ Color 类的作用是什么？

④ Graphics2D 类的作用是什么？绘制 2D 图形时，如何得到 Graphics2D 对象？在用 Graphics2D 对象绘图时，如何设置画笔的样式？如何设置 Graphics2D 绘图颜色和填充图案？

⑤ TexturePaint 和 GradientPaint 类的作用是什么？

⑥ Graphics2D 的 draw 和 fill 方法的参数类型是什么？

⑦ GeneralPath 类的作用是什么？

8.2 编写一个程序，以随机颜色绘制随机长度的直线。

8.3 编写一个程序，以随机颜色绘制 8 个同心圆，圆与圆之间相差 10 个像素。使用 Graphics 类的 drawOval 方法。

8.4 编写一个程序，以随机颜色绘制 8 个同心圆，圆与圆之间相差 10 个像素。使用 Graphics 类的 drawArc 方法。

8.5 编写一个程序，使用 GeneralPath 类绘制一个彩色的五角星。

8.6 编写一个程序，绘制一个彩色的立方体。要求结合虚线和线宽绘制线段。

8.7 使用红—黄—红色渐变来绘制一个填充矩形。

8.8 编写一个程序，利用 GeneralPath 绘制一个正弦曲线。

第9章 GUI组件与用户界面设计

一个程序的图形用户界面(Graphical User Interfaces,GUI)给用户提供了交互式的图形化操作界面,通过GUI接受用户的输入并向用户输出程序运行的结果。本章将要讲述的主要内容有:与设计GUI相关的AWT和Swing中的常用组件、布局管理器以及事件处理机制。并通过大量程序设计例子,介绍GUI程序设计的方法。通过本章的学习,可使用户掌握使用AWT包和Swing包构建图形用户界面。

9.1 AWT和Swing组件概述

Java语言中,为了方便图形用户界面的开发,设计了专门的类库用来生成各种标准图形界面元素和处理图形界面的各种事件。用来生成图形用户界面的类库主要在java.awt包和javax.swing包中。

1. AWT介绍

AWT是Abstract Window Toolkit(抽象窗口工具集)的缩写,java.awt包提供了基本的Java程序的GUI设计工具。包中的主要类或接口之间的继承关系如图9-1所示。

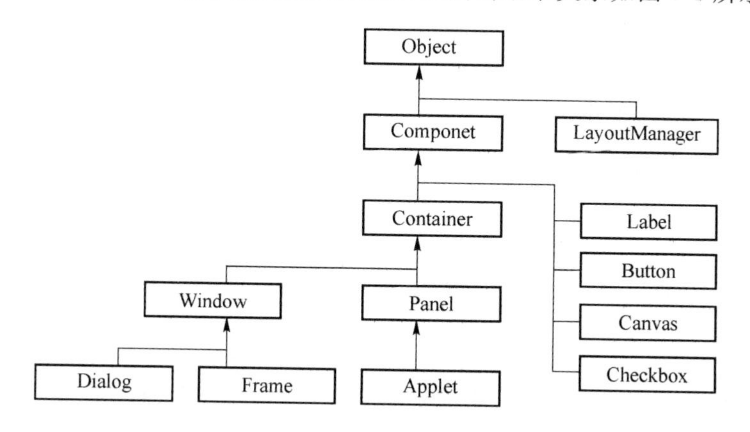

图9-1 AWT包中的组件间继承关系

(1) 组件

组件(Component)是一个以图形化的方式显示在屏幕上并能与用户进行交互的对象,例如一个按钮(Button)、一个标签(Label)等,组件通常被放在容器中。Component类是抽象类,定义了所有组件所具有的通用特性和行为,并派生出其他所有的组件。Component类提供的功能如下。

- 基本的绘画支持。方法repaint()、paint()、update()等用来在屏幕上绘制组件。

• 外形控制。包括字体、颜色等。相应的方法有：getFont()、setFont()、setBackground()、SetForeground()等。
• 大小和位置控制。一个组件的大小和位置主要是由包括该组件的容器所设置的布局决定的，但组件也提供了一些方法来指明合适的大小和位置，相应的方法包括 setSize()、setLocation()等。
• 图像处理。Component 类实现了接口 ImageObserver，其中的方法 imageUpdate()用来进行图像跟踪。
• 组件的状态控制。例如：setEnable()控制组件用来判断是否接收用户的输入，isEnable()、isVisible()、isValid()用来返回组件的状态。

(2) 容器

容器(Container)是 Component 的子类，它具有组件的所有性质，同时又具有容纳其他组件和容器的功能。每个容器用方法 add()向容器添加某个组件，用方法 remove()从容器中删除某个组件。每个容器都与一个布局管理器相连，以确定容器内组件的布局方式。容器可以通过方法 setLayout()设置某种布局。

(3) 布局管理器

布局管理器(LayoutManager)用于管理组件在容器中的布局方式。布局管理器类都实现了接口 LayoutManager。Java 系统提供的标准布局管理器类有：FlowLayout、BorderLayout、GridLayout、CardLayout、BoxLayout 和 GridBagLayout。

2. Swing 介绍

Swing 组件在 javax. swing 包中。它具有如下特点。

(1) Swing 组件是用 100%纯 Java 代码实现的轻量级(light-weight)组件，没有本地代码，不依赖操作系统的支持，这是它与重量级组件 AWT 的最大区别。由于 AWT 组件通过与具体平台相关的对等类(Peer)实现，因此 Swing 比 AWT 组件具有更强的实用性和美观性。

(2) Swing 组件的多样化。Swing 是 AWT 的扩展。Swing 组件都以"J"开头，除了有与 AWT 类似的按钮(JButton)、标签(JLabel)、复选框(JCheckBox)、菜单(JMenu)等基本组件外，还增加了一个丰富的高层组件集合，如表格(JTable)、树(JTree)。大多数 Swing 组件从 JComponent 继承而来，JComponent 是一个抽象类，类继承层次结构如图 9-2 所示，它用于定义所有子类组件的一般方法，例如：使用 setBorder()方法可以设置组件外围的边框；使用 setTooltipText()方法为组件设置对用户有帮助的提示信息。

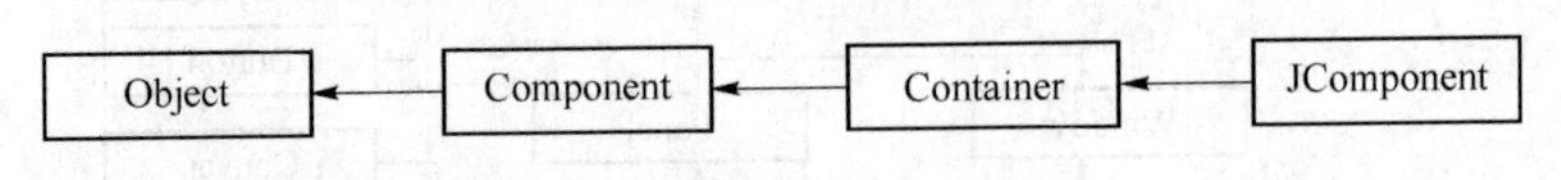

图 9-2　大多数 Swing 组件的超类

(3) 可插入的外观感觉。Swing 使得程序在一个平台上运行时能够有不同的外观。用户可以选择自己习惯的外观。

3. Swing 组件的分类

Swing 组件从功能上分为：

(1) 顶层容器。如 JFrame、JApplet、JDialog。

(2) 普通容器。如 JPanel、JScrollPane、JSplitPane、JToolBar。

(3) 特殊容器。在 GUI 上起特殊作用的中间层。如 JInternalFrame、JLayeredPane。

(4) 向用户显示不可编辑信息的组件。如 JLabel、JProgressBar、ToolTip。

(5) 向用户显示能被编辑信息的组件。如 JTextArea、JTextField、JColorChooser、JFile-

Chooser、JTable。

(6) 实现人机交互的控制组件。如 JButton、JComboBox、JList、JMenu、JSlider。

4. 使用 Swing 的基本规则

与 AWT 组件不同，Swing 组件不能直接添加到顶层容器中，它必须添加到一个与 Swing 顶层容器相关联的内容面板(content pane)上。内容面板是顶层容器包含的一个普通容器。

例如，对 JFrame 顶层容器而言，添加组件有两种方式。

方式 1：用 getContentPane( )方法获得 JFrame 的内容面板，再对其加入组件：

```
frame.getContentPane().add(childComponent);
```

方式 2：建立一个 JPanel 或 JDesktopPane 之类的中间容器，把组件添加到容器中，用 setContentPane()方法把该容器置为 JFrame 的内容面板：

```
JPanel contentPane = new JPanel( );
…//把其他组件添加到 JPanel 中;
frame.setContentPane(contentPane); //把 contentPane 对象设置为 frame 的内容面板
```

9.2 事件处理模型

Java 的图形用户界面是事件驱动的，即当用户与 GUI 组件交互时会引发一个系统预先定义好的事件。一些常见的事件包括：鼠标移动、单击鼠标按钮、鼠标单击按钮(button)、在文本框中进行输入、在菜单中选择菜单项、关闭窗口等。GUI 事件是一个从 java.awt.event.AWTEvent 类扩展的某个子类对象。图 9-3 说明了 java.awt.event 包中的事件类的继承层次，它包含了由 AWT 组件所激发的各类事件的接口和类。本章将在以下各节结合 GUI 组件特性讨论其中的一些事件类。

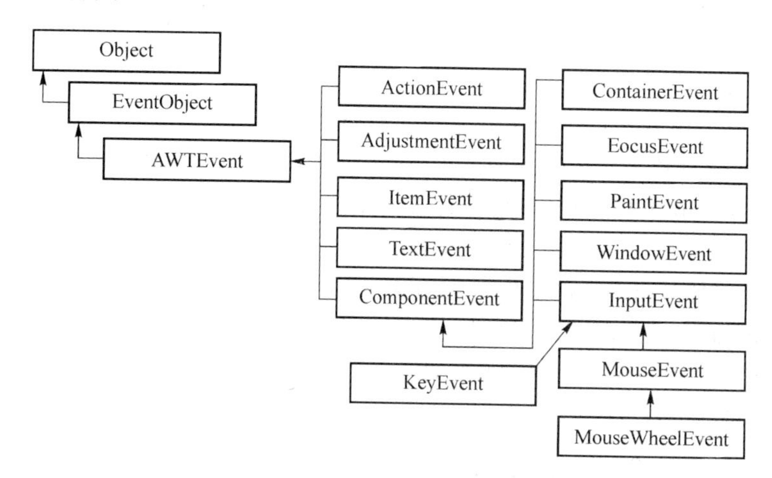

图 9-3 java.awt.event 包中的事件类

1. Java 中的事件处理模型

Java 中的事件处理模型由以下 3 部分组成。

- 事件源(Event source)：是用户交互的各种 GUI 组件。
- 事件对象(Event object)：封装了发生事件的有关信息。
- 事件监听器(Event listener)：当事件发生时被通知到接收事件的事件监听对象，然后调用事件监听对象中对应的方法响应该事件。

一个事件监听对象是实现了系统规定的事件监听接口的类的对象。事件监听接口提供了事件处理的抽象方法的描述。如图 9-4 所示，java. awt. event 包定义了 Java 系统的一组事件监听接口类型，一个监听接口往往声明了一个以上的抽象方法。每个抽象方法对应着要处理的事件动作，由用户实现它。

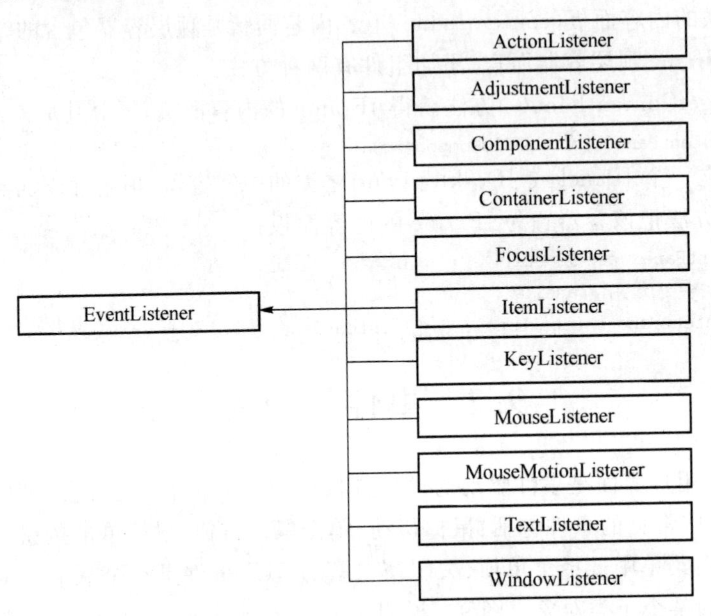

图 9-4　Java 系统的事件监听接口

2. 对组件的响应和处理

当需要对组件的某种事件进行响应和处理时，程序员必须完成两个步骤：

(1) 为组件注册实现规定接口的事件监听器；

(2) 实现事件监听器接口中声明的事件处理抽象方法。

例如，用户用鼠标单击了按钮对象 button，则该按钮 button 就是事件源，而 java 运行时系统会生成 ActionEvent 事件对象，该事件封装了用户单击事件源发生时的一些信息；然后，该事件被传递给事件源 button 通过 addActionListener()方法注册了的监听 ActionEvent 事件的 ButtonHandler 监听器对象。事件发生时自动调用 ButtonHandler 对象中actionPerformed()方法处理该事件。actionPerformed()方法定义了事件处理过程，在 ActionListener 接口中只定义了一个方法 actionPerformed()，它用来接收一个 ActionEvent。Button 的事件注册和事件处理过程如图 9-5 所示。

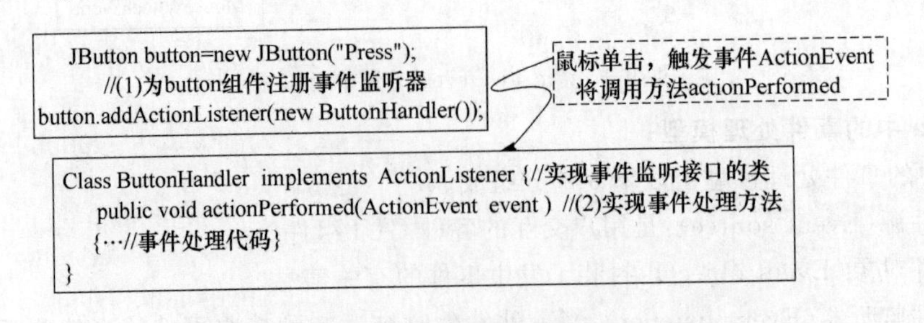

图 9-5　JButton 组件的事件注册和事件处理模型

3. ActionEvent 动作事件

ActionEvent(动作事件)指示发生了组件已定义动作的语义事件。当特定组件(例如 Button)的动作(例如单击按钮)发生时,由组件生成此动作事件。该事件被传递给使用组件的 addActionListener 方法注册的每一个 ActionListener 对象,用以接收这类事件。

能够触发这个事件的动作主要包括:

- 单击按钮(注意要使用键盘在 Button 上触发 ActionEvent,请使用空格键);
- 双击一个列表中的选项;
- 选择菜单项;
- 在文本框中输入回车键。

ActionEvent 事件常用的一些方法如下。

(1) String getActionCommand( ):返回引发事件的动作的命令字符串。例如,对于 Button 按钮返回的是 Button 的标签,对于 JTextField 返回的是文本内容。

(2) String getSource( ):返回引发该事件的组件对象(事件源)的引用。使用 getSource( )方法可以区分产生动作命令的不同事件源。

Java 程序提供了处理事件的 3 种技术:实现接口、使用适配器类(adapter class)和使用内部类(inner class)。有的监听器常常能监听多个相关事件,监听器接口中就提供了多个事件处理的方法 (Method),监听器接收到某个事件后,会自动调用相关的事件处理方法。在本章后面的各节中,介绍各个 GUI 组件时,将结合相应的事件类型和事件监听器接口类型进行程序设计。

9.3 命令按钮

命令按钮(JButton)是一种组件,当用户通过单击按钮时将产生动作事件 ActionEvent,以触发特定的动作代码,从而完成预先指定的功能。Swing 中 JButton 既支持文本按钮,也支持图像按钮,或两者兼有。

1. 创建按钮对象

JButton 类常用的一组构造方法如下。

(1) JButton(String text):创建一个带文本的按钮。

(2) JButton(Icon icon) :创建一个带图标的按钮。

(3) JButton(String text, Icon icon) :创建一个带文本和图标的按钮。

2. 按钮对象的常用方法

(1) getLabel( ):返回按钮的标签字符串。

(2) setLabel(String s):设置按钮的标签为字符串 s。

3. ActionEvent 事件响应

当用户单击一个按钮时就引发了一个动作事件 ActionEvent。如果希望程序能响应按钮引发的动作事件,则通过调用 JButton 的方法 addActionListener 给该按钮注册实现 ActionListener 接口的事件监听器,同时,为这个接口的 actionPerformed(ActionEvent e)方法书写方法体。在方法体中,可以调用 e. getSource( )方法来获取引发动作事件的按钮对象引用,也可以调用 e. getActionCommand( )方法来获取按钮的标签。

例 9-1 文本按钮和图像按钮的程序设计。

```
import java.awt. * ;
import java.awt.event. * ;
import javax.swing. * ;
public class ButtonTest extends JFrame {
    private JButton plainButton, fancyButton;
    public ButtonTest() // set up GUI
    {   super( "JButton 的程序设计" );
        // get content pane and set its layout
        Container container = getContentPane();
        container.setLayout( new FlowLayout() );
        // create buttons
        plainButton = new JButton( "文本按钮" );
        container.add( plainButton );
        Icon bug1 = new ImageIcon( getClass().getResource("bug1.gif" ));
        Icon bug2 = new ImageIcon( getClass().getResource("bug2.gif" ));
        //fancyButton 的默任图标 bug1
        fancyButton = new JButton( "文本图像按钮", bug1 );
        //当鼠标置于该按钮上时,显示图标 bug2
        fancyButton.setRolloverIcon( bug2 );
        container.add( fancyButton );
        // create an instance of class ButtonHandler for button event handling
        ButtonHandler handler = new ButtonHandler();
        fancyButton.addActionListener( handler );
        plainButton.addActionListener( handler );
        setSize( 275, 100 );
        setVisible( true );
    } // end ButtonTest constructor
    public static void main( String args [] )
    {  ButtonTest application = new ButtonTest();
       application.setDefaultCloseOperation( JFrame.EXIT_ON_CLOSE );
    }
    // inner class for button event handling
    private class ButtonHandler implements ActionListener {

        public void actionPerformed( ActionEvent event ) // handle button event
        {JOptionPane.showMessageDialog( ButtonTest.this,
         "你按下的是：" + event.getActionCommand() );
        }
    } // end private inner class ButtonHandler
} // end class ButtonTest
```

程序运行显示结果如图 9-6 所示。当鼠标单击标签为“文本按钮”按钮时,将显示第一个对话框;当鼠标置于标签为“文本图像按钮”按钮时,小图标图像会改变,单击该按钮时,会显示第二个对话框。

图 9-6　例 9-1 程序的运行结果

9.4 标签、单行文本框、多行文本域与滚动条面板

标签(JLabel)、单行文本框(JTextField、JPasswordField)与多行文本域(JTextArea)是搭建 GUI 图形界面使用最多的 Swing 组件,下面分别介绍它们。

9.4.1 标签

Swing 中的 JLabel 是用户不能修改只能查看其内容的文本显示区域,它起到信息说明的作用。JLabel 对象可以显示文本和图像。

使用 JLabel 类的构造方法创建标签对象。JLabel 类常用的构造方法如下。

(1) JLabel(String text):创建具有指定文本 text 的 JLabel 对象。

(2) JLabel(Icon image):创建具有指定图像 image 的 JLabel 对象。

(3) JLabel(String text int horizontalAlignment):创建具有指定文本和水平对齐方式的 JLabel 对象。horizontalAlignment 指定 JLabel 对象在显示区域的水平对齐方式,其值是 SwingConstants 接口中定义的以下静态常量之一:LEFT、CENTER、RIGHT、LEADING 或 TRAILING。

(4) JLabel(String text, Icon icon, int horizontalAlignment):创建具有指定文本、图像和水平对齐方式的 JLabel 对象。该标签在其显示区内垂直居中对齐。文本位于图像的结尾。

设置或读取 JLabel 对象属性的常用方法如下。

(1) setText(String text) 和 String getText():设置或读取 JLabel 显示的文本。

(2) setIcon(Icon image)和 getIcon():设置或读取 JLabel 显示的图像。

(3) setForeground(Color c)和 setBackground(Color c):设置 JLabel 显示文本的前景色和背景色。

(4) setFont(Font f):设置 JLabel 显示的文本字体。

例如,创建文本内容是"文本标签"且颜色是红色的 JLabel 实例的语句:

```
label1 = new JLabel( "文本标签" );
label1.setForeground(Color.RED);
```

又如,创建 JLabel 实例,其文本内容为"文本图像标签",图像是 czims.jpg,水平靠左对齐,语句如下:

```
Icon bug = new ImageIcon( "czims.jpg" );
label2 = new JLabel( "文本图像标签", bug, SwingConstants.LEFT );
```

9.4.2 单行文本框与多行文本域

Swing 中用于文本编辑的基本组件有单行文本框 JTextField、JPasswordField 和多行文本域 JTextArea。它们来自于包 javax.swing,并从 JTextComponent 父类派生。JTextField 和 JPasswordField 允许编辑单行文本,但 JTextField 显示用户输入的内容,而 JPasswordField 隐藏用户实际输入的内容。多行文本域 JTextArea 允许用户编辑多行文本,往往利用 JScrollPane 滚动条面板提供内容水平滚动和纵向滚动。

1. JTextField 和 JPasswordField

JTxtField 常用的构造方法如下。

(1) JTextField():构造一个内容为空的 JTextField 实例。

(2) JTextField(int columns):构造一个内容为空和指定列宽的 JTextField 实例。

(3) JTextField(String text) :构造一个用指定文本的 JTextField 实例。

(4) JTextField(String text, int columns) :构造一个指定文本和列宽的 JTextField 实例。

JPassword 类的构造方法类似于 JTextField,这里不再赘述。

2. JTextArea

JTextArea 常用的一组构造方法如下。

① JTextArea():构造一个内容为空的 JTextArea 实例。

② JTextArea(int rows, int columns):构造一个具有指定行数和列数的内容为空的 JTextArea 实例。

③ JTextArea(String text):构造显示指定文本的 JTextArea 实例。

④ JTextArea(String text, int rows, int columns) :构造具有指定文本、行数和列数的 JTextArea 实例。

3. 常用方法

操作 JTextComponent 对象的常用方法如下。

① void setText(String)和 String getText():设置或取得文本内容。

② void setEditable(boolean b):设置 TextComponent 是否为可编辑的(b 值为 true 时)。

③ String getSelectedText():返回 TextComponent 中包含的选定文本。

4. TextEvent 和 ActionEvent 事件响应

JTextField 和 JTextArea 的事件响应首先由它们的父类 TextComponent 决定,所以先讨论 TextComponent 的事件响应。TextComponent 可以引发一种事件 TextEvent:当用户对文本做编辑修改操作时,将引发 TextEvent 文本改变事件。在此基础上 JTextField 还比 JTextArea 多产生一种事件,当用户在文本框中按回车键时,将引发 ActionEvent 动作事件。JTextArea 却不能产生 ActionEvent 事件,也没有 addActionListener( )这个方法。

9.4.3 滚动条面板

滚动条面板 JScrollPane 是一个容器组件,其作用是通过滚动条在一个较小的区域显示较多的内容。

JScrollPane 常用的构造方法是 JScrollPane(Component view),即创建一个显示指定组件内容的 JScrollPane,只要组件的内容超过视图大小就会显示水平和垂直滚动条。

在例 9-2 中,会将两个 JTextArea 对象分别放入两个 JScrollPane 滚动条面板对象中,是为了在 JTextArea 的文本内容足够多时提供水平滚动条和垂直滚动条。

例 9-2 JLabel、JTextField、JPasswordField 和 JTextArea 的应用。

```
import java.awt.*;
import java.awt.event.*;
import javax.swing.*;
public class LabelandTextTest extends JFrame {
    private JTextField textField1, textField2,textField3;
```

```
    private JTextArea textArea1,textArea2;
    private JPasswordField passwordField;
    private JButton copyButton;
    public LabelandTextTest()                    // set up GUI
    {   super( "JTextField,JPasswordField 和 JTextArea 的使用" );
        Container container = getContentPane();
        container.setLayout( new FlowLayout() );
        JLabel label1 = new JLabel( "用户名:" );
        container.add( label1 );
        textField1 = new JTextField( 10 );
        container.add( textField1 );
        container.add( new JLabel( "口令:" ));
        passwordField = new JPasswordField( 10 );
        container.add( passwordField );
        Icon bug = new ImageIcon( - getClass().getResource( "czims.jpg" ));
        JLabel label2 = new JLabel( "照片:" , bug, SwingConstants.LEFT );
//设置 label2 的文本相对于图片的位置:水平方向居中对齐,垂直方向在图片的底部
        label2.setHorizontalTextPosition( SwingConstants.CENTER );
        label2.setVerticalTextPosition( SwingConstants.BOTTOM );
        container.add( label2 );
        textField3 = new JTextField( "显示用户名或口令", 30 );
        textField3.setEditable( false );        //设置 textField3 不能修改
        container.add( textField3 );
        // register event handlers
        TextFieldHandler handler = new TextFieldHandler();
        textField1.addActionListener( handler );
        passwordField.addActionListener( handler );
        Box box = Box.createHorizontalBox(); //创建一个水平的 Box 对象
        //set up textArea1
        String string = "This is a demo string to\n" +
            "illustrate textarea programming \n" ;
        textArea1 = new JTextArea( string, 5, 12 );
        box.add(new JScrollPane(textArea1));       //将 textArea1 加入 Box
        copyButton = new JButton( "拷贝 >>>" );
        box.add( copyButton );                     //将 copyButton 加入 Box
        copyButton.addActionListener(              // 给 copyButton 注册事件监听器
            new ActionListener() {                 // anonymous inner class
                // set text in textArea2 to selected text from textArea1
                public void actionPerformed( ActionEvent event )
                { textArea2.setText( textArea1.getSelectedText() );
                }
            } // end anonymous inner class
        ); // end call to addActionListener
        // set up textArea2
        textArea2 = new JTextArea( 5, 12 );
        textArea2.setEditable( false );
        box.add(new JScrollPane(textArea2));       //将 textArea2 加入 Box
        // add box to content pane
        container.add( box ); //将 Box 对象加入 container
        setSize( 500, 400 );
        setVisible( true );
```

```
    } // end constructor
    public static void main( String args [] )
    {   LabelandTextTest application = new LabelandTextTest();
        application.setDefaultCloseOperation( JFrame.EXIT_ON_CLOSE );
    }
    // private inner class for event handling for two TextFields
    private class TextFieldHandler implements ActionListener {
        public void actionPerformed( ActionEvent event )        {
            String string = "";
            // user pressed Enter in JTextField textField1
            if ( event.getSource() == textField1 )
              string = "textField1: " + event.getActionCommand();
            // user pressed Enter in JTextField passwordField
            else if ( event.getSource() == passwordField ) {
            string = "passwordField: " +
                new String( passwordField.getPassword() );
            }
            textField3.setText( string );
        } // end method actionPerformed
    } // end private inner class TextFieldHandler
}
```

程序运行输出结果如图 9-7 所示。

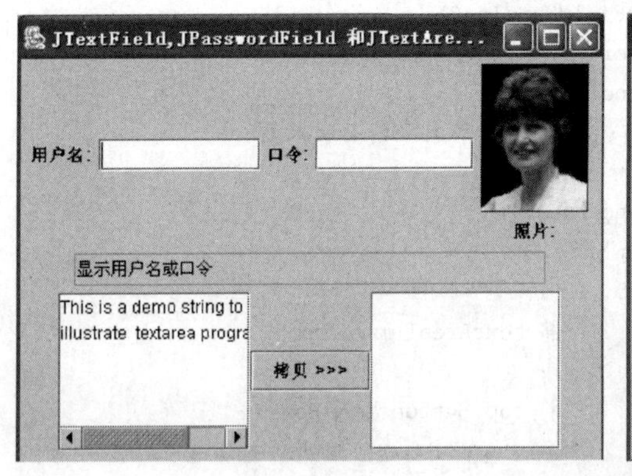

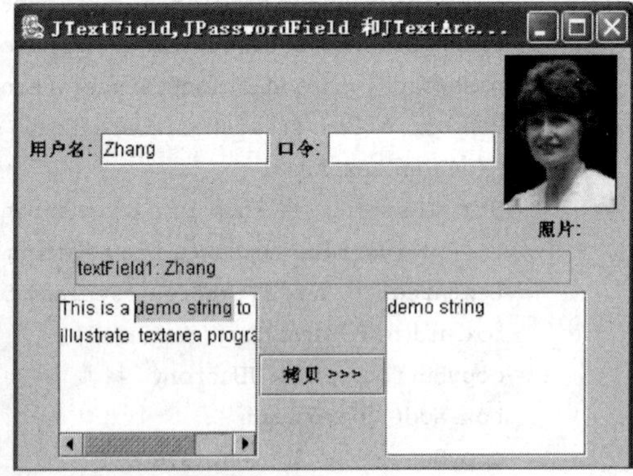

图 9-7 程序运行的部分输出结果

程序的功能如下。

(1) 当用户分别在用户名和口令的文本框中输入内容并按回车键后，将在标签为“显示用户名或口令”的文本框中显示用户输入的内容。

(2) 在图的下半部分定义了一个水平方向的 Box 对象，加入 Box 的组件按水平方向排列。在 Box 对象中加入了 3 个 GUI 组件：两个 JTextArea 和一个 JButton。用户可通过鼠标选择左边的 textArea1 的内容，然后鼠标单击标签为“拷贝”按钮，会将选中的 textArea1 的内容复制到右边 JScrollPane 中。两个 JTextArea 对象分别放入两个 JScrollPane 滚动条面板对象中，是为了在 JTextArea 的文本内容足够多时提供水平滚动条和垂直滚动条。

9.5 复选框按钮和单选按钮

SwingGUI 组件包含 3 种类型的状态按钮：JToggleButton、JCheckBox 和 JRadioButton，这些按钮都有开/关（真/假）两种值，使用户能在给定的数据中进行选择。JCheckBox 和 JRadioButton 是 JToggleButton 的子类。

1. 复选框

复选框（JCheckBox）提供简单的"on/off"开关，旁边显示文本标签。它只有两种状态：选中和未选中。在一组复选框中，可以同时选择多项。

复选框的常用的构造方法如下。

(1) JCheckBox(String text)：创建一个带文本的、最初未被选定的复选框。

(2) JCheckBox(String text, boolean selected)：创建一个带文本的复选框，并指定其最初是否处于选定状态。

(3) JCheckBox(Icon icon, boolean selected)：创建一个带图标的复选框，并指定其最初是否处于选定状态。

(4) JCheckBox(String text, Icon icon, boolean selected)：创建一个带文本和图标的复选框，并指定其最初是否处于选定状态。

2. 单选按钮

单选按钮（JRadioButton）与复选框一样也只有两种状态：选中和未选中。但与 ButtonGroup 对象配合使用可创建一组按钮，一次只能选择其中的一个。

JRadioButton 的常用的构造方法如下。

(1) JRadioButton(String text)：创建一个具有指定文本和状态为未选择的单选按钮。

(2) JRadioButton(String text, boolean selected)：创建一个具有指定文本和选择状态的单选按钮。

(3) JRadioButton(Icon icon)：创建一个初始化为未选择的单选按钮，其具有指定的图像但无文本。

(4) JRadioButton(String text, Icon icon, boolean selected)：创建一个具有指定的文本、图像和选择状态的单选按钮。

单选按钮与 ButtonGroup 对象配合使用时，先创建一个 ButtonGroup 对象并用其 add 方法将 JRadioButton 对象包含在此组中。但要注意的是：ButtonGroup 对象只是将一组单选按钮逻辑上分组，而不是物理上分组，也就是说加入面板容器中的组件不是 ButtonGroup 对象，而是一组单选按钮对象。例如，下面代码创建了性别的两个单选按钮对象并用 ButtonGroup 对象按逻辑上分为一组，这使用户一次只能选择其中的一个按钮。

```
maleButton = new JRadioButton( "男", true );
womenButton = new JRadioButton( "女", false );
radioGroup = new ButtonGroup();
radioGroup.add( maleButton );              //两个单选按钮逻辑上分为一组
radioGroup.add( womenButton );
Container container = getContentPane();  //将两个 JRadioButton 对象加入面板容器中
container.add( maleButton );
container.add( womenButton );
```

3. 复选框或单选按钮对象的常用方法

(1) boolean isSelected():返回复选框的状态,是否被选中。

(2) void setSelected (boolean):设置复选框是否被选中。

4. ItemEvent 事件响应

当用户单击复选框(或单选按钮)使其状态发生变化时就会引发 ItemEvent 事件。如果希望程序能响应复选框(或单选按钮)引发的 ItemEvent 事件,则通过调用复选框(或单选按钮)组件的方法 addItemListener,给复选框(或单选按钮)组件注册实现 ItemListener 接口的事件监听器,同时,为这个接口的抽象方法 void itemStateChanged(ItemEvent event)书写方法体。

例如,使用匿名内部类的复选框的事件监听的程序结构为:

```
复选框对象.addItemListener(new ItemListener() { //给复选框组件注册事件监听器
    public void itemStateChanged(ItemEvent event) {
    …
}});
```

ItemListener 接口负责监听所有与复选框(或单选按钮)有关的事件。当复选框(或单选按钮)的状态发生改变时,在实现方法 itemStateChanged 的方法体中,可以使用 ItemEvent 类的事件对象 event 的方法。

(1) int getStateChange()方法:获得当前复选框(或单选按钮)的状态值。该方法将会返回 ItemEvent. DESELECTED 或者 ItemEvent. SELECTED。

(2) Object getItem()方法:得到当前复选框对象(或单选按钮)的引用。

例 9-3 复选框的应用。

利用两个复选框的选中/未选中状态,控制一个 JTextField 的显示文本的字体风格变化。

```
import java.awt. * ;
import java.awt.event. * ;
import javax.swing. * ;
public class CheckBoxTest extends JFrame {
  private JTextField field;
  private JCheckBox bold, italic;
  // set up GUI
  public CheckBoxTest()
  { super( "JCheckBox 的应用" );
    Container container = getContentPane();
    container.setLayout( new FlowLayout() );
    // set up JTextField and set its font and color
    field = new JTextField( "点击复选框,请观察字体风格的变化", 22 );
    field.setFont( new Font( "Serif", Font.PLAIN, 14 ) );
    field.setForeground(Color.red);
    container.add( field );
    // create checkbox objects
    bold = new JCheckBox( "Bold" );
    container.add( bold );
    italic = new JCheckBox( "Italic" );
    container.add( italic );
    // register listeners for two JCheckBoxes
    CheckBoxHandler handler = new CheckBoxHandler();
    bold.addItemListener( handler );
    italic.addItemListener( handler );
        setSize( 275, 100 );
```

```
        setVisible( true );
    }
    public static void main( String args [] )
    {  CheckBoxTest application = new CheckBoxTest();
        application.setDefaultCloseOperation( JFrame.EXIT_ON_CLOSE );
    }
    // private inner class for ItemListener event handling
    private class CheckBoxHandler implements ItemListener {
        private int valBold = Font.PLAIN;
        private int valItalic = Font.PLAIN;
        public void itemStateChanged(ItemEvent event)   // respond to checkbox events
      {   if(event.getSource() == bold)                  // process bold checkbox events
              valBold = bold.isSelected() ? Font.BOLD : Font.PLAIN;
          if ( event.getSource() == italic )          // process italic checkbox events
              valItalic = italic.isSelected() ? Font.ITALIC : Font.PLAIN;
          // set text field font
          field.setFont( new Font( "Serif", valBold + valItalic, 14 ) );
        } // end method itemStateChanged
    } // end private inner class CheckBoxHandler
} // end class CheckBoxTest
```

程序运行结果如图 9-8 所示。

图 9-8 例 9-3 程序运行结果

例 9-4 创建两个单选按钮对象,使用 ButtonGroup 对象进行逻辑分组。

程序用两个单选按钮提供性别值的选择,规定一次只能选择一个单选按钮。

```
    import java.awt.*;
    import java.awt.event.*;
    import javax.swing.*;
    public class RadioButtonTest extends JFrame {
      private JTextField field;
      private JRadioButton maleButton, womenButton
      private ButtonGroup radioGroup;
      public RadioButtonTest()
      { super( "RadioButton Test" );
        // get content pane and set its layout
        Container container = getContentPane();
        container.setLayout( new FlowLayout() );
        JLabel label1 = new JLabel( "性别:" );
        container.add( label1 );
        // create radio buttons
        maleButton = new JRadioButton( "男"", true );
        container.add( maleButton );
        womenButton = new JRadioButton( "女", false );
        container.add( womenButton );
        // create logical relationship between JRadioButtons
        radioGroup = new ButtonGroup();
```

```
        radioGroup.add( maleButton );
        radioGroup.add( womenButton );
        // register events for JRadioButtons
        maleButton.addItemListener( new RadioButtonHandler( ) );
        womenButton.addItemListener( new RadioButtonHandler( ) );
        field = new JTextField( "性别初值是：男" ); // set up JTextField
        container.add( field );
        setSize( 300, 100 );
        setVisible( true );
    }
    public static void main( String args [] )
    {  RadioButtonTest application = new RadioButtonTest();
       application.setDefaultCloseOperation( JFrame.EXIT_ON_CLOSE );
    }
    // private inner class to handle radio button events
    private class RadioButtonHandler implements ItemListener {
      // handle radio button events
      public void itemStateChanged( ItemEvent event )
      {   if (event.getItem() == maleButton)
           field.setText( "性别值是：男" );
         else if (event.getItem() == womenButton)
         field.setText( "性别值是：女" );
      }
    } // end private inner class RadioButtonHandler
}
```

程序运行结果如图 9-9 所示。

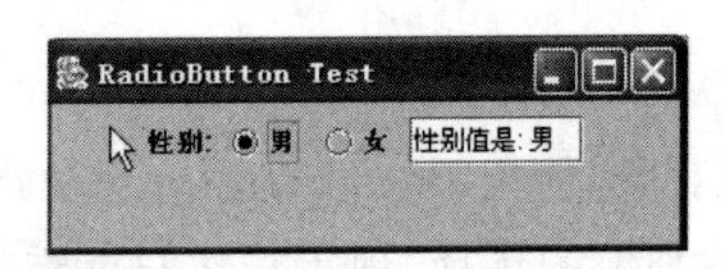

图 9-9　例 9-4 程序运行结果

9.6 组合框

组合框(JComboBox)也叫下拉菜单，用户可以从下拉菜单列表中选择值，下拉列表在用户请求时显示。如果使组合框处于可编辑状态，则组合框将包括用户可在其中键入值的可编辑字段。

1. JComboBox 类的常用的构造方法

JComboBox(Object [] items)：创建包含指定数组中的元素的 JComboBox。

2. JComboBox 对象的常用方法

(1) void setMaximumRowCount()：设置单击组合框时允许显示的最大列表项数。如果组合框的列表项数超过了能显示的最大数目，则 JComboBox 自动提供滚动条。

(2) int getSelectedIndex() ：返回组合框的当前所选项的索引值 (0～n)。

(3) Object getSelectedItem()：返回组合框的当前所选项。

(4) void setEditable(boolean aFlag)：确定 JComboBox 字段是否可编辑。

3. ItemEvent 事件响应

当用户单击组合框时就会引发 ItemEvent 事件。如果希望程序能响应复选框(或单选按钮)引发的 ItemEvent 事件,则通过调用组合框组件的方法 addItemListener,给组合框组件注册实现 ItemListener 接口的事件监听器,同时,为这个接口的抽象方法 void itemStateChanged (ItemEvent event)书写方法体。

例 9-5 组合框示例程序。

用组合框的下拉菜单提供一组图像,用户可选择其中一项,将选中的图像在 JLabel 组件中显示。

```
import java.awt.*;
import java.awt.event.*;
import javax.swing.*;
public class ComboBoxTest extends JFrame {
    private JComboBox imagesComboBox;
    private JLabel label;
    private String names [] =
        { "bug1.gif", "bug2.gif", "travelbug.gif", "buganim.gif" };
    private Icon icons [] = new ImageIcon [names.length];
    // set up GUI
    public ComboBoxTest() {
    super( "用组合框选择一组图像" );
        Container container = getContentPane();
        container.setLayout( new FlowLayout() );
        // set up JComboBox and register its event handler
        imagesComboBox = new JComboBox( names );
        imagesComboBox.setMaximumRowCount( 3 ); //设置允许显示的最大行数
        imagesComboBox.addItemListener(
            new ItemListener() { // anonymous inner class
                // handle JComboBox event
                public void itemStateChanged( ItemEvent event )
                {  // determine whether check box selected
                    if ( event.getStateChange() == ItemEvent.SELECTED )
                       label.setIcon( icons [
                          imagesComboBox.getSelectedIndex() ] );
                }
            } // end anonymous inner class
        ); // end call to addItemListener
        container.add( imagesComboBox );
        for (int i = 0;i<icons.length;i++ )
            icons [i] = new ImageIcon( getClass().getResource(names [ i ]) );
        label = new JLabel( icons [ 0 ] ); // set up JLabel to display ImageIcons
        container.add( label );
        setSize( 350, 100 );
        setVisible( true );
    }
    public static void main( String args [] )
    {  ComboBoxTest application = new ComboBoxTest();
       application.setDefaultCloseOperation( JFrame.EXIT_ON_CLOSE );
    }
}
```

程序运行结果如图 9-10 所示。

图 9-10 例 9-5 程序运行结果

9.7 列 表

列表(JList)组件允许用户从列表中选择一个“单选”或选择多个“多选”对象。JList 不支持直接滚动。要创建一个带滚动条的列表,需要将一个 JList 加入 JScrollPane 对象中。

1. JList 的常用构造方法

JList(Object [] listData):构造一个 JList,使其显示指定数组中的元素。

2. JList 对象的常用方法

(1) void setVisibleRowCount(int visibleRowCount):设置不使用滚动条可以在列表中显示的预定行数。

(2) void setSelectionMode(int selectionMode):确定允许单项选择还是多项选择模式。selectionMode 值允许如下。

- ListSelectionModel. SINGLE_SELECTION:一次只能选择一个列表项。
- ListSelectionModel. SINGLE_INTERVAL_SELECTION:一次可选择多个连续的列表项。
- ListSelectionModel. MULTIPLE_INTERVAL_SELECTION(默认设置):对多项列表选择没有限制。

(3) void setListData(Object [] listData):用 object 数组设置 JList 的列表内容。

(4) int getSelectedIndex():返回所选的第一个索引;如果没有选择项,则返回 -1。

(5) int [] getSelectedIndices():返回所选的全部列表项索引的数组(按升序排列)。

(6) Object getSelectedValue():返回所选的第一个列表项值;如果选择为空,则返回 null。

(7) Object [] getSelectedValues():返回所选一组列表项值。

3. ListSelectionEvent 事件响应

当用户在 JList 中选择列表项时,将产生 ListSelectionEvent 事件。如果希望程序能响应 ListSelectionEvent 事件,则通过调用列表组件的方法 addListSelectionListener,给列表组件注册实现 ListSelectionListener 接口的事件监听器,同时,为这个接口的抽象方法 void valueChanged(ListSelectionEvent event)书写方法体。

例 9-6 列表的应用示例。

程序创建一个带滚动条的 JList 列表对象,JList 列表的内容是放置了一组颜色。当用户选中列表中某颜色项时,将会设置 JFrame 的背景色。

```
import java.awt. * ;
```

```
import javax.swing.*;
import javax.swing.event.*;
public class ListTest extends JFrame {
    private JList colorList;
    private Container container;
    private final String colorNames [] = { "Black", "Blue", "Cyan",
        "Dark Gray", "Gray", "Green", "Light Gray", "Magenta",
        "Orange", "Pink", "Red", "White", "Yellow" };
    private final Color colors [] = { Color.BLACK, Color.BLUE, Color.CYAN,
        Color.DARK_GRAY, Color.GRAY, Color.GREEN, Color.LIGHT_GRAY,
        Color.MAGENTA, Color.ORANGE, Color.PINK, Color.RED, Color.WHITE,
        Color.YELLOW };
    public ListTest() // set up GUI
    {   super( "列表的应用" );
        container = getContentPane();
        container.setLayout( new FlowLayout() );
        // 创建 一个 list 对象,内容来自于数组 colorNames
        colorList = new JList( colorNames );
        //设置在列表中显示的预定行数 5,允许单项选择菜单
        colorList.setVisibleRowCount( 5 );
        colorList.setSelectionMode( ListSelectionModel.SINGLE_SELECTION );
        container.add(new JLabel("请选择颜色,以控制 JFrame 背景色"));
        container.add( new JScrollPane( colorList ) );
        colorList.addListSelectionListener(
          new ListSelectionListener() {
            // handle list selection events
            public void valueChanged( ListSelectionEvent event )
            {   container.setBackground(
                    colors [ colorList.getSelectedIndex() ] );
            } }
    ); // end call to addListSelectionListener
    setSize( 350, 150 ); setVisible( true );
} // end ListTest constructor
public static void main( String args [] )
{   ListTest application = new ListTest();
   application.setDefaultCloseOperation( JFrame.EXIT_ON_CLOSE );
}}
```

程序运行结果如图 9-11 所示。

图 9-11　例 9-6 程序运行结果

9.8　布局管理器

布局管理器用于管理组件在容器中的布局方式。布局管理器类都实现了接口 Layout-

Manager。Java 提供的标准布局管理器类有:FlowLayout、BorderLayout、GridLayout、CardLayout、GridBagLayout 和 BoxLayout,其中前 5 个布局管理器类在 java.awt 包中,而 BoxLayout 在 javax.swing 包中。

Java 的每一个容器都有一个相关联的布局管理器,用来管理组件。设定容器的布局管理器,是用 Container 类的方法:setLayout(LayoutManager layoutmanager)。

每个容器在不设定布局管理器时,将采用默认的布局管理器。如 Frame 容器的默认布局管理器是 BorderLayout,JPanel 和 Applet 容器的默认布局管理器是 FlowLayout,而 getContentPane()容器的默认布局管理器是 BorderLayout。

9.8.1 FlowLayout 布局管理器

FlowLayout 流式布局管理器提供的布局管理:将组件按加入的顺序,将它们从左至右、从上至下(一行放不下,就放入下一行)地放置在容器中,并允许设定组件的纵横间隔和水平对齐方式。FlowLayouts 往往用来布局容器中的一组按钮。

FlowLayout 类的常用构造方法如下。

(1) FlowLayout():构造一个新的 FlowLayout,居中对齐,默认的水平和垂直间隙是 5 个单位。

(2) FlowLayout(int align):构造一个新的 FlowLayout,align 指定对齐方式,默认的水平和垂直间隙是 5 个单位。align 指明每一行的对齐方式,其值为 FlowLayout 类中封装的整型静态常量。

- CENTER:指示每一行组件都应该是居中的。
- LEADING:指示每一行组件都应该与容器方向的开始边对齐。
- LEFT:指示每一行组件都应该是左对齐的。
- RIGHT:指示每一行组件都应该是右对齐的。
- TRAILING:指示每行组件都应该与容器方向的结束边对齐。

(3) FlowLayout(int align, int hgap, int vgap):创建具有指定的对齐方式以及指定的水平和垂直间隙的流布局管理器。

FlowLayout 布局对象的常用方法如下。

(1) void setAlignment(int align):设置此布局的对齐方式。

(2) void layoutContainer(Container target):以 FlowLayout 的对齐方式重新布置容器。

例 9-7 使用 FlowLayout 布局管理器来定位 Frame 组件中的 3 个按钮。

程序在 Frame 组件中加入 3 个按钮,通过用户单击按钮可改变流布局管理器的对齐方式。

```
import java.awt.*;
import java.awt.event.*;
import javax.swing.*;
public class FlowLayoutDemo extends JFrame {
    private JButton leftButton, centerButton, rightButton;
    private Container container;
    private FlowLayout layout;
    public FlowLayoutDemo() { // 设置 GUI 和注册事件处理
        super("FlowLayout 布局管理");
        layout = new FlowLayout();
        container = getContentPane();
```

```
        container.setLayout( layout );
        // set up leftButton and register listener
        leftButton = new JButton( "左对齐" );
        container.add( leftButton );
        leftButton.addActionListener(
          new ActionListener() {
            public void actionPerformed( ActionEvent event )
            {  layout.setAlignment( FlowLayout.LEFT );
                // 重新对齐容器中的组件
                layout.layoutContainer( container );
            }
          } ); // end call to addActionListener
        // set up centerButton and register listener
        centerButton = new JButton( "中心对齐" );
        container.add( centerButton );
        centerButton.addActionListener(
          new ActionListener() {
            public void actionPerformed( ActionEvent event )
            {  layout.setAlignment( FlowLayout.CENTER );
               layout.layoutContainer( container );
            }
        }  );
        // set up rightButton and register listener
        rightButton = new JButton( "右对齐" );
        container.add( rightButton );
        rightButton.addActionListener(
          new ActionListener() {
            public void actionPerformed( ActionEvent event )
            {  layout.setAlignment( FlowLayout.RIGHT );
               layout.layoutContainer( container );
            }
        } );
        setSize( 300, 75 );
        setVisible( true );
    }
    public static void main( String args [] )
    {  FlowLayoutDemo application = new FlowLayoutDemo();
    application.setDefaultCloseOperation( JFrame.EXIT_ON_CLOSE );
    }
}
```

程序运行结果如图 9-12 所示。

图 9-12　例 9-7 程序运行结果

9.8.2 BorderLayout 布局管理器

BorderLayout 边界布局管理器使用地理上的 5 个方向：NORTH（北）、SOUTH（南）、WEST（西）、EAST（东）和 CENTER（中心）来确定组件添加的位置。每个方位区域只能放一个组件。

当使用 BorderLayout 布局管理器将一个组件添加到一个容器中时，使用 BorderLayout 类中 5 个静态常量表示 5 个方位中的一个，例如：

```
container.setLayout(new BorderLayout());
container.add(new Button("one"), BorderLayout.NORTH);
```

BorderLayout 将缺少方位说明的情况解释为 BorderLayout.CENTER。例如：

```
container.add(new TextArea());
```

等价于：

```
container.add(new TextArea(),BorderLayout. CENTER)
```

BorderLayout 布局管理器根据组件的最佳尺寸和容器尺寸的约束条件来对组件进行布局。如果某个方位上无组件时，则其他方位上的组件自动进行缩放占有其位置：NORTH 和 SOUTH 组件可在水平方向进行伸展；EAST 和 WEST 组件可在垂直方向进行伸展；CENTER 组件可在水平和垂直两个方向上伸展，来填充整个剩余空间。

BorderLayout 常用的构造方法如下。

(1) BorderLayout()：构造一个新的边界布局，组件中没有间距。

(2) BorderLayout(int, int)：构造一个边界布局，组件间指定水平间距和垂直间距。

例 9-8 使用 BorderLayout 布局管理器来定位 Frame 组件中的 5 个按钮。

程序在 Frame 组件中加入 5 个按钮，通过用户单击某一按钮可隐藏此方位上的按钮组件。通过程序的运行，请观察如果某个方位上无组件，则其他方位上的组件自动缩放尺寸的规则。

```
import java.awt.*;
import java.awt.event.*;
import javax.swing.*;
public class BorderLayoutDemo extends JFrame implements ActionListener {
    private JButton buttons [];
    private final String names [] = { "North", "South",
        "East", "West", "Center" };
    private BorderLayout layout;
    private Container container;
     public BorderLayoutDemo()               // 设置 GUI 和注册事件处理
     {  super( "BorderLayout 布局管理" );
        layout = new BorderLayout( 5, 5 );   // 设置组件间 5 个像素的间距
        container = getContentPane();
        container.setLayout( layout );
        buttons = new JButton [ names.length ];
        for ( int count = 0; count < names.length; count++ ) {
            buttons [ count ] = new JButton( names [ count ] );
            buttons [ count ].addActionListener( this );
        }
        container.add( buttons [ 0 ], BorderLayout.NORTH );
        container.add( buttons [ 1 ], BorderLayout.SOUTH );
        container.add( buttons [ 2 ], BorderLayout.EAST );
        container.add( buttons [ 3 ], BorderLayout.WEST );
        container.add( buttons [ 4 ], BorderLayout.CENTER );
        setSize( 300, 200 ); setVisible( true );
    }
    //处理按钮事件的方法代码
```

```
    public void actionPerformed( ActionEvent event ) {
        for ( int count = 0; count < buttons.length; count ++ )
            if ( event.getSource() == buttons [ count ] )
                buttons [ count ].setVisible( false );
            else
                buttons [ count ].setVisible( true );
        layout.layoutContainer(container );          // 重新布局容器
    }
    public static void main( String args [] )
    {  BorderLayoutDemo application = new BorderLayoutDemo();
        application.setDefaultCloseOperation( JFrame.EXIT_ON_CLOSE );
    }
}
```

程序运行结果如图 9-13 所示。

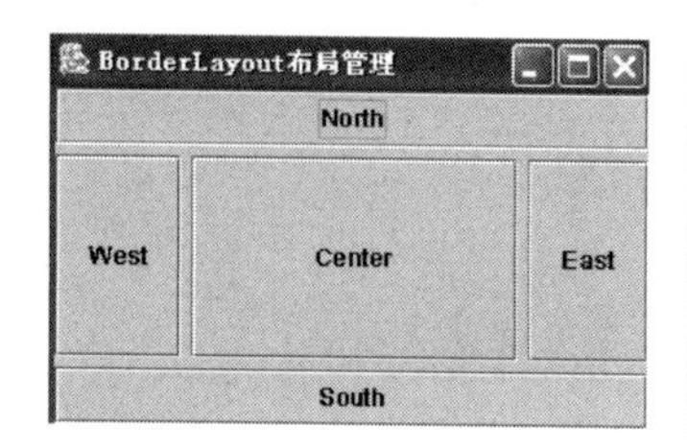

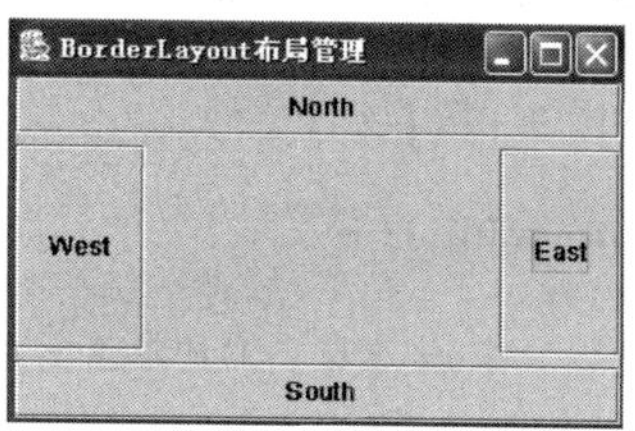

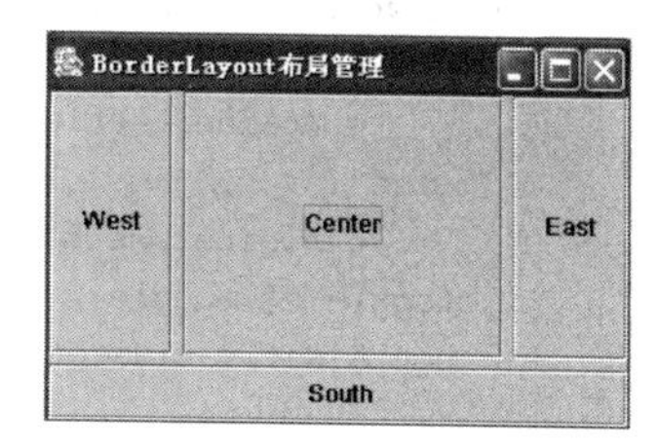

图 9-13　例 9-8 程序运行结果

9.8.3 GridLayout 布局管理器

GridLayout 布局管理器将容器划分为网格，以便在行列中放置组件。添加到 GridLayout 中的组件会依序从左至右、由上至下地填充到每个格子。GridLayout 中的每个 Component 都有相同的宽度和高度。

GridLayout 类的构造方法如下。

(1) GridLayout():创建一个 1 行 1 列的网格布局。

(2) GridLayout(int rows, int cols):用指定的行数和列数创建一个网格布局。

(3) GridLayout(int rows, int cols, int hgap, int vgap):用指定的行数和列数创建一个网格布局，并指定网格单元间水平间距和垂直间距。

例 9-9　使用 GridLayout 布局 Frame 组件中的 6 个按钮。

JFrame 组件中的 6 个按钮被布局为两种网格方式:3 行 2 列网格和 2 行 3 列网格。用户单击任一按钮可切换两种网格布局方式。

程序中使用了 Contanier 的 validate()方法确认这个容器和它的所有子组件。此方法使容器在其包含的组件被添加或修改时，或布局管理发生变化时，对其组件重新进行布局。

```
import java.awt.*;
import java.awt.event.*;
import javax.swing.*;
public class GridLayoutDemo extends JFrame implements ActionListener {
    private JButton buttons [];
    private final String names [] =
        { "one", "two", "three", "four", "five", "six" };
    private boolean toggle = true;
    private Container container;
```

```
    private GridLayout grid1, grid2;
    public GridLayoutDemo()
    {    super( "GridLayout 布局管理的应用" );
        grid1 = new GridLayout( 2, 3, 5, 5 );       //建立 2*3 网格,单元间隔为 5
        grid2 = new GridLayout( 3, 2 );             //建立 3*2 网格,单元间隔为 5
        container = getContentPane();
        container.setLayout( grid1 );
        buttons = new JButton[ names.length ];
        for ( int count = 0; count < names.length; count ++ ) {
            buttons[ count ] = new JButton( names[ count ] );
            buttons[ count ].addActionListener( this );
            container.add( buttons[ count ] );
        }
        setSize( 300, 150 );          setVisible( true );
    }
    public void actionPerformed( ActionEvent event )    //处理按钮事件的代码
    {   if ( toggle )
            container.setLayout( grid2 );
        else
            container.setLayout( grid1 );
        toggle =! toggle;
        container.validate();       //使容器和它的所有子组件有效地显示
    }
    public static void main( String args[] )
    {  GridLayoutDemo application = new GridLayoutDemo();
       application.setDefaultCloseOperation( JFrame.EXIT_ON_CLOSE );
    }
}
```

程序的运行结果如图 9-14 所示。

图 9-14　网格布局示例

9.8.4　CardLayout 布局管理器

CardLayout 布局管理器将容器中的每个组件看成一张卡片,一次只能看到一张卡片,而容器充当卡片的堆栈。当容器第一次显示时,第一个添加到 CardLayout 对象的组件为可见组件。

卡片的顺序由组件对象本身在容器内部的顺序决定。CardLayout 定义了一组方法,这些方法允许应用程序按顺序地浏览这些卡片,或者显示指定的卡片(将一个字符串标识符与给定卡片相关联,以便进行快速随机访问)。

1. CardLayout 类常用的构造方法

(1) CardLayout():创建一个间隙大小为 0 的新卡片布局。

(2) CardLayout(int hgap, int vgap):创建一个具有指定的水平和垂直间隙的新卡片布局。

2. 将指定的组件添加到卡片布局的容器中的方法

```
container.add(String s1,Component comp);
```

其中,s1 是指定的卡片名称。

3. CardLayout 对象的常用方法

(1) void first(Container parent):翻转到容器的第一张卡片。

(2) void last(Container parent):翻转到容器的最后一张卡片。

(3) void next(Container parent):翻转到指定容器的下一张卡片。

(4) void show(Container parent, String name):翻转到已添加到此布局的、具有指定卡片名称是 name 组件的卡片。

例 9-10 CardLayout 布局管理器的举例。

在 JFrame 的 BorderLayout 布局中的 Center 和 East 方位上添加两个 JPanel 组件并命名为 pleft 和 pright。在 pleft 面板的布局方式为 CardLayout,并添加 10 个 button; pright 面板的布局方式为 FlowdLayout, 并添加 1 个 button。程序运行时,通过单击"下一个按钮",可顺序地浏览这些卡片。

```
import java.awt. * ;
import java.awt.event. * ;
import javax.swing. * ;
public class CardLayoutDemo extends JFrame {
    JPanel pleft;
    JPanel pright;
    CardLayout card;
    public CardLayoutDemo() {
     this.setBounds(80,60,600,300);
    Container c = this.getContentPane();
    c.setLayout(new BorderLayout());
    pleft = new JPanel();
    pleft.setBackground(Color.red);
    card = new CardLayout(10,10);
    pleft.setLayout(card);
    JButton [] b = new JButton [10];
    for(int i = 0;i<10;i ++ )
    {   b [i] = new JButton("第" + i +"个 Button");
        b [i].setFont(new Font("Helvetica", Font.PLAIN, 18));
        pleft.add("card" + i,b [i]);
    }
    pright = new JPanel(); pright.setBackground(Color.blue);
    pright.setLayout(new FlowLayout());
    JButton b1 = new JButton("下一个按钮");
    b1.addActionListener(new ActionListener()
    {   public void actionPerformed(ActionEvent e)
     {   card.next(pleft);
     }
    }  );
    pright.add(b1);
    c.add(pleft,BorderLayout.CENTER);
    c.add(pright,BorderLayout.EAST);
  }
  public static void main(String []args)
```

```
{  CardLayoutDemo f = new CardLayoutDemo();
   f.setVisible(true);
   f.setDefaultCloseOperation(JFrame.DISPOSE_ON_CLOSE);
  }
}
```

程序运行结果如图 9-15 所示。

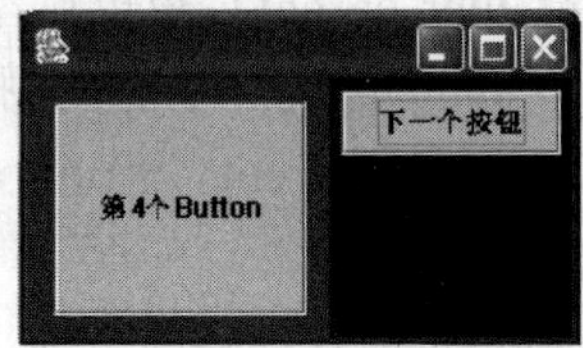

图 9-15　CardLayout 布局示例

9.8.5 BoxLayout 布局管理器

BoxLayout(在 javax.swing 包中)是允许纵向或横向地布置多个组件的布局管理器。Box 类声明了一个以 BoxLayout 作为默认布局管理器的容器,并提供了静态方法来创建带有水平或垂直方向 BoxLayout 的 Box 对象。

例如,创建一个从左到右显示其组件的 Box 对象:

```
Box horizontal1 = Box.createHorizontalBox();
```

创建一个从上到下显示其组件的 Box 对象:

```
Box vertical1 = Box.createVerticalBox();
```

将指定组件追加到 Box 容器对象的尾部的方法:

```
Box 容器对象.add(Component comp);
```

Box 容器提供了如下常用的一组静态方法。

(1) Component createHorizontalGlue():创建一个横向 glue 组件。

(2) Component createHorizontalStrut(int width):创建一个不可见的、固定宽度的 glue 组件。

(3) Component createRigidArea(Dimension d):创建一个总是具有指定大小的不可见组件。

(4) Component createVerticalGlue():创建一个纵向 glue 组件。

(5) Component createVerticalStrut(int height):创建一个不可见的、固定高度的组件。

例 9-11　BoxLayout 布局管理器的举例。

程序完成在 JFrame 的 FlowLayout 布局中添加一个横向 Box 组件和一个纵向 Box 组件,并分别给每一个 Box 添加 3 个 button。

```
import java.awt.*;
import java.awt.event.*;
import javax.swing.*;
public class BoxLayoutDemo extends JFrame {
 public BoxLayoutDemo()
 {  super("Demostrating BoxLayout");
     // create Box containers with BoxLayout
     Box horizontal1 = Box.createHorizontalBox();
     Box vertical1 = Box.createVerticalBox();
     for(int count = 1;count< = 3;count ++ ) // add buttons to Box horizontal1
         horizontal1.add( new JButton( "Button " + count ) );
     // create strut and add buttons to Box vertical1
```

```
        for ( int count = 4; count <= 6; count++ ) {
            vertical1.add( Box.createVerticalStrut( 25 ) );
            vertical1.add( new JButton( "Button " + count ) );
        }
        Container c = getContentPane();
        c.setLayout(new FlowLayout());
        c.add( horizontal1 );
        c.add( vertical1 );
        setSize( 400, 220 );
        setVisible( true );
    }
    public static void main( String args [] )
    {   BoxLayoutDemo application = new BoxLayoutDemo();
        application.setDefaultCloseOperation( JFrame.EXIT_ON_CLOSE );
    }
}
```

程序运行结果如图 9-16 所示。

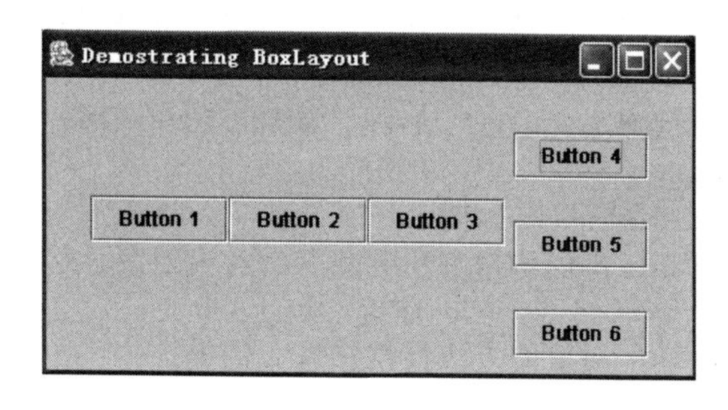

图 9-16　BoxLayout 布局举例

9.8.6 GridBagLayout 布局管理器

GridBagLayout 类似于 GridLayout 的布局管理，也是将容器分成若干行与列组成的单元，但行和列的大小可以不同，且每个组件可占用一个或多个单元(又称为它的显示区域)；组件加入容器的顺序可为任意。

每个由 GridBagLayout 布局的组件都与一个 GridBagConstraints 实例相关联，该对象指定了如何将组件放置到 GridBagLayout 布局中，又称为约束对象。

GridBagConstraints 类封装了一组 public 型的约束变量，设置这组约束变量的值将对布局行为进行约束。常用的一组约束变量定义如下。

(1) gridx,gridy ：放置组件的左上方单元的行号和列号。GridBagLayout 布局中的最左上角单元地址为 gridx 为 0 ，gridy 为 0 。gridx 和 gridy 的值也可设置为 GridBagConstraints 类的常量 RELATIVE (缺省值)和 REMAINDER。RELATIVE 定义此组件放在当前行的最后一个组件的右边(gridx 为 REMAINDER)或当前列的最后一个组件的下面(gridy 为 REMAINDER)。REMAINDER 定义此组件是当前行(或列)的最后一个组件。

(2) gridwidth,gridheight：指定组件占有的行单元数和列单元数。它们的缺省值为 1 。也可设定为 GridBagConstraints 类中的常量 REMAINDER 和 RELATIVE。REMAINDER 指定该组件是当前行(列)的最后一个；RELATIVE 指定该组件放在当前行(列)的最后--个组件的后面。

(3) fill：当组件的显示区域大于组件尺寸时，如何将组件变大。可能值为GridBagConstraints类的常量之一：NONE、HORIZONTAL、VERTICAL和BOTH。NONE（缺省值）表示不改变组件的尺寸。HORIZONTAL使组件的宽度在水平方向上变大以填充它的显示区域，但不改变它的高度。VERTICAL使组件的高度在垂直方向上足够大来填充它的显示区域，但不改变它的宽度。BOTH使组件在水平和垂直方向变大，以填满它的显示区域。

(4) weightx，weighty：给组件指定一个行权值(weightx)和列权值(weighty)，以确定如何将多余空白空间分配给网格单元。当容器调整变大时，则网格单元会水平(weightx非0时)或垂直(weighty非0时)地变大。而当weightx和weighty权值为0(缺省值)时，容器调整变大时，则网格单元不会水平地或垂直地变大；然而，如果组件横跨一列(行)，而此列(行)包含一个weightx(weighty)非0的组件，则weightx(weighty)为0的组件与同一列(行)中其他的组件按相同的比例水平(垂直)地变大。

(5) anchor：当组件没有填满显示区域时，确定如何将组件置于显示区域的何处。可赋予下列GridBagConstraints类的常量之一：NORTH、SOUTH、WEST、EAST、NORTHWEST、NORTHEAST、SOUTHWEST、SOUTHEAST和CENTER（缺省值）。

应用GridBagLayout布局方式的基本步骤为：

① 设置容器的布局方式是GridBagLayout。例如：

```
layout = new GridBagLayout();
container.setLayout( layout );
```

② 用GridBagConstraints类的构造方法创建一个约束对象。例如：

```
GridBagConstraints contarints = new GridBagConstraints()
```

③ 用GridBagConstraints对象的方法设置约束。例如：

```
constraints.gridx = column; contarints.fill = GridBagConstraints.BOTH;
```

④ 用GridBagLayout对象的方法setConstraints(Component comp，GridBagConstraints constraints)，将组件与GridBagConstraints对象相关联。

⑤ 用add(Component comp)方法，将组件加入设置了GridBagLayout的容器中，就完成这一个组件在该容器中的布局。

例9-12 使用BorderLayout布局Frame组件中的9个按钮。

```
import java.awt.*;
import java.awt.event.*;
import javax.swing.*;
public class GridBagDemo extends JFrame {
 private GridBagLayout layout;
 private GridBagConstraints constraints;
 private Container container;
 public GridBagDemo()
 { super( "GridBag布局应用" );
  container = getContentPane();
  layout = new GridBagLayout();
  container.setLayout( layout );
  constraints = new GridBagConstraints();
  constraints.fill = GridBagConstraints.BOTH;
  constraints.weightx = 1.0;     // can grow wider
  makebutton("Button1");
  makebutton("Button2");
  makebutton("Button3");
  constraints.gridwidth = GridBagConstraints.REMAINDER;     //end row
```

```
  makebutton("Button4");
  constraints.weightx = 0.0;      //reset to the default
  makebutton("Button5"); //another row
  constraints.gridwidth = GridBagConstraints.RELATIVE; //next-to-last in row
  makebutton("Button6");
  constraints.gridwidth = GridBagConstraints.REMAINDER; //end row
  makebutton("Button7");
  constraints.gridwidth = 1;      //reset to the default
  constraints.gridheight = 2;
  constraints.weighty = 1.0;      // can grow taller
  makebutton("Button8");
   constraints.weighty = 0.0;      //reset to the default
   constraints.gridwidth = GridBagConstraints.REMAINDER; //end row
   constraints.gridheight = 1;     //reset to the default
   makebutton("Button9");
   makebutton("Button10");
  setSize( 400, 200 );
  setVisible( true );
}  // end constructor
private void makebutton(String name) {   // add a component to the container
    JButton button = new JButton(name);
    button.setFont(new Font("Helvetica", Font.PLAIN, 18));
    layout.setConstraints(button, constraints);
    container.add(button);
}
public static void main( String args [] )
{  GridBagDemo application = new GridBagDemo();
   application.setDefaultCloseOperation( JFrame.EXIT_ON_CLOSE );
}}
```

程序运行结果如图 9-17 所示。图 9-17(b)是 Frame 容器尺寸变大时的结果,请注意观察。

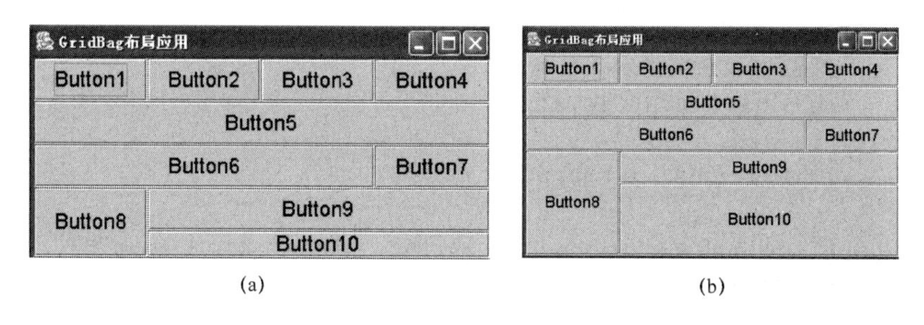

(a) (b)

图 9-17 BorderLayout 布局实例

9.9 面板和窗口

9.9.1 面板

面板(JPanel)是一个轻量容器组件,用于容纳界面元素,以便在布局管理器的设定下可容纳更多的组件，以实现容器的嵌套。JPanel、JScrollPane、JSplitPane 和 JInteralFrame 都属于常用的中间容器，是轻量级组件。JPanel 类从 JComponet 类继承，其缺省布局管理器是 FlowLayout。

JPanel 类的常用构造方法如下。

(1) JPanel()：创建具有双缓冲和流布局的新 JPanel 对象。

(2) JPanel(LayoutManager layout)：创建具有双缓冲和指定布局的新 JPanel 对象。

JPanel 类属于 JComponet 的子类，因此可调用 JComponet 类的方法。一些常用的 JComponet 类的方法如下。

(1) void setBorder(Border border)：设置组件外围的边框。

(2) void setTooltipText(String text)：为组件设置对用户有帮助的提示信息。

(3) protected void paintComponent(Graphics g)：在轻量级 SwingGUI 的环境中进行绘图。调用 repaint()方法可间接调用 paintComponent 方法以达到重新绘图的目的。

(4) void setOpaque(boolean isOpaque)：isOpaque 为 true 时，表示组件为不透明（绘制组件时将清除组件的背景）；否则该组件为透明（绘制组件时不清除组件的背景），默认值为 false。

JPanel 在使用上非常灵活。复杂 GUI 需要将每个组件放置在精确的位置上。它们需要由多个面板组成，而每个面板以特定的布局来排列组件。另外，通过扩展 JPanel 类以创建定制 JPanel 类，以完成创建专用的绘图区域和创建接收鼠标事件的区域。

例 9-13 利用 JPanel 创建一个专用的绘图区域。

程序将图形用户界面的其他部分与绘图区域分开，这有利于 Swing 图形用户界面，否则可能会导致不能正确显示 GUI 组件。

```
import java.awt. * ;
import java.awt.event. * ;
import javax.swing. * ;
//创建 Jpanel 子类用作定制的绘图区域
class CustomPanel extends JPanel {
    private int shape;
    public void paintComponent( Graphics g )// use shape to draw an oval or rectangle
    {    super.paintComponent( g );
         g.setColor(Color.red);
         if ( shape == 1 )
              g.fillOval( 50, 10, 60, 60 );
         else if ( shape == 2 )
              g.fillRect( 190, 10, 60, 60 );
    }
    public void draw( int shapeToDraw )   // set shape value and repaint CustomPanel
    {    shape = shapeToDraw;
    repaint();
    }
} // end class CustomPanel
//在 JPanel 的定制子类上进行绘图
public class CustomPanelTest extends JFrame {
    private JPanel buttonPanel;
    private CustomPanel myPanel;
    private JButton circleButton, squareButton;
    public CustomPanelTest()
    {    super( "创建定制 JPanel 类" );
         myPanel = new CustomPanel();    // create custom drawing area
```

```
        myPanel.setBackground( Color.GREEN );
        squareButton = new JButton( "画正方形" ); // set up squareButton
        squareButton.addActionListener(
            new ActionListener() {
                public void actionPerformed( ActionEvent event ) // draw a square
                {   myPanel.draw(2);
                }
            } );
        circleButton = new JButton( "画圆形" );
        circleButton.addActionListener(
            new ActionListener() {
                public void actionPerformed( ActionEvent event ) // draw a circle
                {   myPanel.draw( 1 );
                }
            }    );
        buttonPanel = new JPanel(); // set up panel containing buttons
        buttonPanel.setLayout( new GridLayout( 1, 2 ) );
        buttonPanel.add( circleButton );
        buttonPanel.add( squareButton );
        // attach button panel & custom drawing area to content pane
        Container container = getContentPane();
        container.add( myPanel, BorderLayout.CENTER );
        container.add( buttonPanel, BorderLayout.SOUTH );
        setSize( 300, 150 ); setVisible( true );
    }    // end constructor CustomPanelTest
    public static void main( String args [] ) {
        CustomPanelTest application = new CustomPanelTest();
        application.setDefaultCloseOperation( JFrame.EXIT_ON_CLOSE );
    }
}
```

程序运行结果如图 9-18 所示。

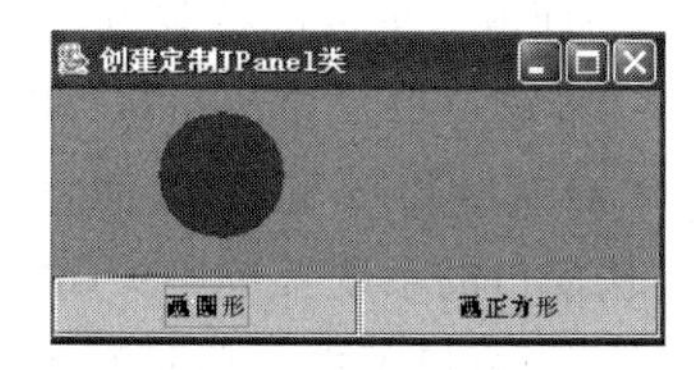

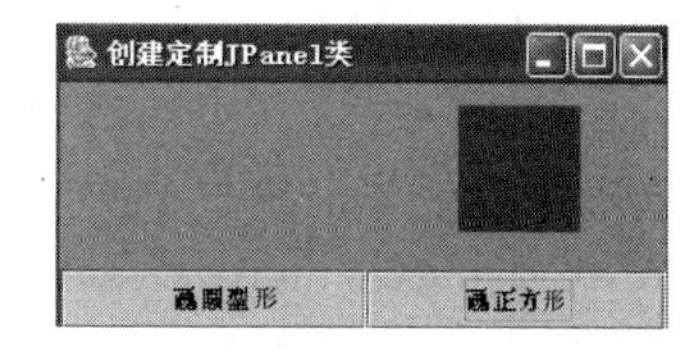

图 9-18　例 9-13 程序运行结果

9.9.2 窗口

设计程序的用户界面往往使用 JFrame 子类的实例作为应用程序的窗口。JFrame(框架)是一个带有标题栏和边框的窗口。JFrame 类的继承结构如图 9-19 所示，它是 Swing GUI 中少数几个不属于轻量级组件中的一个。

java.awt.Component←java.awt.Container←java.awt.Window←java.awt.Frame←javax.swing.JFrame

图 9-19　JFrame 类的继承结构

JFrame 对象常用的方法(注：有些是超类的方法)有如下几种。

(1) Container getContentPane()：获得内容面板。

(2) void setContentPane(Container)：设置内容面板。

(3) void setMenuBar(JMenuBar)：设置菜单条。

(4) void setTitle(String title):将窗口标题设置为指定的字符串。

(5) void setVisible(boolean b):根据参数 b 的值显示或隐藏此组件。

(6) void setSize(Dimension d):设置窗口的大小,宽度 d. width,高度 d. height。

(7) void pack():窗体调整到足够显示所有组件。

(8) void validate():验证此容器及其所有子组件。

(9) void setLocation(int x, int y):将组件移到新位置。

(10) void setResizable(boolean):设置窗体是否允许拖动以调整大小。

(11) void setDefaultCloseOperation(int operation):设置用户在此窗体上发起"close"时默认执行的操作。Operation 的值必须指定为以下 WindowConstants 接口(javax. swing 包)中的常量值选项之一。

- WindowConstants. DO_NOTHING_ON_CLOSE:不执行任何操作。要求程序在已注册的实现 ActionListener 接口的事件监听对象中的 windowClosing 方法中处理该操作。
- WindowConstants. HIDE_ON_CLOSE(默认值):调用已注册的实现 ActionListener 接口的事件监听对象后自动隐藏该窗体。
- WindowConstants. DISPOSE_ON_CLOSE:调用已注册的实现 ActionListener 接口事件监听对象后自动隐藏并释放该窗体。相当于调用窗口对象的 dispose 方法。
- JFrame . EXIT_ON_CLOSE:调用 System. exit()方法退出应用程序。

在默认情况下,如果程序不以 true 作为参数调用窗口的 setVisible 方法,则该窗口不会显示在屏幕上,此外应调用 setSize 方法来设置窗口的大小。在应用程序不需要某个窗口对象时,应及时地释放(dispose)窗口对象。

在用户操作窗口时,窗口组件将产生 WindowEvent 窗口事件。调用 Window 类的方法 add WindowListener()给窗口组件注册实现 WindowListener 接口的事件监听器。

WindowListener 接口中提供了以下 7 个用于处理窗口事件的抽象方法。

(1) void windowActivated(WindowEvent e) :将窗口设置为活动窗口时调用。

(2) void windowClosed(WindowEvent e) :因对窗口调用 dispose 而将其关闭时调用。

(3) void windowClosing(WindowEvent e) :从窗口的系统菜单中关闭窗口时调用。

(4) void windowDeactivated(WindowEvent e):当窗口不再活动时调用。

(5) void windowDeiconified(WindowEvent e):窗口从最小化状态变为正常状态时调用。

(6) void windowIconified(WindowEvent e):窗口从正常状态变为最小化状态时调用。

(7) void windowOpened(WindowEvent e):窗口首次变为可见时调用。

9.10 鼠标事件处理

触发鼠标事件的事件源通常是一个 GUI 容器,当鼠标进入、离开容器或者在容器中单击鼠标、拖动鼠标等操作时,都会发生鼠标事件。鼠标事件对应的类是 MouseEvent。Java 中的鼠标事件处理由两个事件监听接口 MouseListener 和 MouseMotionListener 描述。MouseListener 接口负责接收和处理鼠标的 press(按下)、release(释放)、click(单击)、enter(进入)、exit(移出)动作触发的事件;而 MouseMotionListener 接口负责接收和处理鼠标的 move(移动)和 drag(拖动) 动作触发的事件。表 9-1 列出了这两个接口中相应的方法描述。

表 9-1　鼠标事件对应的监听器接口中的方法描述

接口名	接口中的方法	方法调用描述信息
MouseListener 鼠标不移动	mousePressed(MouseEvent)	鼠标光标位于组件上方时,按下鼠标按钮时将调用该方法
	mouseReleased(MouseEvent)	鼠标释放时调用该方法。发生在 mousePressed 方法调用之后
	mouseClicked(MouseEvent)	鼠标光标在组件单击时(按下再释放)调用该方法
	mouseEntered(MouseEvent)	鼠标进入组件边界时将调用该方法
	mouseExited(MouseEvent)	鼠标离开组件边界时将调用该方法
MouseMotion Listener 鼠标移动	mouseDragged(MouseEvent)	鼠标光标位于组件上方时,拖动时将调用该方法。发生在 mousePressed 方法调用之后
	mouseMoved(MouseEvent)	鼠标光标位于组件上方并移动鼠标时,将调用该方法

MouseEvent 事件对象包含了鼠标事件发生时的信息,如事件发生时鼠标的 x、y 坐标位置等。MouseEvent 对象提供了如下常用方法。

(1) int GetX()和 int getY():获取鼠标的 x 坐标和 y 坐标。

(2) Point getPoint():获取鼠标的位置。

(3) int getClickcount():获取鼠标单击次数。

如果程序通过组件提供的方法 addMouseListener()与 addMouseMotionListener(),为 Component 组件注册了合适的事件监听器对象,则在鼠标与 Component 进行交互时,将调用 MouseListener 和 MouseMotionListener 接口中实现的方法。程序员在实现鼠标事件接口时,必须实现接口中的所有方法。

例 9-14　鼠标事件处理举例。

该程序在 JFrame 组件上监听鼠标事件。程序给 JFrame 组件分别注册了实现接口 MouseListener 和 MouseMotionListener 的事件监听对象 this(this 指向当前的 JFrame 对象)。每个鼠标事件的动作处理,都会在窗口底部的 JLabel 对象中显示一个字符串,以指明当前用户鼠标操作类型和鼠标光标位置。

```
import java.awt. * ;
import java.awt.event. * ;
import javax.swing. * ;
public class MouseTracker extends JFrame
        mplements MouseListener, MouseMotionListener {
    private JLabel statusBar;
    // set up GUI and register mouse event handlers
    public MouseTracker()
    {     super( "鼠标事件处理应用" );
        statusBar = new JLabel("",SwingConstants.CENTER);
        //使用 JFrame 的默认布局管理器 BorderLayout,将 JLabel 对象放入容器底部
        getContentPane().add( statusBar, BorderLayout.SOUTH );
        addMouseListener( this );              // listens for own mouse and
        addMouseMotionListener( this );        // mouse-motion events
        setSize( 275, 100 );
        setVisible( true );
    }
    // 实现 MouseListener 事件监听接口
    public void mouseClicked( MouseEvent event )    //鼠标单击事件的处理方法
```

```
{      statusBar.setText("鼠标单击[" + event.getX() + "," + event.getY() + "]");
}
  public void mousePressed( MouseEvent event ) // 鼠标按下事件的处理方法
{      statusBar.setText("鼠标按下[" + event.getX() + "," + event.getY() + "]");
}
public void mouseReleased( MouseEvent event )  // 鼠标释放事件的处理方法
{      statusBar.setText("鼠标释放[" + event.getX() + "," + event.getY() + "]");
}
public void mouseEntered( MouseEvent event )  // 鼠标进入事件的处理方法
{      statusBar.setText("鼠标进入[" + event.getX() + "," + event.getY() + "]");
       getContentPane().setBackground( Color.GREEN );
}
public void mouseExited( MouseEvent event )   // 鼠标移出事件的处理方法
{      statusBar.setText( "鼠标移出" );
       getContentPane().setBackground( Color.WHITE );
}
// 实现 MouseMotionListener 监听接口
public void mouseDragged( MouseEvent event )  // 鼠标拖动事件的处理方法
{      statusBar.setText( "鼠标拖动 [" + event.getX() +
         "," + event.getY() + "]" );
}
public void mouseMoved( MouseEvent event )     //// 鼠标移动事件的处理方法
{      statusBar.setText("鼠标移动[" + event.getX() + "," + event.getY() + "]");
}
public static void main( String args [] )
{      MouseTracker application = new MouseTracker();
       application.setDefaultCloseOperation( JFrame.EXIT_ON_CLOSE );
}
}
```

程序运行结果如图 9-20 所示。

图 9-20　例 9-14 程序运行结果

9.11　适配器类

Java 的许多事件监听接口中含有多个抽象方法时,程序员必须实现该接口中声明的所有方法,如上面讨论的 MouseListener 和 MouseMotionListener 接口。但有时也许只需要实现事件监听接口的某些方法,而不是实现接口中的全部方法。例如:一个程序可能只需要 MouseListener 的 mouseClicked 处理方法,或 MouseMotionListener 接口的 mouseDragged 处理方法。为此,对于包含多个方法的一些监听接口,java.awt.event 和 javax.swing.event 包提供了事件适配器类。一个适配器类实现了一个事件监听接口,它为该接口的所有方法提供了默认实现(空方法体)。java.awt.event 包中定义的事件适配器类包括以下几个:

① ComponentAdapter 组件适配器(实现 ComponentListener 接口)

② ContainerAdapter 容器适配器(实现 ContainerListener 接口)

③ FocusAdapter 焦点适配器（实现 FocusListener 接口）

④ KeyAdapter 键盘适配器（实现 KeyListener 接口）

⑤ MouseAdapter 鼠标适配器（实现 MouseListener 接口）

⑥ MouseMotionAdapter 鼠标运动适配器（实现 MouseMotionListener 接口）

⑦ WindowAdapter 窗口适配器（实现 WindowListener 接口）

在处理事件时，程序通过扩展适配器类，只需重写需要的事件处理方法，而不是对应的原先事件监听接口中的所有方法，这样缩短了程序代码。

例 9-15 鼠标适配器类 MouseMotionListener 的应用举例。

程序实现功能：通过拖动鼠标在窗口组件上绘制任意图形。应用程序在响应鼠标事件时，给 Frame 组件注册的事件监听器是一个扩展适配器类 MouseMotionAdapter 的子类——匿名内部类的对象。在子类中重写了 mouseDragged 方法，在方法体中获取鼠标拖动坐标，存放到 pointCount 数组中，并调用 repaint 方法，以间接地调用 paint 方法。在 paint 方法中，遍历 pointCount 的每一个数组元素，在每一个拖动坐标处绘制一个小椭圆，多个小椭圆组成了用鼠标在窗口上绘制图形的功能。

```
import java.awt.*;
import java.awt.event.*;
import javax.swing.*;
public class MouseDrawShape extends JFrame {
    private int pointCount = 0;
    private Point points [] = new Point [ 1000 ]; // java.awt.Point
    // set up GUI and register mouse event handler
    public MouseDrawShape()
    {   super( "通过拖动鼠标在窗口上绘制任意的图形" );
        getContentPane().add( new JLabel( "拖动鼠标绘图",
                SwingConstants.CENTER), BorderLayout.SOUTH );
        addMouseMotionListener(
            new MouseMotionAdapter() {
                //获取鼠标拖动坐标，调用 repaint 方法
                public void mouseDragged( MouseEvent event ) {
                    if ( pointCount < points.length ) {
                        points [ pointCount ] = event.getPoint();
                        ++ pointCount;
                        repaint();
                    } }
        } // end anonymous inner class
      ); // end call to addMouseMotionListener
      setSize( 300, 150 ); setVisible( true );
    } // end MouseDrawShape constructor
    //在窗口上每一个鼠标拖动坐标处绘制一个小椭圆
    public void paint( Graphics g )
    {  super.paint( g ); //以背景色清除绘画区域
       for ( int i = 0; i < points.length && points [ i ] != null; i++ )
          g.fillOval( points [ i ].x, points [ i ].y, 4, 4 );
    }
    public static void main( String args [] )
    {   MouseDrawShape application = new MouseDrawShape();
        application.setDefaultCloseOperation( JFrame.EXIT_ON_CLOSE );
    }
} // end class MouseDrawShape
```

程序运行结果如图 9-21 所示。

图 9-21　例 9-15 程序运行结果

9.12　键盘事件处理

在用户按下或松开键盘上的某键时，会产生键盘事件 KeyEvent。Java 中的键盘事件处理由事件监听接口 KeyListener 描述。

1. KeyListener 接口中声明了 3 个抽象方法

(1) void keyPressed(KeyEvent e)：为响应某键按下时，即响应"按下键"(KEY_PRESSED)事件将调用此方法。

(2) void keyReleased(KeyEvent e)：为响应某键被释放时，即响应"释放键"(KEY_RELEASED)事件将调用此方法。

(3) void keyTyped(KeyEvent e)：为响应某键被键入时，即响应"键入键"(KEY_TYPED) 事件将调用此方法。

对于不生成 Unicode 字符的键是不会生成键入键事件的(如动作键、组合键等)。只有按下任何非动作键(动作键包括箭头键、Home 键、End 键、PageDown 键和 Pause 键等)时，才生成键入键事件，将调用 keyTyped 方法。在 keyPressed 或 keyTyped 事件之后释放键时，将调用 keyReleased 方法。

2. KeyEvent 键盘事件对象的常用方法

(1) char getKeyChar()：返回与此事件中的键相关联的 Unicode 字符或 CHAR_UNDEFINED(仅对 KEY_TYPED 事件有意义)。

(2) int getKeyCode()：返回与此事件中的键相关联的虚拟键码 keyCode(仅对按下键和释放键事件有意义)。对于按下键和释放键事件，getKeyCode 方法返回该事件的 keyCode。对于键入键事件，getKeyCode 方法总是返回 VK_UNDEFINED。虚拟键码用于报告按下了键盘上的哪个键，而不是通过一个或多个击键组合所生成的字符(如"A"是由 shift+"a"生成的)。

例如，按下 Shift 键会生成 keyCode 为 VK_SHIFT 的 KEY_PRESSED 事件，而按下"a"键将生成 keyCode 为 VK_A 的 KEY_PRESSED 事件。释放"a"键后，会激发 keyCode 为 VK_A 的 KEY_RELEASED 事件。另外，还会生成一个 keyChar 值为 "A" 的 KEY_TYPED 事件。

(3) static String getKeyText(int keyCode)：返回描述 keyCode 的 String，如"HOME"、"F1"或"A"。

(4) boolean isActionKey()：返回此事件中的键是否为"动作"键。

3. 给组件注册 KeyListener 接口实现的事件监听器

通过调用组件的 addKeyListener 方法，注册 KeyListener 接口实现的事件监听器。KeyListener 接口对应的键盘适配器是 KeyAdapter。

例 9-16　键盘事件处理举例。

在一个 Applet 容器中显示一小方块，用户可通过键盘的上、下、左、右 4 个移动键控制小方块的移动，并且通过键入字母键“Y”和“R”能改变小方块的填充色。

程序 MyPanel.java 定义了定制 Panel 类，在其上将添加画图和注册键盘处理事件对象。程序 keyEventDemo.java 是小应用程序，在 init 方法中将创建 MyPanel 对象，并将对象放置在 Applet 上。

MyPanel.java 代码如下：

```
//MyPanel.java
import javax.swing.*;
import java.awt.*;
import java.awt.event.*;
public class MyPanel extends JPanel implements KeyListener
{    final int SIZE = 20;              //小正方形的边长
    int x,y;
    int x0,y0;
    Color fillColor = Color.red;   //小正方形的填充颜色
    Color bg = Color.CYAN;         //面板的背景色
    public MyPanel(Dimension d)    //d 为 panel 所在父容器的 size
    {
        setBackground(bg);
        x = d.width/2 - SIZE/2;
        y = d.height/2 - SIZE/2;
        x0 = x; y0 = y;
        this.setFocusable(true);
        this.addKeyListener(this);
    }
    public void paintComponent(Graphics g) {
    super.paintComponent(g);
    g.setColor(bg);
    g.fillRect(x0,y0,SIZE,SIZE); //清除原先位置上的小正方形
    g.setColor(fillColor);
    g.fillRect(x,y,SIZE,SIZE);    //画移动以后的小正方形
    }
    // 按移动键头的处理
    public void keyPressed(KeyEvent e)
    {
        if(e.getKeyCode() == KeyEvent.VK_UP)
          {  y0 = y; x0 = x;
             y = y - 8;
            if(y<= 3) y = 3;
            repaint();
          }
        else if(e.getKeyCode() == KeyEvent.VK_DOWN)
        {   y0 = y; x0 = x;
            y = y + 8;
            if(y>= getSize().height - 3 - SIZE) y = getSize().height - 3 - SIZE;
            repaint();
        }
        else if(e.getKeyCode() == KeyEvent.VK_LEFT)
        {   x0 = x; y0 = y;
            x = x - 8;
            if(x<= 3) x = 3;
            repaint();
        }
```

```
        else if(e.getKeyCode() == KeyEvent.VK_RIGHT)
            {   x0 = x; y0 = y;
                x = x + 8;
                if(x>= getSize().width - 3 - SIZE) x = getSize().width - 3 - SIZE;
                repaint();
            }
    }
    // 键入是字母键的处理
    public void keyTyped(KeyEvent e) {
    char ch = e.getKeyChar();
    if (ch == 'Y'||ch == 'y') {
        fillColor = Color.yellow;
        repaint();
    }
    else if (ch == 'R'||ch == 'r') {
        fillColor = Color.red;
        repaint();
    }
  }
  public void keyReleased(KeyEvent e) {}
}
```

文件 keyEventDemo.java 的代码如下：

```
//keyEventDemo.java
import javax.swing.*;
import java.awt.*;
import java.awt.event.*;
public class KeyEventDemo extends JApplet
{
    public void init()
  {   Container container = getContentPane();
    Dimension d = this.getSize(); // d 为 applet 区域的大小
    MyPanel panel = new MyPanel(d);
    panel.setLayout(new BorderLayout());
    container.add(panel);
  }
}
```

HTML 文件 keyEventDemo.HTML 代码如下：

```
<html>
<applet code = "KeyEventDemo.class" width = "400" height = "400">
</applet>
</html>
```

程序运行结果如图 9-22 所示。

图 9-22　例 9-16 程序运行结果

9.13 菜　单

菜单是编程中经常用到的一种控件，Swing 菜单控件具有的显著特性为：能在菜单中使用图标；菜单项可以为单选按钮或者复选框；可以为菜单项指定加速键和快捷键，能提供菜单分隔线、弹出式菜单和子菜单等。与菜单相关的类继承关系如图 9-23 所示。

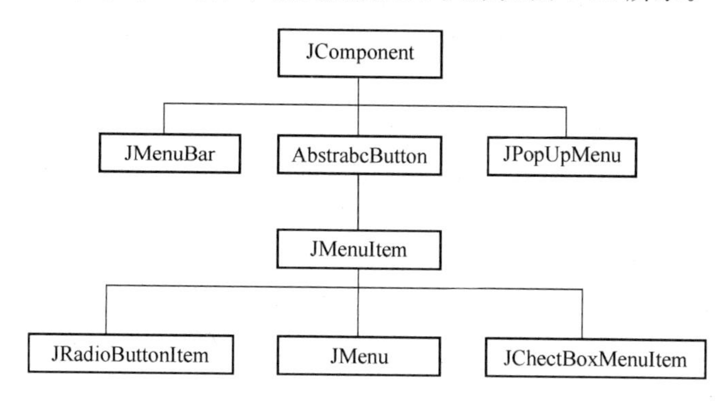

图 9-23　菜单相关类的继承关系

9.13.1 顶层菜单

顶层菜单位于应用程序的主窗口或顶层窗口中，与 JMenuBar 相关。

JMenuBar 类是 JComponent 的子类，程序设计时，一般将多个 JMenu（菜单）加入到 JMenuBar（菜单栏）中，而一个 JMenu 又可包含子菜单，或多个包含 JMenuItem（菜单项）。当用户单击菜单时，该菜单会展开并显示菜单项的列表。当用户单击菜单项时会产生动作事件。在 Java 中，响应事件的监听器应注册在菜单项上。

1. 设计菜单的编程步骤

（1）创建菜单栏 JMenuBar。创建一个空的菜单栏的语句如下：

```
JMenuBar bar = new JMenuBar();
```

创建一个带有菜单的菜单栏语句如下：

```
JMenuBar bar = new JMenuBar(new JMenu("File"));
```

创建完菜单栏以后，可以使用 JDialog、JApplet 或 JFrame 类的 setJMenuBar 方法将菜单栏添加到窗体中。以下代码给出了如何将前面创建的 bar 添加到 JFrame 中。

```
JFrame myJFrame = new JFrame();
myJFrame.setJMenuBar(bar);
```

（2）创建多个 JMenu 菜单并加入到菜单栏中。例如：

```
JMenu fileMenu = new JMenu( "File" );
bar.add(fileMenu);
```

（3）为每个菜单（JMenu）创建其所包含的子菜单或菜单项，并把子菜单或菜单项加入到菜单中。例如，下列语句创建 2 个 JMenuItem 对象并加入菜单 fileMenu 中。

```
JMenuItem aboutItem = new JMenuItem( "About..." );
JMenuItem exitItem = new JMenuItem( "Exit" );
fileMenu.add( exitItem );
fileMenu.add( aboutItem );
```

Swing 中除了提供菜单项 JMenuItem 外，还提供了另外的两种菜单项：JCheckBoxMenu-

Item(复选菜单项)与 JRadioButtonMenuItem(单选菜单项)。

JCheckBoxMenuItem 是实现了复选框功能的菜单项,具有 JCheckBox 的一切特征。JCheckBoxMenuItem 可以实现同时选中几个菜单的功能。

JRadioButtonMenuItem 类是实现了单选按钮功能的菜单项,其具有 JRadioButton 类的一切特征。使用 JRadioButtonMenuItem 主要是为了实现多选一的功能。

(4) 给每一个菜单项注册事件处理监听器。菜单项触发的主要事件有 ActionEvet,其在菜单项被选中时触发。通过调用 addActionListener 给菜单项注册实现 ActionListener 接口的事件监听器。

2. 添加快捷键和加速器

为菜单项添加快捷键和加速器,用户在使用时可以提高工作效率。选择菜单项,可用鼠标进行,还可为菜单项定义一个键盘快捷键,这样用键盘可以选择菜单项。定义了快捷键的菜单项在显示时,会在菜单项标签字符串中与快捷键字母相同的第一个字母上加下画线。除了在菜单项中使用快捷键外,开发人员还可以使用 setAccelerator 方法为菜单项指定一个加速器。加速器和快捷键的区别是:快捷键用来从当前打开的菜单中选择一个子菜单或者菜单项,而加速器可以在不打开菜单的情况下选择菜单项。

javax.swing.AbstractButton 的所有子类都可以使用快捷键。添加快捷键的方法有两种:

(1) 在创建菜单项的同时定义快捷键。例如:

```
JMenuItem exitItem = new JMenuItem("Exit",(int)'x');   //指定该菜单项的快捷键 x
```

(2) 为已经存在的菜单项定义快捷键。例如:

```
JMenu fileMenu  =  new JMenu( "File" );
fileMenu.setMnemonic( 'F' );          //设置该菜单项的快捷键 F
```

3. 使用分隔线

有时希望在菜单项之间增加一条横向分隔线,以便把菜单项分成几组。加入分隔线的方法是 JMenu 的方法 addSeparator( ),使用时要注意该语句的位置。菜单项是按照加入的先后顺序排列在菜单中的,希望把分隔线加在哪里,就要把分隔线语句放在哪里。

例如,在"Font"菜单中的子菜单项 1 后加入一条分隔线,语句如下:

```
JMenu fontMenu  =  new JMenu( "Font" );
fontMenu.add (子菜单项 1);
fontMenu.addSeparator();   //加一条横向分隔线
```

例 9-17 菜单应用举例。

在菜单栏加入两个菜单:File 和 Format。File 菜单中加入两个菜单项用于显示对话框信息和中断程序运行。Format 菜单加入一组单选菜单项与一组复选菜单项用于控制 JFame 的内容面板上 Label 内容的颜色和字体。

```
import java.awt.*;
import java.awt.event.*;
import javax.swing.*;
public class MenuTest extends JFrame {
    private final Color colorValues[] = {Color.black,Color.blue,Color.red,Color.green};
    private JRadioButtonMenuItem colorItems [], fonts [];
    private JCheckBoxMenuItem styleItems [];
    private JLabel displayLabel;
    private ButtonGroup fontGroup, colorGroup;
    private int style;
```

```
public MenuTest()                                   // set up GUI
{    super( " 使用 Swing 中的菜单" );
    // 创建菜单栏并置于 MenuTest 窗体中
    JMenuBar bar = new JMenuBar();
    setJMenuBar( bar );
    // 创建菜单 File 和它的菜单项
    JMenu fileMenu = new JMenu( "File" );
    fileMenu.setMnemonic( 'F' );
    JMenuItem aboutItem = new JMenuItem("About…");   // 创建菜单项 About…
    aboutItem.setMnemonic( 'A' );                     //设置快捷键
    fileMenu.add( aboutItem );
    aboutItem.addActionListener(      //给菜单项 aboutItem 注册事件监听器
        new ActionListener() {
            // 当用户选中 About…时,显示对话框
            public void actionPerformed( ActionEvent event )
            {JOptionPane.showMessageDialog( MenuTest.this,"这是菜单应用的例子",
                    "About", JOptionPane.PLAIN_MESSAGE );
            }
    } );
    // 创建菜单项 Exit,并设置快捷键
    JMenuItem exitItem = new JMenuItem( "Exit", (int)'x');
    fileMenu.add( exitItem );
    exitItem.addActionListener(
        new ActionListener() {
            // 当用户选中 exitItem,程序中断运行
            public void actionPerformed( ActionEvent event )
            { System.exit( 0 );
            }
      });
    bar.add( fileMenu );                              //将菜单 fileMenu 加入菜单栏
    // 创建菜单 Format 和它的子菜单,菜单项
    JMenu formatMenu = new JMenu( "Format" );
    formatMenu.setMnemonic( 'r' );
    // 创建子菜单 Color
    String colors [] = { "Black", "Blue", "Red", "Green" };
    JMenu colorMenu = new JMenu( "Color" );
    colorMenu.setMnemonic( 'C' );
    colorItems = new JRadioButtonMenuItem [ colors.length ];
    colorGroup = new ButtonGroup();
    ItemHandler itemHandler = new ItemHandler();
    // 创建 Color 的菜单项 radiobutton
    for ( int count = 0; count < colors.length; count ++ ) {
        colorItems [ count ] = new JRadioButtonMenuItem( colors [ count ] );
        colorMenu.add( colorItems [ count ] );
        colorGroup.add( colorItems [ count ] );
        colorItems [ count ].addActionListener( itemHandler );
    }
    colorItems [ 0 ].setSelected( true );
    formatMenu.add( colorMenu ); //将菜单 format 加入菜单栏
    formatMenu.addSeparator();
    // 创建 Font 子菜单
    String fontNames [] = { "Serif", "Monospaced", "SansSerif" };
    JMenu fontMenu = new JMenu( "Font" );
    fontMenu.setMnemonic( 'n' );
    fonts = new JRadioButtonMenuItem [ fontNames.length ];
```

```
        fontGroup = new ButtonGroup();
        // 创建 Font 的菜单项 radio button
        for ( int count = 0; count < fonts.length; count ++ ) {
            fonts [ count ] = new JRadioButtonMenuItem( fontNames [ count ] );
            fontMenu.add( fonts [ count ] );
            fontGroup.add( fonts [ count ] );
            fonts [ count ].addActionListener( itemHandler );
        }
        fonts [ 0 ].setSelected( true ); // select first Font menu item
        fontMenu.addSeparator();
        // 创建 style 菜单项
        String styleNames [] = { "Bold", "Italic" };
        styleItems = new JCheckBoxMenuItem [ styleNames.length ];
        StyleHandler styleHandler = new StyleHandler();
        for ( int count = 0; count < styleNames.length; count ++ ) {
            styleItems [ count ] =
                new JCheckBoxMenuItem( styleNames [ count ] );
            fontMenu.add( styleItems [ count ] );
            styleItems [ count ].addItemListener( styleHandler );
        }
        formatMenu.add( fontMenu ); // 将子菜单 Font 加入菜单 Format 中
        bar.add( formatMenu ); // 将菜单 formatMenu 加入菜单栏 bar
        // 在 Frame 的内容面板上加入一 JLabel 对象,以显示 text
        displayLabel = new JLabel("Demonstrate JMenu",SwingConstants.CENTER);
        displayLabel.setForeground( colorValues [ 0 ] );
        displayLabel.setFont( new Font( "Serif", Font.PLAIN, 50 ) );
        getContentPane().setBackground( Color.CYAN );
        getContentPane().add( displayLabel, BorderLayout.CENTER );
        setSize( 500, 150 );
        setVisible( true );
    }
    public static void main( String args [] )
    {   MenuTest application = new MenuTest();
        application.setDefaultCloseOperation( JFrame.EXIT_ON_CLOSE );
    }
    // 定义处理 radio button 菜单项上的事件 Action events 的内部类
    private class ItemHandler implements ActionListener {
        // 选中 color 和 font 时的处理代码
        public void actionPerformed( ActionEvent event )
        {   // 选中 color 的处理代码
            for ( int count = 0; count < colorItems.length; count ++ )
                if ( colorItems [ count ].isSelected() ) {
                    displayLabel.setForeground( colorValues [ count ] );
                    break;
                }
            // 选中 font 时的处理代码
            for ( int count = 0; count < fonts.length; count ++ )
                if ( event.getSource() == fonts [ count ] ) {
                    displayLabel.setFont(
```

```
                        new Font( fonts [ count ].getText(), style, 50 ) );
                    break;
                }
            } // end method actionPerformed
        } // end class ItemHandler
        // 定义处理 check box 菜单项上的事件 Item events 的内部类
        private class StyleHandler implements ItemListener {
            // process font style selections
            public void itemStateChanged( ItemEvent e )
            {   style = 0;
                if ( styleItems [ 0 ].isSelected() ) // 如果 bold 被选中
            style + = Font.BOLD;
        if ( styleItems [ 1 ].isSelected() ) // 如果 italic 被选中
            style + = Font.ITALIC;
        displayLabel.setFont(
            new Font( displayLabel.getFont().getName(), style, 50 ) );
        }
    } // end class StyleHandler
}
```

程序运行的部分输出结果如图 9-24 所示。

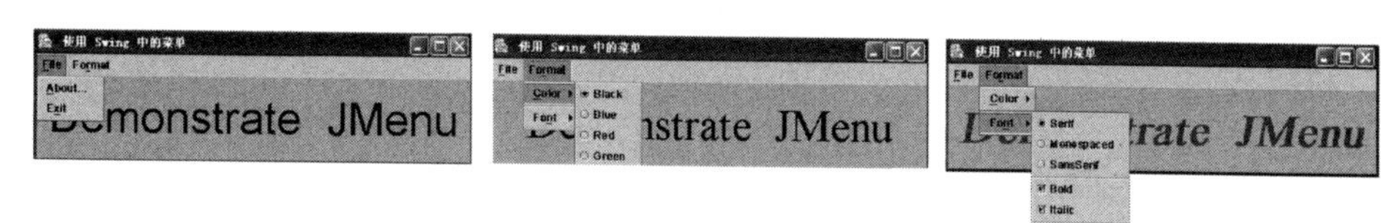

图 9-24　例 9-17 程序运行输出结果

9.13.2　弹出式菜单

弹出式菜单(JPopupMenu)是一个可弹出并显示一系列选项的小窗口,它并不固定在菜单栏中,而是能够自由浮动。JPopupMenu 具有很好的环境相关(context-sensitive)特性,每一个 JPopupMenu 都与相应的控件相关联,该控件被称做调用者(invoker)。

1. 创建弹出式菜单

创建 JPopupMenu 对象的常用构造方法如下。

JPopupMenu() :构造一个不带"调用者"的 JPopupMenu。

在弹出式菜单创建后,程序员可使用 add 方法和 insert 方法向 JPopupMenu 中添加或者插入 JMenuItem 与 JComponent。

2. 弹出式菜单对象常用的方法

(1) void add(JMenuItem menuItem):将指定菜单项追加到此菜单的末尾。

(2) void addSeparator():将新分隔符追加到菜单的末尾。

(3) void show(Component invoker, int x, int y) :在控件调用者的坐标空间中的位置(x,y)处,显示弹出菜单。

3. 弹出式菜单触发事件

用户在支持弹出式菜单的控件上,单击鼠标右键时产生弹出式触发事件(popup trigger event)。程序员应该检查所有的 MouseEvent 事件,看其是否是弹出式菜单触发器,然后显示弹出式菜单。下面的 checkForTriggerEvent 方法在收到适当的触发器事件时就会在当前控件的鼠标位置处显示弹出式菜单,代码如下:

```
void checkForTriggerEvent( MouseEvent event )
{   if ( event.isPopupTrigger() )
   popupMenu.show(event.getComponent(),event.getX(), event.getY() );
}
```

例 9-18 利用弹出式菜单中的选项控制 JFrame 主窗口的背景颜色。

```
// PopupTest.java
import java.awt.*;
import java.awt.event.*;
import javax.swing.*;
public class PopupTest extends JFrame {
    private JRadioButtonMenuItem items[];
    private final Color colorValues[] =
        { Color.GREEN, Color.YELLOW, Color.RED };
    private JPopupMenu popupMenu;
    public PopupTest() // set up GUI
    {   super( "弹出式菜单的简单使用" );
        // 设置弹出式菜单和它的菜单项
        popupMenu = new JPopupMenu();
        String colors[] = { "Green", "Yellow", "Red" };
        ButtonGroup colorGroup = new ButtonGroup();
        items = new JRadioButtonMenuItem[colors.length];
        ItemHandler handler = new ItemHandler();
        // 创建弹出式菜单中的各个菜单项,并给每个菜单项注册事件监听器
        for ( int count = 0; count < items.length; count++ ) {
            items[ count ] = new JRadioButtonMenuItem( colors[ count ] );
            popupMenu.add( items[ count ] );
            colorGroup.add( items[ count ] );
            items[ count ].addActionListener( handler );
        }
        getContentPane().setBackground( Color.WHITE ); //设置窗口背景色初值
            this.addMouseListener( //向主窗口注册监听器
          new MouseAdapter() {
            // handle mouse press event
            public void mousePressed( MouseEvent event )
            {   checkForTriggerEvent( event );
            }
            // handle mouse release event
            public void mouseReleased( MouseEvent event )
            {   checkForTriggerEvent( event );
            }
            // 检查 MouseEvent 事件,看其是否是弹出式菜单触发器
            private void checkForTriggerEvent( MouseEvent event )
            { if ( event.isPopupTrigger() )
                popupMenu.show (event.getComponent(),
                                event.getX(),event.getY() );
            }
        } ); // end call to addMouseListener
      setSize( 300, 200 );
```

```
            setVisible( true );
        } // end constructor PopupTest
        public static void main( String args [] )
        {    PopupTest application = new PopupTest();
             application.setDefaultCloseOperation( JFrame.EXIT_ON_CLOSE );
        }
        // 菜单项选中时的事件处理代码
        private class ItemHandler implements ActionListener {
             public void actionPerformed( ActionEvent event )
             { // 判断哪个菜单项被选中
               for ( int i = 0; i < items.length; i ++ )
                 if ( event.getSource() == items [ i ] ) {
                      getContentPane().setBackground( colorValues [ i ] );
                 }
             }
        } // end class ItemHandler
}
```

程序运行结果如图 9-25 所示。

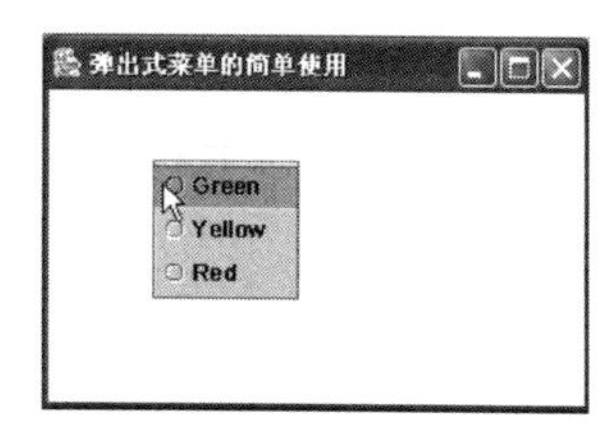

图 9-25 弹出式菜单应用例子

9.14 选项卡面板

选项卡面板(JTabbedPane)控件是一个容器,它允许用户通过单击给定标题或图标的选项卡,在一组组件之间进行切换。JTabbedPane 类是 JComponent 的子类。

1. JTabbedPane 常用的构造方法

(1) JTabbedPane():创建一个空的 TabbedPane 对象。

(2) JTabbedPane(int tabPlacement):创建一个空的 JTabbedPane 对象,并指定选项卡布局(摆放位置)。tabPlacement 可为 JTabbedPane 类中的 4 个静态常量之一:TOP(缺省值)、BOTTOM、LEFT、RIGHT。

(3) JTabbedPane(int tabPlacement, int tabLayoutPolicy):创建一个空的 TabbedPane,使其具有指定的选项卡布局和选项卡布局策略。布局策略定义了当在一次运行中不能放入所有的选项卡时,放置选项卡的策略。tabLayoutPolicy 可为 JTabbedPane 类中的两个静态常量之一:

① WRAP_TAB_LAYOUT(缺省值):选项卡一行放不下时,将换行。

② SCROLL_TAB_LAYOUT:选项卡将不再换行,而是停留在一行,并将箭头被添加到尾部。

2. JTabbedPane 对象常用的方法

(1) void addTab(String title, Component componcnt):添加一个具有指定选项卡标题且

没有图标的组件。

(2) void addTab(String title, Icon icon, Component component):添加一个具有指定选项卡标题 title 和/或 icon 表示的组件,其任意一个都可以为 null。

(3) void addTab(String title, Icon icon, Component component, String tip):添加由 title 和/或 icon 表示的 component 和 tip,其中任意一个都可以为 null。

(4) int getSelectedIndex() :返回当前选择的选项卡标签的索引。JTabbedPane 上的每个标签都有索引值(index),一般若没有加以设置,索引值从左到右依次是 0,1,2,…。

3. ChangeEvent 事件响应

JTabbedPane 以处理 ChangeEvent(在 javax.swing.event 包中)事件为主。每当用户在 JTabbedPane 上选换标签时,都会产生 ChangeEvent 事件。因此要处理选换标签所对应的操作,程序员就必须通过调用 addChangeListener(ChangeListener e) 方法给选项卡面板组件注册实现 ChangeListener 接口的监听器,并实现接口中的 stateChanged(ChangeEvent e)方法。

例 9-19 该程序在 JFrame 窗口中创建一个带有 3 个选项标签的 JTabbedPane 对象。每个标签显示一个 JPanel 对象:panel1、panel2 和 panel3。

```
//JTabbedPaneDemo.java
import java.awt.*;
import javax.swing.*;
public class JTabbedPaneDemo extends JFrame {
    public JTabbedPaneDemo() // set up GUI
    {    super( "JTabbedPane 简单应用 " );
        JTabbedPane tabbedPane = new JTabbedPane(); // create JTabbedPane
        // 创建 panel1 并加入 JTabbedPane
        JLabel label1 = new JLabel( "panel 1", SwingConstants.CENTER );
        JPanel panel1 = new JPanel();
        panel1.add( label1 );
        tabbedPane.addTab( "Tab 1", null, panel1, "First Panel" );
        //创建 panel2 并加入 JTabbedPane e
        JLabel label2 = new JLabel( "panel 2", SwingConstants.CENTER );
        JPanel panel2 = new JPanel();
        panel2.setBackground( Color.YELLOW );
        panel2.add( label2 );
        tabbedPane.addTab( "Tab 2", null, panel2, "Second Panel" );
        //创建 panel3 并加入 JTabbedPane
        JLabel label3 = new JLabel( "panel 3" ,SwingConstants.CENTER );
        JPanel panel3 = new JPanel();
        panel3.setLayout( new BorderLayout() );
        panel3.add( new JButton( "北" ), BorderLayout.NORTH );
        panel3.add( new JButton( "西" ), BorderLayout.WEST );
        panel3.add( new JButton( "东" ), BorderLayout.EAST );
        panel3.add( new JButton( "南" ), BorderLayout.SOUTH );
        panel3.add( label3, BorderLayout.CENTER );
        tabbedPane.addTab( "Tab 3", null, panel3, "Third Panel" );
        getContentPane().add( tabbedPane ); // add JTabbedPane to container
        setSize( 250, 200 ); setVisible( true );
    }
    public static void main( String args [])
    { JTabbedPaneDemo tabbedPaneDemo = new JTabbedPaneDemo();
      tabbedPaneDemo.setDefaultCloseOperation( JFrame.EXIT_ON_CLOSE );
    }
}
```

程序运行结果如图 9-26 所示。

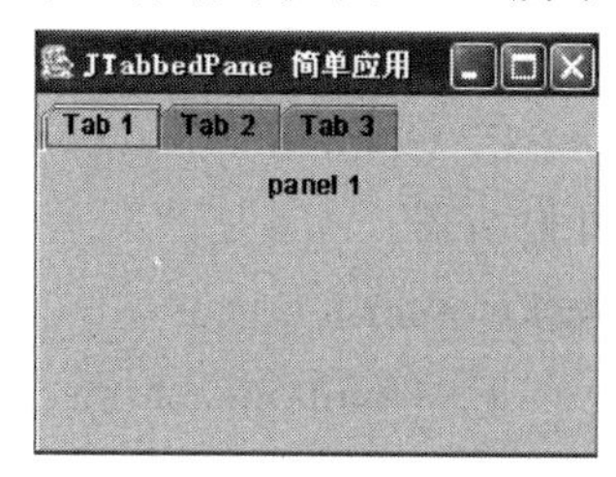

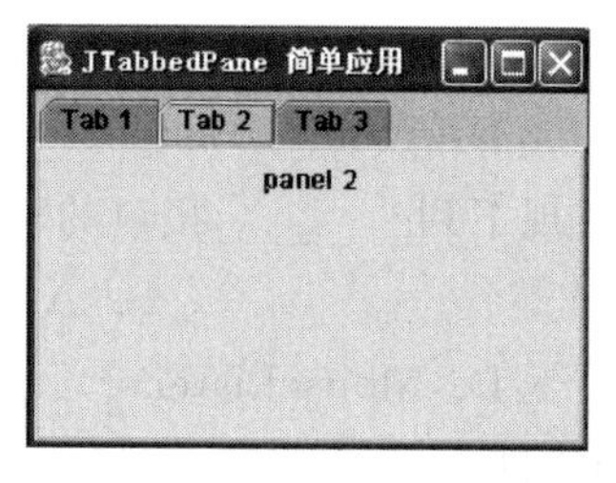

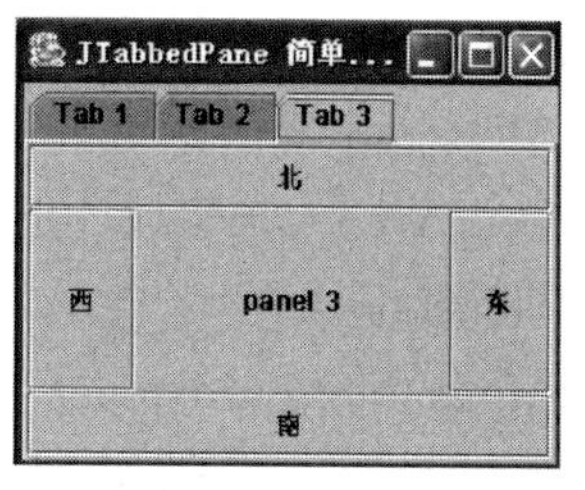

图 9-26 JTabbedPane 应用示例

9.15 小 结

GUI 编程涉及 AWT 包和 Swing 包中的大量的类接口。

java.awt 包提供了基本的 Java 程序的 GUI 设计工具，而 Swing 是 AWT 的扩展，Swing 组件都以"J"开头。

事件处理机制能够让图形界面响应用户的操作。Java 中的事件处理模型三要素为事件源、事件对象、事件监听器。当需要对组件的某种事件进行响应和处理时，程序员必须完成两个步骤：一是为该组件注册事件监听器；二是实现事件监听器接口中声明的事件处理方法。事件对象封装了发生事件的有关信息，常用的事件对象有 ActionEvent、KeyEvent、ItemEvent 和 MouseEvent 等。

布局管理器用于管理组件在容器中的布局方式，Java 系统提供的标准布局管理器类有 FlowLayout、BorderLayout、GridLayout、CardLayout、BoxLayout 和 GridBagLayout。其中 GridBagLayout 最灵活、最复杂。

JFrame、JApplet、JDialog 是 Java 的顶层容器，Swing 组件必须添加到一个与 Swing 顶层容器相关联的内容面板(content pane)上。

搭建图形用户界面时，要理解各种 GUI 组件特性和使用方法。如 JButton、JLabel、JTextField、JTeaxArea、JCheckBox、JRadioButton、JComboBox、JList、JScrollPane、JPanel、JTabbedPane、JPopMenu 等。

用于声明菜单的类有 JMenuBar、JcheckBoxMenuItem、JradioButtonMenuItem、JMenu、JMenuItem。

习 题

9.1 每个容器用什么方法向容器添加某个组件？用什么方法从容器中删除某个组件？使用什么方法可为容器指定布局管理器？添加 GUI 组件的 add 方法是哪个类的方法？

9.2 Java 布局管理器有哪些？各具有什么特点？

9.3 在 Java 事件处理模型中，程序员如何对 GUI 组件的某种发生事件进行响应？

9.4 选择题

① 按钮可以产生 ActionEvent 事件，实现________接口可处理此事件。

A. A、FocusListener B. ComponentListener C. WindowListener D. ActionListener

② 通过实现下列________接口可以对 TextField 对象的事件进行监听和处理。

A. ActionListener　　B. FocusListener

C. MouseMotionListener　　D. WindowListener　　E. ContainerListener

③ TextArea 对象可以注册实现下列________接口的事件监听器。

A. TextListener　　B. ActionListener

C. MouseMotionListener　　D. MouseListener　　E. ComponentListener

④ 下列________属于容器的构件。

A. JFrame　　B. JButton　　C. JPanel　　D. JApplet

9.5　填空题。

① 传递给实现了 java.awt.event.MouseMotionListener 接口的类中 mouseDragged()方法的事件对象是________类型的。

② 当用户在 TextField 中输入一行文字后，按回车键，实现______接口可实现对事件的响应。

③ 内容面板的默认布局管理器是________。

④ JFrame 类的________方法用来指定用户关闭窗口时发生的操作。

⑤ 方法________将 JMenubar 添加到 JFrame 中。

9.6　编写一个 Applet，利用两个文本框对象 input1 和 input2，接受用户从键盘输入的两个整型数。当用户单击“计算”按钮时，可进行算术运算，并输出运算结果；运算结果放在多行文本域 JTtextArea 组件中。GUI 界面参考如下：

9.7　编写一个图形界面的 Application 程序，包含一个带图标的 JButton 对象。当用户单击这个按钮时，能将 Frame 的背景色进行重新设置。

提示：使用 javax.swing.JColorChoos 组件选择各种颜色。

9.8　编写一个 Application 程序，其 GUI 参考如左下图。用户可选择复合框的前景颜色和背景颜色。

9.9　编写一个 Application 程序，其 GUI 参考如右下图。

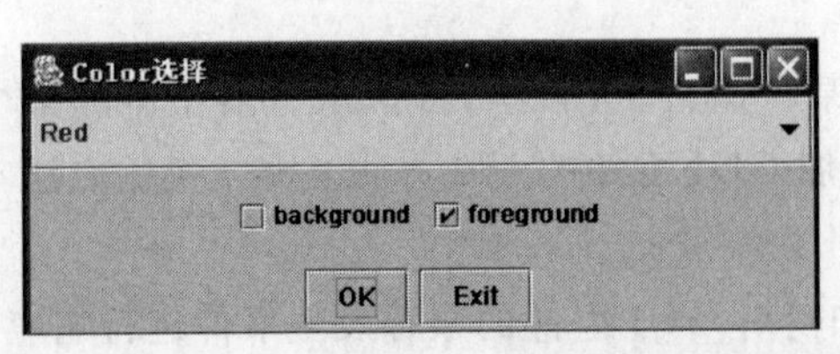

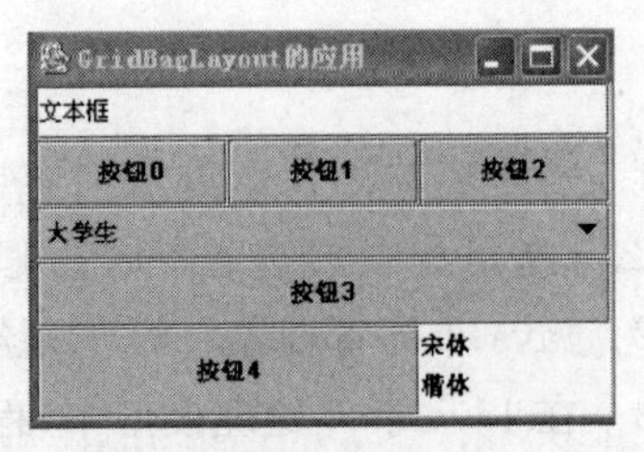

9.10　对学生类(包括学号、姓名、性别、年龄、爱好等)完成 GUI 的设计，使能对学生对象进行添加、修改、删除、查询操作。请使用菜单技术。

第10章　异常处理

Java 通过面向对象的方法进行异常处理，把各种不同的异常事件进行分类，体现了良好的层次性，提供了良好的接口，这种机制为具有动态运行特性的复杂程序提供了强有力的控制。本章介绍异常的概念、Java 异常的分类、异常处理块的形式，并通过实例介绍了异常抛出、异常重新抛出和异常捕获处理的过程。最后介绍如何定义用户自己的异常，并给出了应用的实例。

10.1　异常处理概述

所谓异常(Exception)，是指程序在运行过程中出现的不正常情况或错误。比如除零溢出、数组越界、文件找不到等，这些错误的发生将阻止程序的正常运行。为了加强程序的健壮性(robust)，程序设计时必须考虑程序在运行时可能发生的异常情况，并做出相应的处理。在用传统的语言编程时，程序员只能通过函数的返回值来发出错误信息，这易于导致很多错误。而在 Java 语言中，通过面向对象的方法来处理程序运行时异常。当出现异常情况时，一个 Exception 对象就产生了，并交由异常处理程序处理，异常处理程序是使程序得以恢复正常运行的处理过程。

Java 的异常处理机制描述如下：在一个方法的运行过程中，如果发生了异常，

- 则这个方法(或者是 Java 虚拟机)生成一个代表该异常的对象(它包含了异常的详细信息)，并把它交给运行时系统，运行时系统寻找相应的代码来处理这一异常。把生成异常对象并把它提交给运行时系统的过程称为抛出(throw)一个异常。
- 运行时系统寻找相应的代码来处理这一异常，系统在方法的调用栈中查找，从产生异常的方法开始进行回溯，沿着被调用的顺序往前寻找，直到找到包含相应异常处理的方法为止。其过程如图 10-1 所示。这一过程称为捕获(catch)一个异常。
- 如该异常未进行成功捕获，则程序将终止运行。

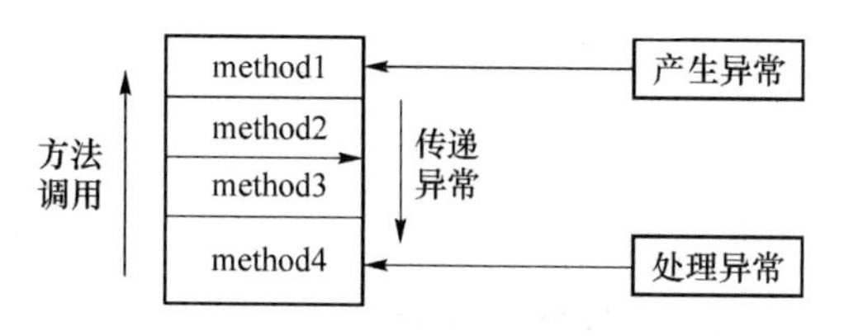

图 10-1　捕获异常

下面看一个未被捕获的异常的例子。下面是一段故意包括“被零除错误”的程序代码，如果不处理它将会有什么样的情况？

例 10-1 一个未被捕获的异常——被零除。

```
//Exc0.java
class Exc0 {
    public static void main(String args[])
    {
     int d = 0;
     int a = 42 / d;
    }
}
```

运行时系统检查到"被零除"的情况,它构造一个新的异常对象,然后抛出该异常。这导致 Exc0 的执行停止,因为一旦一个异常被抛出,它必须被一个异常处理程序捕获并且被立即处理。例 10-1 中,由于没有提供任何自己的异常处理程序,所以异常被 Java 运行时系统的默认处理程序捕获。任何不是被自己程序捕获的异常最终都会被系统的默认处理程序处理。默认处理程序显示一个描述异常的字符串,打印异常发生处的堆栈信息并且终止程序。如图 10-2 所示是由 JDK 运行时解释器执行该程序所产生的输出信息。

```
Exception in thread "main" java.lang.ArithmeticException: / by zero
        at Exc0.main(Exc0.java:5)
```

图 10-2 一个未被捕获的异常——被零除

其中,抛出的异常类型是 ArithmeticException,需要注意类名 Exc0、方法名 main、文件名 Exc0.java 和第 5 行是被包括在一个简单的堆栈使用轨迹中。

将例 10-1 中的程序修改为提供捕获"被零除"的异常处理,只需要把所要监控的代码放进一个 try 块就可以了。紧跟着 try 块的,是包括一个说明捕获的错误类型的 catch 子句。

例 10-2 的程序包含一个处理因为被零除而产生的 ArithmeticException 异常的 try 块和一个 catch 子句。

例 10-2 一个被捕获的异常——被零除。

```
//Exc2.jaa
class Exc2 {
  public static void main(String args[])
  { int d, a;
    try { // monitor a block of code.
      d = 0;
      a = 42 / d;
    }
    catch (ArithmeticException e){ //catch divide-by-zero error
      System.out.println("Division by zero.");
      }
    System.out.println("After catch statement.");
  }
}
```

程序运行将正常结束,并显示出如图 10-3 所示信息。

```
Division by zero.
After catch statement.
```

图 10-3 一个被捕获的异常——被零除

10.2　异常分类

Java 的异常是一个对象，所有的异常都直接或间接地继承 Throwable 类。其类的继承层次结构如图 10-4 所示。

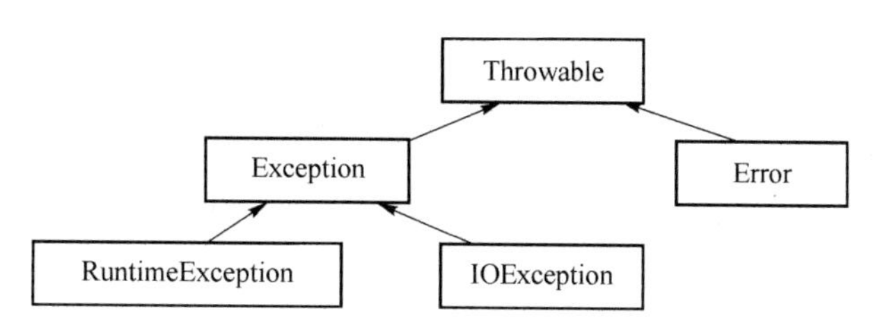

图 10-4　Throwable 类的继承层次

Throwable 类是所有异常的超类。它有两个子类 Error 和 Exception。

Error 类及其子类由 Java 虚拟机生成并抛出，包括动态链接失败、虚拟机错误等。Java 程序不应捕获这些异常。

Exception 类及其子类(RuntimeException 和 IOException)代表 Java 程序中可能发生的异常，并且应用程序可以捕获这些异常。

Java 将异常分为两种类型：编译时检查性异常(checked exception)和非检查性异常(unchecked exception，又称运行时异常)。对于检查性异常，在程序中必须对其进行处理，否则编译器会指出错误。对于非检查性异常，程序中可以不做处理，直接由运行时系统来处理。编译器要求 Java 程序必须捕获或声明所有的编译时异常(或称非运行时异常)。

Runtime Exception 异常是 Java 虚拟机在运行时生成的异常，如被零除等系统错误、数组下标超范围等，其产生比较频繁，处理麻烦，对程序可读性和运行效率影响很大。因此由系统检测，用户可不做处理，系统将它们交给缺省的异常处理程序。当然，必要时，用户可对其进行处理。

在 Java 类库的每个类包中都定义了异常类，这些异常类分成两大类——Error 类及 Exception 类，后者是 Java 程序中需要大量处理的。

除了 Java 类库所定义的异常类之外，用户也可以通过继承已有的异常类(一般从 Throwable 类或 Exception 类继承)来定义自己的异常类，并在程序中使用(利用 throw 产生，catch 捕捉)。

常见的异常如下。

- ArithmeticException：算术运算溢出。如除数为零。
- ArrayIndexOutOfBandsException：数组下标越界异常。
- ArrayStoreException：数组储存异常。数组复制时，若源数组和目标数组的类型不一致时，导致异常。
- NullPointerException：空指针异常。
- NumberFormatException：数据格式异常。将字符串(含有非数字)直接转换成数值时产生。
- OutOfMemoryException：内存溢出异常。在创建一个对象时，由于申请的内存空间过大，导致系统内存不够。

• IOException：输入/输出中的异常。
• FileNotFoundException：文件找不到异常。
• NoClassDefFoundException：没有找到类定义时的异常。

10.3 异常的捕获处理

Java 的异常处理是通过 3 个关键词来实现的：try、catch、finally。为防止和处理一个运行时错误，只需要把所要监控的代码放进一个 try 块，如果出现异常，系统抛出异常对象，可以通过它的类型来捕捉并处理它，或最后由 finally 进行最终处理。下面是异常处理块的一般形式：

```
try{
    //接受监视的程序块，在此区域内发生的异常，由 catch 中指定的程序处理；
}
catch(要处理的异常类型和标识符){
    //捕获(catch)一个异常并进行异常处理；
}
catch(要处理的异常类型和标识符){
    //捕获(catch)一个异常并进行异常处理；
}
…

finally{
    //最终处理；
}
```

1. try 语句

捕获异常的第一步就是用 try {…}语句指定一段代码，该段代码就是一次捕获并处理异常的范围。在执行过程中，该段代码可能会产生并抛出一个或多个异常，因此，它后面的 catch 语句进行捕获时也要做相应的处理。

2. catch 语句

每个 try 语句必须伴随一个或多个 catch 语句，用于捕获 try 代码块所产生的异常，并做相应的处理。catch 语句有一个形式参数，用于指明其所能捕获的异常类型，运行时系统通过参数值把被抛出的异常对象传递给 catch 语句。

程序设计中要根据具体的情况来选择 catch 语句的异常处理类型，一般应该按照 try 代码块中异常可能产生的顺序及其真正类型进行捕获和处理。

在某些情况下，由单个代码段可能引起多个异常。处理这种情况，可以定义两个或更多的 catch 子句，每个子句捕获一种类型的异常。当异常被引发时，每一个 catch 子句被依次检查，第一个匹配异常类型的子句执行。当一个 catch 语句执行以后，其他的子句被旁路，执行从 try/catch 块以后的代码开始继续。因此，当用多个 catch 语句时需要记住，异常子类必须在它们任何父类之前使用，这一点是很重要的。因为运用父类的 catch 语句将捕获该类型及其所有子类类型的异常。这样，如果子类在父类后面，子类将永远不会到达。

下面例子中，try 语句块执行时，可能产生两种异常类型：FileNotFoundException 和 IOException，这两个异常中，FileNotFoundException 类是 IOException（一般类）的特殊类，所以，捕获 FileNotFoundException 异常类型的 catch 语句应写在捕获 IOException 异常类型的 catch 语句之前。

```
public class ExceptionDemo
{
  public static void main(String args[])
  {
    try
      {
        FileInputStream fis = new FileInputStream("test1.txt");
        int b;
        while( (b = fis.read())!= -1 )
        {
          System.out.print(b);
        }
        fis.close();
      }
      catch(FileNotFoundException e){
        ...
      }
      catch(IOException e){
        ...
      }
    }
}
```

3. finally 语句

无论 try 所指定的程序块中抛出或不抛出异常，也无论 catch 语句的异常类型是否与所抛出的异常类型一致，finally 所指定的代码都要被执行，它提供了统一的出口。

通常在 finally 语句中可以进行资源的清除工作，如关闭打开的文件等。finally 在文件处理时非常有用，例如：

```
try {
    对文件进行处理的程序;
}catch(IOException e) {
    //对文件异常进行处理;
}finally {
    //不论是否发生异常，都关闭文件;
}
```

要注意：在使用能够产生异常的方法而没有捕获和处理，程序将不能通过编译。

10.4 重新抛出异常

10.4.1 异常对象的生成

Java 中异常对象生成的方式为：

(1) 由 Java 虚拟机生成；

(2) 由 Java 类库中的某些类生成；

(3) 在程序中生成自己的异常对象，即异常可以不是出错产生，而是人为地抛出。不论哪种方式，生成异常对象都是通过 throw 语句实现的。例如：

```
throw new ThrowableObject();
ArithmeticException e = new ArithmeticException();
throw e;
```

注意:抛出的异常必须是 Throwable 或其子类的实例。

例 10-3 的程序,抛出了 ArithmeticException 、ArrayIndexOutOfBoundsException 和 StringIndexOutOfBoundsException 3 个异常类型的对象,并通过 catch 语句对 3 种异常对象进行了捕获处理。

例 10-3 抛出 3 个异常类型的对象并捕获处理。

```
//JavaThrow.java
class JavaThrow
{    public static void main(String args[])
{try
  {   throw new ArithmeticException();
  }catch(ArithmeticException ae){
      System.out.println(ae);
  }
  try
  {    throw new ArrayIndexOutOfBoundsException();
  }catch(ArrayIndexOutOfBoundsException ai){
      System.out.println(ai);
  }
  try
  { throw new StringIndexOutOfBoundsException();
  }catch(StringIndexOutOfBoundsException si){
      System.out.println(si);
  }
}
```

程序运行结果如图 10-5 所示。

```
java.lang.ArithmeticException
java.lang.ArrayIndexOutOfBoundsException
java.lang.StringIndexOutOfBoundsException
```

图 10-5 3 个 try 块的异常处理

10.4.2 重新抛出异常对象

重新抛出异常对象首先必须生成异常。如果在一个方法中生成了异常,但是该方法并不处理它自己产生的异常,而是沿着调用层次向上传递,由调用它的方法或方法栈来处理这些异常,叫重新抛出异常。

通常的情况是在该方法中并不确切知道如何对这些异常进行处理,比如 FileNotFound-Exception 异常,它由 FileInputStream 的构造方法产生,但在其构造方法中并不清楚如何处理它。是终止程序的执行还是新生成一个文件,这需要由调用它的方法来处理。

重新抛出异常对象的方法是在产生异常的方法名后面加上要抛出的异常列表:

```
returnType methodName([paramlist]) throws exceptionList
```

例如,类 FileInputStream 中的 read()方法是这样定义的:

```
public int read() throws IOException
{ … }
```

throws 子句中可以同时指明多个异常,说明该方法将不对这些异常进行处理,而是声明抛出它们。

需要强调的是：对于非运行时异常，在程序中必须要作处理，或者捕获，或者声明重新抛出。而对于运行时异常，程序中则可不处理。

例 10-4 重新抛出异常对象。

```
import java.io. * ;
class JavaThrows {
  public int compute(int x) throws ArithmeticException
  {  int z = 100/x;     //可能抛出异常类型 ArithmeticException 的对象
     return z;
  }
  public void method1()
  {  int x;
     try{
       x = System.in.read();      //可能抛出异常类型 IOException 的对象
       x = x - 48;
       x = compute(x);     //抛出异常类型 ArithmeticException 的对象
       System.out.println(x);
     }
     catch(IOException ioe){     //捕获异常类型 IOException 的对象
       System.out.println("read error");
     }
     catch(ArithmeticException e){ //捕获类型 ArithmeticException 的对象
       System.out.println("devided by 0");
     }
     }
     public static void main(String args[]) {
      JavaThrows t1 = new JavaThrows();
       t1.method1();
     }
}
```

例 10-4 程序执行如图 10-6 所示。要求用户从键盘输入一个字符号。当输入'0'时，程序执行结果如图 10-6(a)所示；当输入非'0'字符时，程序执行结果如图 10-6(b)所示。

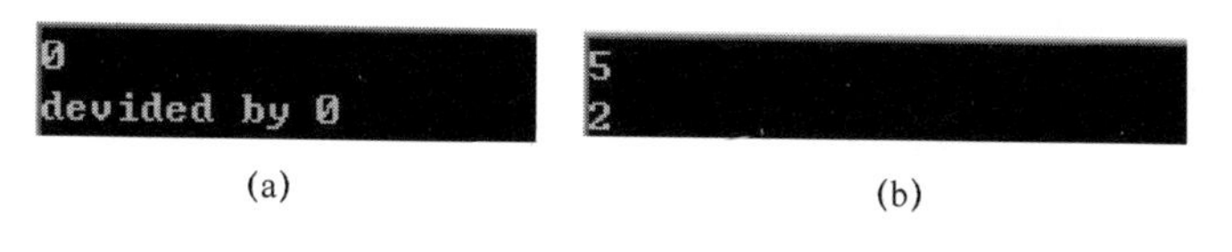

图 10-6 重新抛出异常

(1) 在 try 语句块中，执行语句

```
x = System.in.read();
```

要求用户从键盘输入一个字符，并将此字符的 Unicode 数值放入变量 x 中。此语句执行时可能会抛出异常类型 IOException 的对象。IOException 类在 Java.io 子包中，所以程序的第一条语句 import java.io. * ;用于导入 IOException 类。

(2) 执行语句

```
x = x - 48;
```

是将 x 减去 48('0'的 ASCII 值)后，x 被转换成数字值。例如，'0'被转换数值 0，'5'被转换成数值 5。

(3) 执行语句

```
compute(x);
```

是调用方法 compute()。在被调用方法 compute 中可能会抛出 ArithmeticException 型异常。当输入'0'时,会抛出 ArithmeticException 型异常;当输入非'0'字符时,则不会抛出 ArithmeticException 型异常。而在方法 compute 中并未对此异常对象进行捕获处理,但通过 compute 方法声明中的子句 throws ArithmeticException 将重新抛出此异常对象,此异常对象被传递给调用方法 method1。

(4) 在方法 method1 中,分别对 IOException 型异常和 ArithmeticException 型异常进行了捕获处理。

这里应注意 3 点:①当在 try 块中调用一个方法时,被调用方法中抛出了异常,则此被调用方法被中断执行,try 块也被立即中断执行,这可以从上述程序的执行结果中验证该结论;②处理异常的 catch 块必须紧跟在 try 块之后;③如果一个 catch 块中定义的异常类型和程序抛出的异常类型匹配,则该 catch 块被执行;如果没有任何一个 catch 块中定义的异常类型和程序抛出的异常类型匹配,则该程序立即被中止执行。

10.5 定义新的异常类型

在设计自己的类时,应尽最大的努力为用户提供最好的服务。当程序出现某些异常事件时,希望程序足够健壮,足以从程序中恢复,这时就需要用到异常。

在选择异常类型时,可以使用 Java 类库中已经定义好的类,也可以自定义新的异常类。自定义异常类是由用户自己定义的新的异常类。

自定义异常类必须继承自 Throwable 或 Exception 类,建议用 Exception 类。一般不把自定义异常作为 Error 的子类,因为 Error 通常被用来表示系统内部的严重故障。

当自定义异常是从 RuntimeException 及其子类继承而来时,该自定义异常是运行时异常,即程序中可以不捕获和处理它。当自定义异常是从 Throwable、Exception 及其子类继承而来时,该自定义异常是编译时异常,也即程序中必须捕获并处理它。

自定义异常同样要用 try-catch-finally 形式捕获处理,但异常对象必须由用户自己抛出。自定义异常的一般形式为:

```
class MyException extends Exception
{
…
}
```

例 10-5 计算两个数之和,当任意一个数超出范围(10,20)时,抛出自己的异常。

```
//NewException.java
//声明一个新的异常
class NumberRangeException extends Exception
{  public NumberRangeException(String msg)
   {    super(msg);
   }
}
//捕获异常的主类
public class NewException {
// 在方法 CalcAnswer 中可能产生异常,其中 NumberRangeException 异常在此方法中捕获,而异常
NumberRangeException被重新抛出给调用方法
//throws 重新抛出异常 NumberRangeException
public int CalcAnswer(String str1, String str2) throws NumberRangeException
```

```
{   int int1, int2;
    int answer = -1;
    try
    {   int1 = Integer.parseInt(str1); //可能产生异常 NumberFormatException
        int2 = Integer.parseInt(str2); //可能产生异常 NumberFormatException
            if( (int1 < 10) || (int1 > 20) || (int2 < 10) || (int2 > 20) )
            {   NumberRangeException e = new NumberRangeException
                    ("Numbers are not within the specified range.");
                throw e; //抛出自定义异常对象 NumberRangeException e
            }
            answer = int1 + int2;
    }
    catch (NumberFormatException e){ //捕获异常对象 NumberRangeException e
        System.out.println( e.toString() );
    }
    return answer;
  }
  public void getAnswer() //在调用方法 getAnswer 中捕获异常
  {String answerStr;
  try
  {  int answer = CalcAnswer("12", "5"); //抛出异常 NumberRangeException
     answerStr = String.valueOf(answer);
  }
   catch (NumberRangeException e){ //捕获异常 NumberRangeException e
     answerStr = e.getMessage();
   }
   System.out.println(answerStr);
   }
  public static void main(String args[]) {
   NewException t1 = new NewException();
     t1.getAnswer();
  }
}
```

程序运行输出结果如图 10-7 所示。

```
Numbers are not within the specified range.
```

图 10-7　声明一个新的异常

10.6　小　结

Java 通过面向对象的方法进行异常处理，把各种不同的异常事件进行分类，体现了良好的层次性，提供了良好的接口，这种机制对于具有动态运行特性的复杂程序提供了强有力的控制。

应该根据具体的情况选择在何处处理异常。或者在方法内捕获并处理，或者用 throws 子句把它交给调用栈中上层的方法去处理。

对非运行时异常必须捕获或声明。对运行时异常，若不能预测它何时发生，程序可不作处理，而是交给 Java 运行时系统来处理；若能预知它可能发生的地点和时间，则应在程序中处理，而不应简单地交给运行时系统。

异常可以人为地抛出，用 throw new 语句。异常可以是系统已经定义好的，也可以是用户自己定义的。用户自己定义的异常一定继承自 Throwable 或 Exception 类。

在自定义异常类时，如果它所对应的异常事件通常总是在运行时产生，而且不容易预测它将在何时何地发生，可以将它定义为运行时异常，否则应定义为非运行时异常。

应该使用 finally 语句为异常处理提供统一的出口。

习　题

10.1　什么叫异常？列出 5 个异常的例子。

10.2　Java 的异常处理块的形式是什么样的？

10.3　Java 中关键字 try、catch、finally、throw 和 throws 各有何作用？

10.4　如下代码输出结果是什么？为什么？

```
public class test {
    public static void main(String args[]) {
        int i = 1, j = 1;
        try {
            i++;
            j--;
            if(i/j > 1)
                i++;
        }
        catch(ArithmeticException e) {
            System.out.println(0);
        }
        catch(ArrayIndexOutOfBoundsException e) {
            System.out.println(1);
        }
        catch(Exception e) {
            System.out.println(2);
        }
        finally {
            System.out.println(3);
        }
        System.out.println(4);
    }
}
```

10.5　从键盘输入两个数，进行相加，显示和。当输入串中含有非数字时，通过异常处理机制使程序能正确运行。

10.6　编写一个程序，完成在银行的取款和存款操作。在定义银行类时，若取款数大于余额则作为异常处理。

第11章 多线程

以往开发的程序，大多是单线程的，即一个程序只有一条从头至尾的执行路线。例如，对于一个 Java 的 Application 程序，其主线程就是 main( )方法执行的路线；对于一 Applet，主线程指挥浏览器加载并执行 Java 小程序。然而，在实际应用中，一个程序往往需要同时并发地处理多个任务。比如，用户在从万维网下载大型文件时，可能需要同时浏览网页。

Java 语言支持多线程机制，它使得编程人员可以很方便地开发出能同时处理多个任务的功能强大的应用程序。本章内容将主要介绍：线程的概念、线程的生命周期、线程的创建与执行、线程的优先级与调度策略、线程的同步机制等，并介绍多线程的程序设计实例。

11.1 线程的概念

程序是一段静态的代码，它是应用软件执行的蓝本；而进程是程序一次动态执行的过程，它对应着从代码加载、执行到执行完毕的一个完整过程，这个过程也是进程本身从产生、发展到消亡的过程。每一个进程都有自己独立的一块内存空间、一组系统资源。在进程概念中，每一个进程的内部数据和状态都是完全独立的。并且，进程是操作系统中的概念，由操作系统调度，通过多进程使操作系统能同时运行多个任务程序。

线程是进程中可独立执行的子任务，一个进程可以含有一个或多个线程，每个线程都有一个唯一的标识符。进程和线程的区别为：进程空间大体分为数据区、代码区、栈区、堆区，多个进程的内部数据和状态都是完全独立的；而线程共享进程的数据区、代码区、堆区，只有栈区是独立的，所以线程切换比进程切换的代价小。

多线程技术使得单个程序内部可以在同一时刻执行多个代码段，完成不同的任务，这种机制称为多线程。如图 11-1 所示是一个程序中含有两个线程的情形。当在一个程序中需要同时执行几段代码时，应使用多线程程序设计来实现，多线程由程序负责管理。

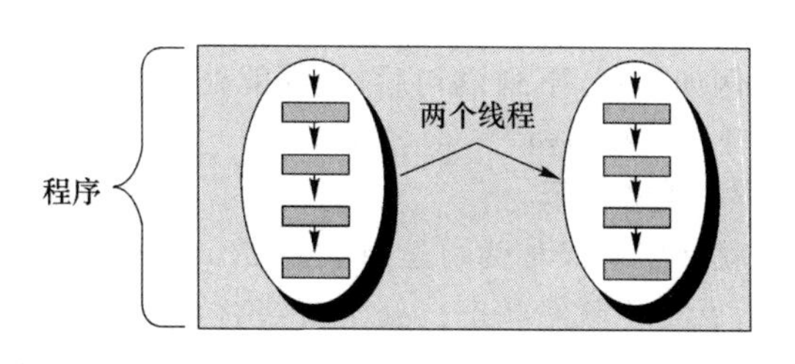

图 11-1 一个程序中的两个线程

值得注意的是，上面提到的“同时”并不是真正意义上的同时，而是指多个线程轮流占用 CPU 的时间片。比如说，某 CPU 每秒可执行 100 万次运算，现有 4 个线程在“同时”运行，事实上每个线程每秒钟只执行 25 万次运算，它们“平分”了 CPU 的时间片轮流执行。对于最终用户而言，由于线程对 CPU 的控制切换的非常快，用户看到的是有 4 个线程在同时运行。

11.2 线程的状态与生命周期

每个 Java 程序都有一个缺省的主线程，对于 Application，主线程是 main( )方法执行的线索；对于 Applet，主线程指挥浏览器加载并执行 Java 小程序。要想实现多线程，必须在主线程中创建新的线程对象。线程有它自身的产生、存在和消亡的过程，是一个动态的概念。一个线程完整的生命周期通常要经历 5 个状态：创建状态(Born)、就绪状态(Ready)、运行状态(Running)、阻塞状态(Blocked，Waiting，Sleeping)和死亡状态(Dead)。图 11-2 表示了 Java 线程的生命周期：线程的不同状态以及各状态之间转换的过程。

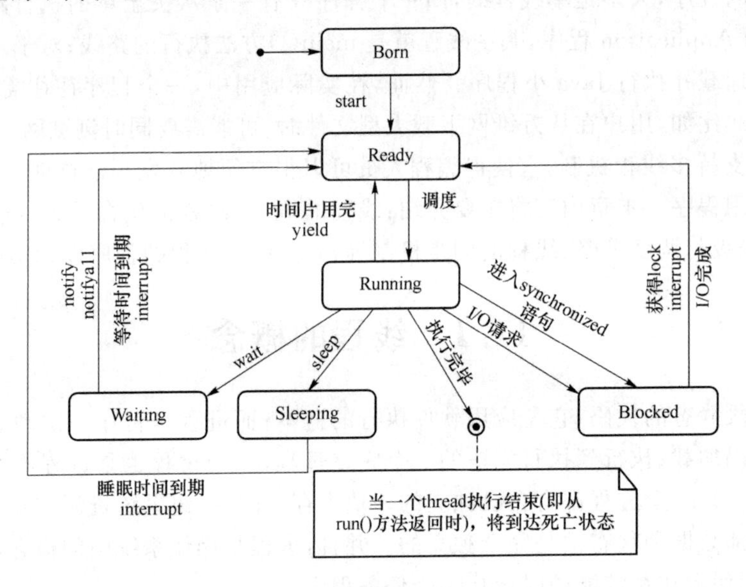

图 11-2　线程的生命周期

1. 创建状态

Java 语言使用 Thread 类及其子类的对象来表示线程。当一个 Thread 类或其子类的对象被声明并创建时，新生的线程对象就处于新建状态。此时它已经有了相应的内存空间和其他资源，并已被初始化。

例如，执行下列语句后，线程就处于创建状态。

```
Thread myThread = new Thread( );
```

2. 就绪状态

就绪状态又称为可运行状态(Runnable)。处于新建状态的线程，通过调用 start()方法执行后，就处于就绪状态。处于就绪状态的线程，将进入线程队列排队等待分配 CPU 时间片，此时它已经具备了运行的条件。一旦轮到它来享用 CPU 资源时，就可以脱离创建它的主线程独立开始自己的生命周期了。另外，原来处于阻塞状态的线程被解除阻塞后，也将进入就绪状态。

例如，执行下列语句后，一个线程就处于就绪状态。

```
Thread myThread = new Thread( );
myThread.start( );
```

3. 运行状态

当就绪状态的线程被调度并获得处理器资源时，便进入运行状态。每一个 Thread 类及其子类的对象都有一个重要的 run( )方法，当线程对象被调度执行时，它将自动调用此线程对象的 run( )方法。run( )方法定义了这一线程的操作功能。

处于运行中的线程，因调用了方法 yield()会自动放弃 CPU 而进入就绪状态，从而使其他就绪的线程有运行的机会，这用于非时间片运行的系统中。

4. 阻塞状态

一个正在运行的线程在某些特殊情况下，例如被人为挂起或需要执行费时的输入/输出操作时，将让出 CPU 并暂时中止自己的执行，进入阻塞状态。阻塞时该线程不能进入排队队列。只有当引起阻塞的原因被消除时，线程才可以转入就绪状态，重新进入线程队列中排队等待 CPU 资源，以便从原来终止处开始继续运行。

处于运行状态的线程进入阻塞状态的原因有如下几条。

(1) 线程调用了 sleep()方法，而进入睡眠状态(sleeping)；睡眠时间到期或线程调用了方法 interrupt()，使睡眠状态结束，进入就绪状态。

(2) 为等待访问某个共享对象的信号变量，当前线程调用了 wait()方法而进入等待状态，直到另一线程为共享对象调用了方法 notify()或 notifyAll()为止，使该线程等待状态结束而进入就绪状态。处于等待状态的线程，因调用了方法 interrupt()，也会使等待状态结束，进入就绪状态。

(3) 线程进入同步(synchronized)语句时，因要对共享的对象加排斥锁而未获成功，也将进入阻塞状态，直到获得排斥锁为止。

(4) 线程因输入输出请求，导致线程阻塞(如等待用户输入)，输入/输出操作完成时，使线程进入就绪状态。

处于阻塞状态的线程，因调用了方法 interrupt()，也会使阻塞状态结束，进入就绪状态。

5. 死亡状态

处于死亡状态的线程不具有继续运行的能力。线程死亡的原因往往是正常运行的线程完成了它的全部工作，即执行完 run( )方法的最后一个语句并退出。

11.3 线程优先级与线程调度策略

1. 线程的优先级

在 Java 的多线程系统中，每个线程都有一个优先级(priority)，其数值在 Thread. MIN_PRIORITY(常量值 1)到 Thread. MAX_PRIORITY(常量值 10)范围内，其中，10 是最高优先级，1 是最低优先级。具有较高优先级的线程，完成的任务较紧急，因而应优先于较低优先级的线程分配处理器时间。在默认情况下，线程的优先级是 Thread. NORM_PRIORITY(常量值 5)，每个新创建线程均继承创建线程的优先级。

设置线程的优先级用 Thread 对象的方法 setPriority(int priority)；用方法 getPriority()获得线程的优先级。

2. 线程的调度策略

线程调度器(thread scheduler)支持一种抢先式的调度策略：即当前线程执行过程中有较高优先级的线程进入就绪状态，则高优先级的线程立即被调度执行。而具有相同优先级的所有线程采用轮转的方式来共同分配 CPU 时间片，这是大多数 Java 系统支持的分时概念。

而在一些不支持分时技术的 Java 系统中，则具有相同优先级的每个线程将一直占用

CPU 运行直到完成任务为止；或因为某个原因使该线程进入阻塞状态；或因为某个优先级较高的线程准备就绪而中断。而就绪队列中的其他线程只能等待。

在非分时系统中，处于运行中的线程，因调用了方法 yield()会自动放弃 CPU 而进入就绪状态，从而使其他就绪的相同优先级线程有运行的机会。当调用 yield()方法时，如果其他处于就绪状态的线程的优先级都比当前线程低，则 yield()操作将被忽略。

11.4 线程的创建和执行

Java 的线程相关的类在软件包 java. lang 中。在程序中实现多线程有两种方式：创建 Thread 类的子类或实现 Runnable 接口。这两种方式都要进行两个关键性的操作：

(1) 定义用户线程的操作，即实现线程的 run( )方法的方法体；

(2) 构造 Thread 类对象，实现线程的建立和运行控制。

下面将具体介绍创建线程的两种方式。

11.4.1 Runnable 接口和 Thread 类介绍

1. Runnable 接口介绍

Runnable 接口只有一个方法 run( )，所有实现 Runnable 接口的类必须实现这个方法。它定义了线程体的具体操作。run( )方法能被运行系统自动识别和执行。具体地说，当线程被调度并转入运行状态时，它所执行的就是 run( )方法中规定的操作。

创建实现 Runnable 接口的类，在此类中实现 Runnable 接口中的 run()方法，创建此类的对象，并将此对象作为参数传递给 Thread 类的构造方法，构造 Thread 对象并启动。只要一段代码在单独的线程中运行，则可以采用继承 Runnable 接口的方式，并将该段代码放在 Runnable 接口的 run()方法中。

2. Thread 类介绍

Thread 类是一个具体的类，它封装了一个线程需要拥有的属性和方法。Thread 类实现 Runnable 接口中的 run 方法，但方法体为空。

3. Thread 类的构造方法

用于创建一个线程对象的 Thread 类的构造方法有多个，常用的构造方法如下。

(1) Thread (String threadName)：创建一个线程类的对象，并为新创建的线程对象指定一个字符串名称 threadName。

(2) Thread ( )：创建一个线程类的对象。线程对象的名称由系统指定为“Thread-”连接一个数值，如“Thread-1”、“Thread-2”等。

(3) Thread (Runnable target)：以实现 Runnable 接口的 target 对象中所定义的 run( )方法，来初始化或覆盖新创建的线程对象的 run( )方法。

(4) Thread (Runnable target, String ThreadName)：实现上述(2)和(3)两个构造函数的功能。

(5) Thread(ThreadGroup group,Runnable target,String name)：实现上述(2)和(3)两个构造函数的功能，并将新创建的线程对象加入线程组中。Java 中对一个线程组的操作，将转换为组中的多个线程进行操作。

利用构造函数创建新线程对象之后，这个对象中的有关数据被初始化，从而进入线程的生命周期的第一个状态——新建状态。接着调用线程的 start( )方法，将启动线程对象，使之从新建状态转入就绪状态并进入就绪队列排队。

4. Thread 类的常用方法

(1) Thread 类的静态方法

① static Thread currentThread() :返回当前正在运行的线程的引用。

② static void yield() :使当前正在运行的线程暂时中断，变为就绪状态，以让其他线程有运行的机会。

③ static sleep(int millsecond) :以 millsecond(毫秒)为单位设置当前线程休眠时间。sleep 要抛出异常，必须捕获。

④ static sleep(int millsecond, int nanosecond) : 以 millsecond(毫秒)+nanosecond(纳秒)为单位的休眠时间。

(2) Thread 类的非静态方法

① void start():启动已创建的线程对象。

② void run(): 由线程调度器调用，当从 run()返回时，该进程运行结束。

③ final void setName(String name) :设置线程的名字。

④ final String getName():返回线程的名字。

⑤ interrupt():中断线程对象所处的状态。如果线程处于阻塞状态，因调用了方法 interrupt(),会使阻塞状态结束而进入就绪状态。此方法将抛出 InterruptedException 异常。

⑥ final boolean isAlive() :判断线程是否被启动，若被启动则返回 true。

⑦ void join(): 使当前线程暂停运行，等调用 join 方法的线程运行结束，当前线程才继续运行。例如，mt.join()出现在当前线程的线程体中时，当前线程运行到此语句时会暂停运行，等待 mt 线程运行结束后才继续运行。

11.4.2 通过继承 Thread 的子类创建线程

创建用户定制的 Thread 类的子类，并在子类中重新定义自己的 run( )方法，这个 run( )方法中包含了用户线程的操作。这样在用户程序需要建立自己的线程时，只需要创建一个已定义好的 Thread 子类的实例就可以了。

例 11-1 通过定制的 Thread 的子类，创建多线程。

该程序是一个 Java 的 Application 程序，主线程是 main( )方法执行的路线。在主线程中创建 3 个线程，其名称是 thread1、thread2 和 thread3，每个线程的优先级均为默认的 Thread.NORM_PRIORITY。每个线程启动后，由系统执行 run()方法，运行将显示信息:进入睡眠的线程名称和要休眠的时间(睡眠时间为 0～5 000 ms 之间的随机数)；接着通过调用 Thread.sleep()方法使当前线程进入睡眠状态；在各个线程睡眠结束醒来时，将显示其线程名称和已完成睡眠的信息；然后线程正常运行结束，即执行完了 run( )方法的最后一个语句并退出进入死亡状态。

```
// ThreadTester.java
public class ThreadTester {
    public static void main( String [] args )
    {  // 创建和命名三个线程
        PrintThread thread1 = new PrintThread( "thread1" );
        PrintThread thread2 = new PrintThread( "thread2" );
```

```
        PrintThread thread3 = new PrintThread( "thread3" );
        System.err.println( "主线程将要启动三个线程" );
        thread1.start();    // 启动 thread1,进入就绪状态
        thread2.start();    //启动 thread2,进入就绪状态
        thread3.start();    //启动 thread3,进入就绪状态
        System.err.println( "三个线程启动完成，主线程运行结束\n" );
        }
} // end class ThreadTester
// 创建 Thread 类的子类 PrintThread ,以控制线程的运行
class PrintThread extends Thread {
    private int sleepTime;
        public PrintThread( String name )
    {super( name ); // 通过调用父类构造方法给 thread 命名
    sleepTime = ( int ) ( Math.random() * 5001 ); // 设置睡眠时间 0～5 秒
    }
    // 设置线程运行的线程体
    public void run()
    {   // put thread to sleep for sleepTime amount of time
        try {
          System.err.println(
            getName() + " 进入睡眠状态,睡眠时间是:" + sleepTime );
            Thread.sleep( sleepTime );
        }
        // thread 在睡眠被打断时,抛出异常
        catch ( InterruptedException exception ) { }
        System.err.println( getName() + " 睡眠醒来" ); // 显示线程名称
        } // end method run
} // end class PrintThread
```

程序运行结果如图 11-3 所示。

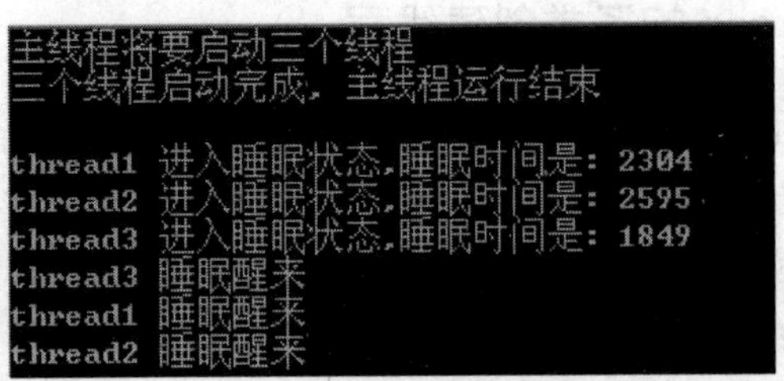

图 11-3　例 11-1 程序运行结果

11.4.3　通过实现 Runnable 接口创建线程

创建实现 Runnable 接口的类,在此类中实现 Runnable 接口中的 run()方法,创建此类的对象,并将此对象作为参数传递给 Thread 类的构造方法,构造 Thread 对象并启动它。只要一段代码在单独线程中运行,则可以继承 Runnable 接口,并将该段代码放在该接口的 run()方法中。

例 11-2　通过实现 Runnable 接口创建线程。

该例子实现一个时钟功能,时钟数据每隔 1 秒就变化一次。程序通过每隔 1 秒执行线程的刷新画面功能,显示当前的时间。

类 Clock 继承了 JApplet，并实现接口 Runnable。JApplet 的方法 init 是 Applet 容器在加载 Applet 时调用的方法，在此方法中创建时钟线程对象并启动它；在时钟线程对象启动后，运行 run()方法；在 run()方法中，安排时钟线程睡眠 1 秒钟，1 秒钟到期时，通过调用 repaint()方法，又间接调用 paint()方法，用以在 Applet 图形界面显示当前时间；在用户关闭页面时通过调用 interrupt()方法中断时钟线程睡眠状态，从 run 方法返回，时钟线程运行结束。

在 JApplet 的方法 init 中，语句

```
clockThread = new Thread(this, "Clock-thread");
```

Thread 构造方法的第一个参数是实现 Runnable 接口的 Clock 的当前对象 this；第二个参数是线程名字，构造 Thread 对象，接着启动该线程对象。

Clock.java 代码如下：

```
//Clock.java
import java.awt.*;
import javax.swing.*;
import java.util.*;      //import Date and Locale
import java.text.SimpleDateFormat;
public class Clock extends JApplet implements Runnable {
    Thread clockThread;
    //在 JApplet 的方法 init 中，创建一线程命名为 Clock-thread
    public void init() {
        clockThread = new Thread(this, "Clock-thread");
        clockThread.start();
    }
    //实现 Runnable 接口中的方法 run，每隔一秒刷新时钟画面
    public void run() {
      boolean flag = true;
      while (flag) {
            repaint();
            try {
              clockThread.sleep(1000);
            }
            catch (InterruptedException e)
            { flag = false; }
        }  //end of while
    }  //end of run
    //JApplet 的方法 paint，显示当前时钟对象的值
      public void paint(Graphics g){
        super.paint(g);
        SimpleDateFormat formatter = new SimpleDateFormat (" hh:mm:ss ",
                Locale.getDefault()); //设置时钟显示格式
        Date currentDate = new Date(); //获得当前时钟对象
        String lastdate = formatter.format(currentDate); //转换格式
        g.drawString(lastdate ,5, 10); //显示时钟对象
      }
    }
```

ClockApplet.html 代码如下：

```
<html>
<applet code = "Clock.class" width = "300" height = "45">
</applet>
</html>
```

程序运行结果如图 11-4 所示。

图 11-4　例 11-2 程序运行结果

11.5　线程同步

多个同时运行的线程之间的通信，往往需要通过共享数据块去完成。而共享数据的读写操作往往封装在一个对象中，此对象称为共享对象。这样，多个同时运行的线程往往需要操作同一个共享的对象。例如，如果有些线程读取共享对象，同时又有一个以上的线程修改这个共享对象，此时如果对共享对象不能有效地管理，则不能保证共享对象的正确性。例如，线程 1 在更新该对象时，线程 2 也试图更新该对象，则该对象将会反映出线程 2 第二次更新的结果，而第一次线程 1 更新的结果被丢失。

11.5.1　synchonized 同步关键字

为了保证共享对象的正确性，Java 语言中，使用关键字 synchonized 修饰对象的同步语句(synchonized statement)或同步方法(synchonized statement)。synchonized 的一般使用格式有：

```
synchonized (对象)
{ … }        //定义对象的同步代码块或同步语句
```

或

```
synchonized 方法声明头
{ … }        //定义对象的同步方法
```

一个对象上可定义多个同步语句(synchonized statement)或同步方法。Java 系统只允许一个线程执行对象的一个同步语句(synchonized statement)或一个同步方法(synchonized statement)。一个线程在进入同步语句时要给对象加互斥锁(又叫获得锁)，一个对象只能加一把互斥锁，加锁成功时才能执行同步语句；而其他所有试图对同一个对象执行同步语句的线程，因加锁不成功都将处于阻塞状态。在一条同步语句完成执行时，同步对象上的锁被解除，并让最高优先级的阻塞线程处理它的同步语句。同步方法相当于一条封装了方法体的同步语句。

synchronized 的使用注意事项如下。

- 一个对象中的所有 synchronized 方法都共享一把锁，这把锁能够防止多个方法对共用内存同时进行的写操作。
- synchronized static 方法，可在一个类范围内被相互间锁定起来。
- 对于访问某个关键共享资源的所有方法，都必须把它们设为 synchronized，否则就不能正常工作。
- 假设已知一个方法不会造成冲突，最明智的方法是不要使用 synchronized 修饰，这样能提高些性能。

• synchronized 不能继承，父类的方法是 synchronized，那么其子类重载方法中就不会继承“同步”。

11.5.2 wait 方法和 notify 方法

在线程获得对象的锁的情况下，如果该线程确定只有在满足一些条件的情况下才能继续对该对象执行线程任务，这时该线程可调用 Object 类的 wait()方法。线程调用 wait 方法会解除对象的锁，并使该线程处于等待状态，让出 CPU 资源，使其他线程将尝试执行该对象的同步语句。wait()和 sleep()的区别是：wait()方法被调用时会解除锁定，但是只能在一个同步的方法或代码块内使用。

在某个线程执行完同步语句，或该线程使另一个线程所等待的条件满足时，它将调用 Object 类的 notify()方法，以允许一个处于等待状态的线程再次进入就绪状态。这时，从等待状态进入就绪状态的线程，将再次尝试获得该对象的锁。Object 类的 notifyAll()方法将使因该对象处于等待状态的全部线程进入就绪状态。要注意的是，wait 方法和 notify 方法一般配合使用。

下面将以生产者和消费者的关系模型，说明如何应用上面介绍的 Java 同步机制进行多线程的程序设计。

11.5.3 多线程同步的程序设计举例

在生产者和消费者的关系模型中，应用程序中的生产者将产生数据，而消费者将使用生产者产生的数据。生产者线程生产的数据放在共享缓冲区中，消费者线程从这个共享缓冲区中读取数据。为了简单起见，程序假设共享缓冲区只用一个整型数据。应用程序中生产者和消费者可以是两个同时运行的线程。生产者线程生产的数据在放入共享缓冲区时，要检查缓冲区是否空，若空，则将数据写入缓冲区，并调用 notify 方法使处于等待状态的消费者线程转为就绪状态；若不空，则调用 wait 方法使自己等待。消费者线程在从共享缓冲区读数据时，应检查缓冲区是否已有数据存在，若有数据存在，则从缓冲区读数据，并调用 notify 方法使处于等待状态的生产者线程转为就绪状态；若无数据存在，则调用 wait 方法使自己等待。程序使用条件变量 occupiedBufferCount 控制缓冲区 buffers 读写，当 occupiedBufferCount 为 0 时，可写但不能读，当 occupiedBufferCount 为 1 时，可读但不能写。

例 11-3 生产者和消费者线程同步的程序设计。

程序由 4 个源文件组成：Buffer.java 定义了共享缓冲区的读写管理类；Producer.java 定义了生产者类；Consumer.java 定义了消费者类；SharedBufferTest.java 定义了测试类。在 main 方法中，创建了一个生产者线程和一个消费者线程，并分别启动了它们。

源文件 Buffer.java 的代码如下：

```
public class Buffer {
  // buffer 缓冲区被 producer 和 consumer 线程共享
  private int buffer = -1;
  // 控制缓冲区 buffers 读写的条件边量
  private int occupiedBufferCount = 0;
  public synchronized void set( int value )                //写 buffer 的方法
  {  String name = Thread.currentThread().getName();       //得到线程的名称
```

```
    while ( occupiedBufferCount == 1 ) {                    //当 buffer 不空时
        try {
            wait(); }
        catch ( InterruptedException exception ) {
          exception.printStackTrace();
        }
    }                                                        // end while
  buffer = value;                                            // 写 buffer
  ++occupiedBufferCount;
  System.err.println(name + " writes " + buffer);
  notify();
  } // end method set; releases lock on Buffer
  public synchronized int get()                              //读 buffer 的方法
  {  String name = Thread.currentThread().getName();
    // while no data to read, place thread in waiting state
    while ( occupiedBufferCount == 0 ) {                     //当 buffer 空时
     try {
           wait(); }
      catch ( InterruptedException exception ) {
        exception.printStackTrace();
        }
     } // end while
     --occupiedBufferCount;
     System.err.println( name + " reads " + buffer );
     notify();
     return buffer;
   } // end method get; releases lock on Buffer
}
```

源文件 Producer.java 的代码如下：

```
public class Producer extends Thread {
    private Buffer sharedLocation;       // 引用指向共享对象
    public Producer( Buffer shared )
    {  super( "Producer" );
       sharedLocation = shared;
    }
    public void run()                    // 将值 1～4 存放到 sharedLocation
    {  for ( int count = 1; count <= 4; count++ ) {
        sharedLocation.set( count );
       }
      System.err.println( getName() + " done producing." + " Terminating " );
    }
} // end class Producer
```

源文件 Consumer.java 的代码如下：

```
public class Consumer extends Thread {
    private Buffer sharedLocation;       // 引用指向共享对象
    public Consumer( Buffer shared )
    {  super( "Consumer" );
       sharedLocation = shared;
    }
    public void run()                    // 从 sharedLocation 读取值四次并累加到 sum
    {  int sum = 0;
       for ( int count = 1; count <= 4; count++ ) {
          sum += sharedLocation.get();
       }
```

```
            System.err.println( getName() + " read values totaling: " + sum +
                " and Terminating" );
        }
    }
```

源文件 SharedBufferTest.java 的代码如下：

```
public class SharedBufferTest {
    public static void main( String [] args )
    {   Buffer sharedLocation = new Buffer();
        Producer producer = new Producer( sharedLocation );
        Consumer consumer = new Consumer( sharedLocation );
        producer.start();      // start producer thread
        consumer.start();      // start consumer thread
    } // end main
} // end class SharedBufferTest2
```

程序运行结果如图 11-5 所示。

```
Producer writes 1
Consumer reads 1
Producer writes 2
Consumer reads 2
Producer writes 3
Consumer reads 3
Producer writes 4
Producer done producing. Terminating
Consumer reads 4
Consumer read values totaling: 10  and Terminating
```

图 11-5　例 11-3 程序运行结果

11.6　Daemon 线程

Daemon(守护)线程是为其他线程提供服务的线程，它的优先级是最低的。典型的守护线程例子是 JVM 中的系统资源自动回收线程，它始终在低级别的状态中运行，用于实时监控和管理系统中的可回收资源。当系统中运行的只有 Daemon 线程，则系统会自动清除 Daemon 线程，退出 Java 解释器。守护线程与非守护线程(或称用户线程)的区别是：一旦所有非 Daemon 线程完成，程序会中止运行；相反，假若有任何非 Daemon 线程仍在运行，则程序的运行不会中止。

将一个线程设置为守护线程的方式是在线程对象启动之前调用线程对象的 public final void setDaemon(boolean on)方法。对于一个 Daemon 线程，它创建的任何线程也会自动具备 Daemon 属性。

通过 final boolean isDaemon()来判断一个线程是否是 Daemon 线程。

11.7　死　锁

多线程在使用互斥机制实现同步的同时不仅会带来阻塞，还会带来“死锁”(deadlock)的潜在危险。如下列情形将触发死锁的发生：线程 A 锁住了线程 B 等待的资源，而且线程 B 锁住了线程 A 等待的资源，即线程 B 一直在等待线程 A 释放锁，线程 A 也是如此。因此，对于多线程的应用而言，程序员编程时应注意死锁问题，尽量避免之。

Java 技术既不能发现死锁也不能避免死锁。对大多数的 Java 程序员来说，防止死锁是一种较好的选择。最简单的防止死锁的方法是：

- 对竞争的资源引入序号，如果一个线程需要几个资源，那么它必须先得到小序号的资源，再申请大序号的资源；

• 在程序中不提倡使用线程的停止/销毁线程方法 stop()/destroy()，挂起/恢复线程的方法 suspend()/resume()。

11.8 小 结

本章介绍了 Java 的线程概念，线程的生命周期包含 5 个状态：创建状态(Born)、就绪状态(Ready)、运行状态(Running)、阻塞状态(Blocked，Waiting，Sleeping)和死亡状态(Dead)。

每个线程都有一个优先级。Java 支持抢先式的线程调度策略。

在程序中实现多线程有两种方式：创建 Thread 类的子类或实现 Runnable 接口。

多线程操作同一个共享资源时，Java 语言使用同步互斥机制，保证共享资源的确定性，中。互斥机制会带来死锁。

习 题

11.1 什么是进程、线程？它们有什么区别？

11.2 Java 的线程的生命周期是怎样的？

11.3 Java 的线程是如何调度的？

11.4 Java 中的多线程有什么特点？

11.5 多线程中的同步和互斥的原理是如何实现的？

11.6 说说 Java 中创建线程的方式。在程序中实现多线程有两种方式：创建 Thread 类的子类或实现 Runnable 接口。

11.7 说说 wait 方法和 notify(notifyAll)方法的作用。

11.8 编程题。

(1) 创建两个线程，每个线程的工作是在自己界面的 TextField 区域中从左到右动态地显示一个字符串；可通过界面按钮启动和中止每个线程的运行。GUI 显示界面可参考下图。

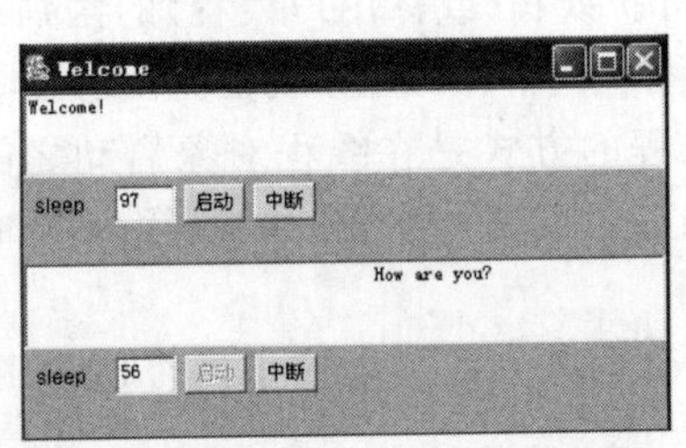

(2) 创建 3 个线程，每个线程的工作是在自己的界面随机地显示 26 个字母中的一个；可通过界面按钮，临时挂起或恢复各线程的运行。GUI 显示界面参考左下图(要求用 Runnable 接口方式创建线程)。

(3) 编写 3 个线程：各线程分别显示各自的运行时间，第一个线程每隔 1 秒钟运行一次，第二个线程每隔 5 秒钟运行一次，第三个线程每隔 10 秒钟运行一次。如右下图所示。

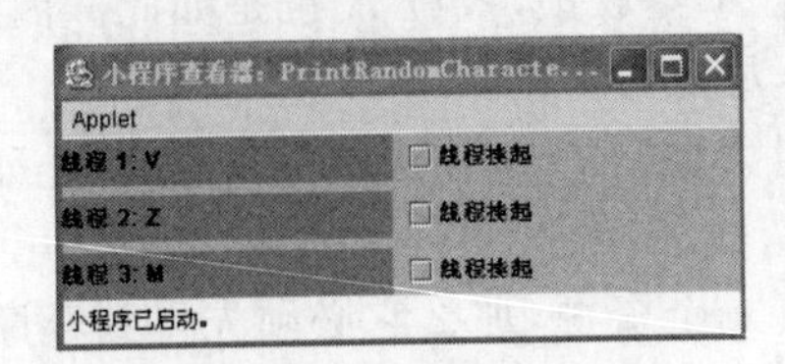

第12章　输入和输出流处理

本章首先介绍了 Java 中与输入和输出处理相关的概念，包括输入流和输出流、字节流和字符流；与处理输入和输出流相关的系统的类所在的包以及类的继承层次结构。接着详细讨论了用于获取磁盘中文件或目录信息的 File 类，基于字节的输入类和输出类，基于字符的输入类和输出类，并结合应用实例给出了标准输入/输出设备的读写方法，以及顺序文件和随机文件的多种读写方法。

12.1　输入和输出流概述

本节概述了 Java 输入和输出处理的特点，输入流和输出流、字节流和字符流的概念，处理输入和输出流相关的系统的类所在的包以及类的继承层次结构，为学习后续的各节打下了基础。

12.1.1　输入流和输出流

大多数程序往往需要输入/输出处理，比如从键盘读取数据、向屏幕输出数据、从文件读数据或者向文件中写数据、在一个网络连接上进行读写操作等。在 Java 中，把这些不同类型的输入/输出源抽象为流(Stream)，而其中输入/输出的数据则称为数据流(Data Stream)，用统一的接口来表示，从而使程序设计简单明了。数据流是指一组有顺序的、有起点和终点的字节集合。

Java 的流分为两类：输入流(Input Stream)和输出流(Output Stream)，输入流是能够读取字节序列的对象；而输出流是能够写字节序列的对象。如一个文件，当向其中写数据时，它就是一个输出流；当从其中读取数据时，它就是一个输入流。例如，键盘是一个输入流，屏幕是一个输出流。

12.1.2　字节流和字符流

Java 定义了两种类型的流：字节流和字符流。最初设计的输入/输出类是面向字节流的，即以 B(8 bit)为基本处理单位的流，这些类分别由抽象类 InputStream 和 OutputStream 派生的类层次。但是随着对国际化支持的需求出现，面向字节的流不能很好地处理使用 Unicode (每个字符使用 2 B)的数据，因此引入了面向字符的流，是从抽象类 Reader 和 Writer 派生的类层次，用于读写双字节的 Unicode 字符，而不是单字节字符。

12.1.3　输入和输出类的继承层次结构

在 Java 开发环境中，主要是由包 java.io 中提供的一系列的类和接口来实现输入/输出处理。诸如 FileInputStream 类(从文件中读取字节)、FileOutputStream 类(向文件输出字节)、FileReader 类(从文件中读取字符)、FileWriter 类(向文件输出字符),这些流类分别派生于 InputStream、OutPutStream、Reader 和 Writer 类,并通过创建它们的对象来打开文件。若要执行具有数据类型的输入/输出,可以将 ObjectInputStream、ObjectOutputStream、DataInputStream、DataOutputStream 类的对象与基于字节的文件流类 FileInputStream 和 FileOutputStream 捆绑使用。java.io 包中的类层次结构如图 12-1 所示(虚线用来表示实现接口)。

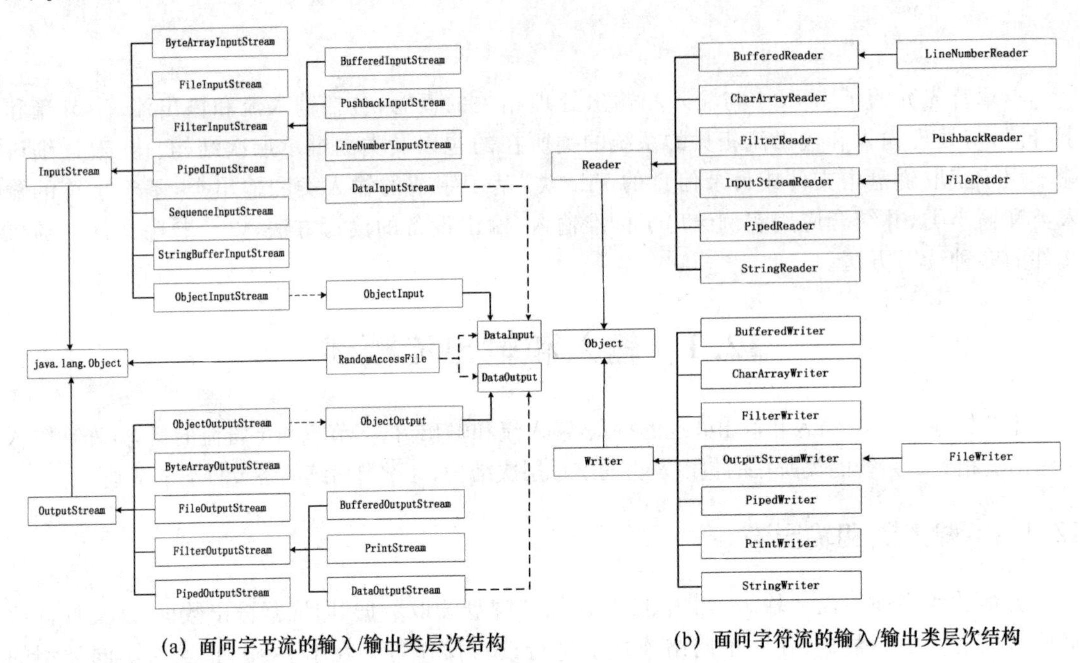

(a) 面向字节流的输入/输出类层次结构　　(b) 面向字符流的输入/输出类层次结构

图 12-1　Java.io 包的类层次结构

在 Java 中,输入/输出流处理时,往往抛出非运行时异常(checked),在程序中必须进行输入/输出异常处理。

12.2 File 类

File 类主要用于获取磁盘中文件或目录的各种信息。File 类的对象并不打开文件,也不提供任何文件处理功能。然而,java.io 包中的其他类经常使用 File 对象来指定要操作的文件或目录。

通过构造函数来构造 File 类的对象。File 类提供的构造函数

```
public Fle(String pathName)
```

指定与 File 对象关联的文件或目录的名称。pathName 可以是包含路径信息的文件名或目录名。路径可为绝对路径或为相对路径。绝对路径是包含从根目录到指定文件或目录间的所有目录;相对路径是仅包含当前目录到指定文件或目录间的部分目录。

例如:

```
File f = new File("data\temp.dat");              //相对路径
File g = new File("d:\java\data\temp.dat");  //绝对路径
```

File 类的一些常见方法如下。

- String getName():返回文件名;
- String getParent():返回文件所在目录名;
- String getPath():返回文件路径;
- String getAbsolutePath():返回绝对路径;
- boolean exists():文件是否存在;
- boolean canWrite():文件是否可写、读;
- boolean isFile():是否为文件;
- boolean isDirectory():是否为目录;
- long length():返回文件的长度;
- boolean delete():删除文件或目录,删除成功返回 true,否则返回 false;
- boolean mkdir():创建一个目录,目录创建成功,返回 true,否则返回 false;
- String[]list():返回一个代表目录下的所有文件的字符串数组。

例如,用上面引用变量 f 调用相应的方法,得到的值如下。

- f.getName():返回 temp.dat;
- f.getParent():返回 data;
- f.getPath():返回 data\temp.dat;
- f.getAbsolutePath():返回 d:\java\data\temp.dat;
- f.exists():若 data\temp.dat 文件存在返回 true,否则返回 false。

File 类另一构造函数

```
public File(URI uri)
```

使用给定的 URI(Uniform Resource Identifier)对象来定位文件。统一资源标识符 URI 是一种较为常用的形式,它通常用于定位分布于 Internet 的网络资源。

例如,http://www.sytu.edu.cn/是一个 Web 站点的 URI。

12.3 基于字节的输入和输出类及应用实例

12.3.1 抽象类 InputStream 和 OutputStream

Java 中每一种字节流的基本功能依赖于基类 InputStream(输入流类)和 OutputStream(输出流类),它们都是抽象类,不能直接使用。InputStream 和 OutputStream 类分别声明了基于字节的输入和输出方法。

属于 InputStream 类的一些常用方法如下。

(1) read():从流中读入数据;

(2) skip():跳过数据流中指定数量的字节不读,返回值表示实际跳过的字节数;

(3) available():返回流中可用字节数;

(4) mark():在流中当前位置标记一个位置;

(5) reset():返回上一个标记过的位置,用于从这一位置读取;

(6) markSupport():是否支持标记和复位操作;

(7) close():关闭流。

在 InputStream 类中,方法 read()提供了 3 种从流中读数据的方法。

(1) int read():从输入流中读一个字节,形成一个0～255之间的整数返回;

(2) int read(byte b[]):读多个字节到数组中,填满整个数组;

(3) int read(byte b[], int off, int len):从输入流中读取长度为 len 的数据,写入数组 b 中从索引 off 开始的位置,并返回读取得字节数。

对于上面的3个方法,若返回－1,表明流结束。

属于 OutputStream 类的方法如下。

(1) write(int b):将一个整数输出到流中(只输出低位字节,为抽象方法);

(2) write(byte b[]):将字节数组中的数据输出到流中;

(3) write(byte b[], int off, int len):将数组 b 中从 off 指定的位置开始,长度为 len 的数据输出到流中;

(4) flush():刷空输出流,并将缓冲区中的数据强制送出;

(5) close():关闭流。

12.3.2 FileInputStream 类和 FileOutputStream 类

FileInputStream(文件输入流类)和 FileOutputStream(文件输出流类)是基于字节的被广泛用于操作顺序文件的两个类。

FileInputStream 类用于打开一个输入文件。若要打开的文件不存在,则会产生异常 FileNotFoundException,这是一个非运行时异常,必须在方法中捕获或向上传递抛出异常。

FileOutputStream 类用于打开一个输出文件。若要打开的文件不存在,则会创建一个新的文件,否则原文件的内容会被新写入的内容所覆盖。如果试图打开一个只读文件,将会引发一个 IOException 异常。

在进行文件的读/写操作时,会产生非运行时异常 IOException,必须在方法中捕获或向上传递抛出异常。

创建输入文件流对象和输出文件流对象的语句如下:

```
File f1 = new File("file1.txt");          //创建 File 对象
File f2 = new File("file2.txt");
FileInputStream in = new FileInputStream(f1);
                    //向 FileInputStream 构造函数传递一个 File 对象
FileOutputStream out1 = new FileOutputStream(f2);
                    //向 FileOutputStream 构造函数传递一个 File 对象
FileOutputStream out2 = new FileOutputStream("file3.txt");
```

输入流的构造函数的参数是用于指定输入的文件名,输出流的构造函数参数则是用于指定输出的文件名。

例 12-1 基于字节流的顺序文件的读写。

完成将一文件 file1.txt 的内容复制生成另一新文件 file2.txt。

```
package filestream;
import java.io.*;
class filestream
  { public static void main(String args[])
    { try
      { File inFile = new File("file1.txt");
```

```
            File outFile = new File("file2.txt");
            FileInputStream fis = new FileInputStream(inFile);
            FileOutputStream fos = new FileOutputStream(outFile);
            int c;
            while((c = fis.read())!= -1)
                fos.write(c);
            fis.close();
            fos.close();
        }
        catch(FileNotFoundException e) {
            System.out.println("FileStreamsTest1: " + e);
        }catch(IOException e)
        {     System.err.println("FileStreamsTest2: " + e);
        }
    }
}
```

程序中的语句解释如下。

(1) `import java.io.*;`

从java.io包导入所需的输入/输出类。

(2)
```
File inFile = new File("file1.txt");
File outFile = new File("file2.txt");
FileInputStream fis = new FileInputStream(inFile);
FileOutputStream fos = new FileOutputStream(outFile);
```

对输入文件file1.txt建立输入流对象inFile，对输出文件file2.txt建立输出流对象outFile。这两个流对象被下面的语句使用来完成文件读写操作和关闭文件操作。

(3)
```
while((c = fis.read())!= -1)
    fos.write(c);
```

完成文件读写。操作过程：从输入流对象inFile读一字节，写入输出流对象outFile中，此过程循环地进行，直到输入文件的结束〔即read()方法到达文件尾时返回值为－1〕。要注意read()方法的返回类型是int.

(4)
```
fis.close();
fos.close();
```

输入流对象调用close方法关闭输入文件；输出流对象调用close方法关闭输出文件。

(5)
```
catch(FileNotFoundException e) {
      System.out.println("Can not open files: " + e);
}
catch(IOException e) {
      System.err.println("File IO error :" + e);
}
```

是对try代码块中产生的异常进行捕获处理。

12.3.3 随机访问文件类

类RandomAccessFile可用于随机文件的读写。对于InputStream和OutputStream来说，它们的实例都是顺序访问流，即只能进行顺序文件的读写，顺序文件的每个连续的输入/输出请求将读取或写入文件的下一段连续数据。而随机文件可将每个连续的读写请求定位到文件的任何部分，并且两个读写请求所获得的数据段在文件中可以相差甚远。类RandomAccessFile则允许对文件内容同时完成读和写操作，它直接继承object，并且同时实现了接口

DataInput 和 DataOutput。

DataInput 接口中描述了用于从输入流中读取基本类型值的一组方法,即从输入流中成组地读取字节,并将其视为基本类型值。而 DataOutput 接口中描述了将基本类型值写入输出流中的一组方法。

类 RandomAccessFile 提供的支持随机文件操作的方法如下。

(1) XXX Read XXX():XXX 对应基本数据类型。如 int readInt()、long readLong()、byte readByte()、char readChar()、unsignedByte readUnsignedByte()等;

(2) void write XXX():XXX 对应基本数据类型,如 writeInt、writeLong、writeByte;

(3) String ReadLine():从输入流中读取下一行;

(4) int skipBytes(int n):将指针向下移动 n 个字节;

(5) long length():返回文件长度;

(6) long getFilePointer():返回指针当前位置;

(7) void seek(long pos):将指针调到所需位置。

在生成一个随机文件对象,构造函数的参数时,除了要指明 File 对象或文件名外,还需要指明访问文件的模式。例如:

```
File f = new File("file.txt");
new RandomAccessFile(f, "r");
new RandomAccessFile(f, "rw");
new RandomAccessFile("file1.txt", "w");
new RandomAccessFile("file2.txt", "rw");
```

其中"r"是只读模式,"w" 是仅允许写模式,"rw"是可读可写模式。

例 12-2 随机文件的读写。

将整型数组中的 10 个数写入随机文件名 temp.dat 中,并以与写入顺序相反的次序从此文件中读入数据,显示在屏幕上。

```
import java.io.*;
public class random_file
{  public static void main(String args[])
   {  int data_arr[] = {12,31,56,23,27,1,43,65,4,99};
      try
      {  File f = new File("temp.dat");
         RandomAccessFile randf = new RandomAccessFile(f,"rw");
         for (int i = 0;i< = data_arr.length - 1;i ++ )
            { randf.writeInt(data_arr[i]);}
         for(int i = data_arr.length - 1;i> = 0;i -- )
         {  randf.seek(i * 4);
            System.out.println (randf.readInt());
         }
         randf.close();
      }catch (IOException e){
         System.out.println("File access error: " + e);
      }
   }
}
```

上面程序执行输出结果如下所示。

```
99
65
…
31
12
```

程序说明：

(1) `int data_arr[] = {12,31,56,23,27,1,43,65,4,99};`

声明整型数组 data_arr，并通过赋初始值，使数组元素的个数为 10。

(2)
```
File f = new File("temp.dat");
RandomAccessFile randf = new RandomAccessFile(f,"rw");
```

将一个 RandomAccessFile 流对象（由引用 randf 指向）与一个文件 temp.dat 关联，此流对象设置的访问文件模式为可读可写。

(3)
```
for (int i = 0;i< = data_arr.length - 1;i + + )
    { randf.writeInt(data_arr[i]);}
```

通过循环语句将数组 data_arr 的 10 个元素的值写入 RandomAccessFile 对象中，一共 10 次调用 writeInt 方法写入数据类型 int 的值。因为 int 类型的数值占有 4 B，所以调用 1 次 writeInt 方法将写入 4 B 到 RandomAccessFile 对象中，写入 10 次时一共写入 400 B 到 RandomAccessFile对象中。

(4)
```
for(int i = data_arr.length - 1;i> = 0;i - - )
{
    randf.seek(i * 4);
    System.out.println (randf.readInt());
}
```

完成从 RandomAccessFile 对象读取 10 个数据类型 int 的值（4 B）。每次循环中，通过 seek 方法定位指针到读取数值的位置，然后调用 readInt 读取 int 型的数值，并显示在屏幕上。第一次循环，i 的值为 9，指针定位的位置为 $9\times4=36$，是 RandomAccessFile 流的最后一个 int 型的数值（即 99）位置。下一次循环时，指针定位的位置是上一次的位置－4（即指向前一个 int 型的位置），所以循环完成时将以反序输出 RandomAccessFile 流的 int 型的数值。

12.3.4 过滤字节流

过滤字节流由 Java 的类 FilterInputStream 和 FilterOutputStream 提供，它们分别对其他输入/输出流（缓冲流、数据流等）进行特殊处理。过滤仅仅意味着可以提供更多的功能，诸如缓冲、监视行数或将数字字节结合成有意义的基本数据类型，它们在读写数据的同时可以对数据进行特殊处理，另外还提供了同步机制，使得某一时刻只有一个线程可以访问一个输入/输出流。这两个类都是抽象类，因此由它们的子类提供额外的功能。

1. 带缓冲的过滤字节流

类 BufferedInputStream（FilterInputStream 的子类）和 BufferedOutputStream（FilterOutputStream 的子类）是带缓冲的两个过滤字节流。它提供了缓冲机制，把任意的 I/O 流"捆绑"到缓冲流上，可以提高该 I/O 流的读取效率。

通过构造函数创建缓冲字节流时，除了要将缓冲流与 I/O 流相连接之外，还可以指定缓冲区的大小。缺省时用 32 B 大小的缓冲区，最优的缓冲区大小常依赖于主机操作系统、可使用的内存空间以及机器的配置等，一般缓冲区的大小为内存页或磁盘块的等整数倍，如 8 9 12 B或更小。例如：

```
FileInputStream in = new FileInputStream("file1.txt");
FileOutputStream out = new FileOutputStream ("file2.txt");
BufferedInputStream bin = new BufferedInputStream(in,256)
BufferedOutputStream bout = new BufferedOutputStream(out,256);
int len;
byte bArray[] = new byte[256];
len = bin.read(bArray); //len 输入长度, bArray 数据
```

对于 BufferedOutputStream，只有缓冲区满时，才会将数据真正送到输出流，但可以使用 flush()方法人为地将尚未填满的缓冲区中的数据送出。

2. 数据过滤流

数据过滤流类 DataInputStream 和 DataOutputStream，提供了一种较为高级的数据输入/输出方式。它们分别实现了接口 DataInput 和 DataOutput，与 RandomAccessFile 类似，也属于过滤字节流，但为顺序流。

利用类 DataInputStream 和 DataOutputStream 处理字节或字节数组；实现对文件的基本数据类型的数值的读写。提供的读写方法包括 read()、readInt()、readByte()、writeChar()、writeBoolean()、write()等。

通过构造函数创建数据流时，要将缓冲流与 I/O 流相连接。创建数据流的例子如下：

```
FileInputStream fis = new FileInputStream("file1.txt");
FileOutputStream fos = new FileOutputStream("file2.txt");
DataInputStream dis = new DataInputStream(fis);
DataOutputStream dos = new DataOutputStream(fos);
```

例 12-3 利用数据过滤流读写顺序文件。

先创建输出数据过滤流，并将输出文件流和输出数据过滤流相连接；向输出数据流顺序写入 8 个基本类型的数据后，关闭输出数据过滤流(因为是顺序流)。接着，先创建输入数据过滤流，并将输入文件流和输入数据过滤流相连接；从输入数据流顺序读取 8 个基本类型的数据送显示器显示后，关闭输入数据过滤流。

```
//datainput_output.java
import java.io.*;
class datainput_output
{  public static void main(String args[]) throws IOException
   {  FileOutputStream fos = new FileOutputStream("a.txt");
      DataOutputStream dos = new DataOutputStream (fos);
      try
      {  dos.writeBoolean(true);
         dos.writeByte((byte)123);
         dos.writeChar('J');
         dos.writeDouble(3.141592654);
         dos.writeFloat(2.7182f);
         dos.writeInt(1234567890);
         dos.writeLong(998877665544332211L);
         dos.writeShort((short)11223);
      }
      finally
      {  dos.close();
      }
      FileInputStream fin = new FileInputStream("a.txt");
```

```
    DataInputStream sin = new DataInputStream (fin);
    try
    {   System.out.println(sin.readBoolean());
        System.out.println(sin.readByte());
        System.out.println(sin.readChar());
        System.out.println(sin.readDouble());
        System.out.println(sin.readFloat());
        System.out.println(sin.readInt());
        System.out.println(sin.readLong());
        System.out.println(sin.readShort());
    }
    finally
    {   sin.close();
    }
  }
}
```

3. 其他过滤流

(1) InputStream：主要用于对文本文件的处理，提供了行号控制功能。

(2) PrintStream：其作用是将 Java 语言中的不同类型的数据以字节表示形式输出到相应的输出流中去。

12.3.5 标准输入/输出流

Java 的标准输入/输出处理由包 java. lang 中提供的类 System 实现。在程序开始执行时，Java 会自动创建 3 个与设备关联的对象：System. in、System. out 和 System. err 对象。System. in 对象（标准输入流对象）是从 InputStream 中继承而来，用于从标准输入设备中获取输入数据（通常是键盘）。System. out 对象（标准输出流对象）是从 PrintStream 中继承而来，把输出送到缺省的显示设备（通常是显示器）。System. err（标准错误流对象）也是从 PrintStream 中继承而来，把错误信息送到缺省的显示设备（通常是显示器）。每种流都可以被重新定向。System 类提供方法 setIn、setOut 和 setErr，分别对标准输入流、标准输出流和标准错误流进行重新定向。

System 类含有标准输入流的静态成员变量 in，可以调用它的 read 方法来读取键盘数据。read 方法有 3 种格式：

① public abstract int read()

② public int read(byte[] b)

③ public int read(byte[] b, int off, int len)

例 12-4 利用标准输入/输出流，从键盘输入一行字符串并显示在屏幕上。

```
import java.io.*;
public class ReadFromKeyboard //ReadFromKB.java
{   public static void main(String args[])
    {   try {
        byte bArray[] = new byte[24];
        String str;
        System.out.print("Enter something Using Keyborad:");
        /* 从键盘读入多个字符(一个字节表示)到数组 bArray 中，直到遇到回车字符为止或字符填满整
           个 */
        System.in.read(bArray);数组为止。
```

```
        str = new String(bArray, 0,bArray.length);
        System.out.print("You entered:");
        System.out.println(str);
        }
        catch(IOException ioe)
        {  System.out.println(ioe.toString());
        }
    }
}
```

程序执行结果如图 12-2 所示。

```
Enter something Using Keyborad:I am a student
You entered:I am a student
```

图 12-2　标准输入/输出流处理

12.3.6　对象流与 Serializable 接口

Java 中的数据流不仅能对基本数据类型的数据进行操作，还提供了把一个完整的对象写入文件或从文件中读出的功能。对象流类 ObjectOutputStream 和 ObjectInputStream 分别实现了接口 ObjectOutput 和 ObjectInput，并且能够在文件(或其他类型的流)中读取和写入一个完整的对象。

对象的持续性(persistence)是指能将对象的状态记录以便将来再生的能力。对象的可串行化(Serialization)是指对象通过写出描述自己状态的数值来记录自己的过程。串行化的主要任务是写出对象实例变量的数值。所有基本类型的变量都是可串行化的，如果变量是另一个对象的引用，则引用的对象也要可串行化。这个过程是递归的。

在 Java 中，只有可串行化的对象才能通过对象流进行传输。只有实现 Serializable 接口的类才能被串行化，Serializable 接口是一个标记接口(tagging interface)，该接口不包含任何方法。当一个类声明实现 Serializable 接口时，只是表明该类加入对象串行化协议。

建立对象输入流时，通常将 FileInputStream 对象封装在类 ObjectInputStream 中，用 readObject()方法直接从输入流中读取一个对象，此方法的返回类型为 Object，往往要转换成当初写入时的对象类型。建立对象输出流时，通常将 FileOutputStream 对象封装在类 ObjectOutputStream 中，用 writeObject()方法可以直接将对象写入到输出流中。

例 12-5　使用对象流对顺序文件进行读写。

程序由两个类 Student.java 和 ObjectSeriWrite.java 组成。学生类 Student 包含的信息为：学号、姓名、年龄和所在系，含有一个构造函数和一个显示学生信息的方法。类 Student 实现了 Serializable 接口，是一个可串行化的类，该类的对象是一个可串行化的对象。

```
    //Student.java 定义可串行化的类
  import java.io.Serializable;
  public class Student implements Serializable
  {  int id;
     String name;
     int age;
     String departmernt;
     public Student(int id, String name, int age, String departmernt)
     {  this.id = id;
```

```
        this.name = name;
        this.age = age;
        this.departmernt  = departmernt;
    }
    void show() { //显示对象
        System.out.println(id + "," + name + "," + age + "," + departmernt);
    }
}

//建立对象的输入输出流,并对可串行化的对象进行读写
//ObjectSeriWrite.java
import java.io. * ;
class ObjectSeriWrite
{ public static void main(String args[])
 { //可串行化的对象
 Student stu1 = new Student(200501, "Li Ming", 16, "Compute Science");
 Student stu2 = new Student(200502, "Zhang Bi", 17, "Chinese");
 try
  { FileOutputStream fo  =  new FileOutputStream("student.dat");
   ObjectOutputStream so  =  new ObjectOutputStream(fo); //建立对象输出流
    so.writeObject(stu1); // 写对象
    so.writeObject(stu2);
    so.close();
    FileInputStream fi  =  new FileInputStream("student.dat");
    ObjectInputStream si  =  new ObjectInputStream(fi); //建立对象输入流
    stu1 = (Student)si.readObject(); //读对象
    stu2 = (Student)si.readObject();
    si.close();
    stu1.show(); //显示对象
    stu2.show();
    }
    catch(Exception e)
    { System.out.println(e) ;
    }
 }
}
```

类 ObjectSeriWrite 中,先创建 Student 类的两个对象并分别赋给引用 stu1 和 stu2。

在 try 代码块中,语句

```
FileOutputStream fo  =  new FileOutputStream("student.dat");
```

首先创建文件输入流,并与输入文件 student.dat 相关联。

```
ObjectOutputStream so  =  new ObjectOutputStream(fo);
```

建立对象输入流,并与文件输入流相关联。

```
so.writeObject(stu1);
so.writeObject(stu2);
so.close();
```

将两个可串行化的对象写入对象输入流中。因为对象流是顺序流,必须先关闭输入文件,然后用对象输出流重新打开文件,进行读取对象。

```
FileInputStream fi  =  new FileInputStream("student.dat");
ObjectInputStream si  =  new ObjectInputStream(fi);
```

建立对象输入流与文件输入流的关联。

```
stu1 = (Student)si.readObject();
stu2 = (Student)si.readObject();
```

从输入流中读两个读对象,关闭文件。跟在 try 代码段之后的 catch 语句块是对 try 代码段中可能发生的异常进行捕获处理。

最后要注意的是:串行化只能保存对象的非静态成员变量,而不能保存任何成员方法和静态成员变量,并且保存的只是变量的值,对于变量的任何修饰符都不能保存。对于某些类型的对象,其状态是瞬时的,这样的对象是无法保存其状态的,如 Thread 对象或流对象。对于这样的成员变量,必须用 transient 关键字标明,否则编译器将报错。任何用 transient 关键字标明的成员变量,都不会被保存。

另外,串行化可能涉及将对象存放到磁盘上或在网络上发送数据,这时会产生安全问题。对于一些需要保密的数据,不应保存在永久介质中(或者不应简单地不加处理地保存下来),为了保证安全,应在这些变量前加上 transient 关键字,以指出它们不是 Serializable 类型的变量,在串行化的过程中应忽略它们。

12.3.7 管道流

管道是用来把一个程序、线程和代码块的输出连接到另一个程序、线程和代码块的输入。java.io 中提供了类 PipedInputStream 和 PipedOutputStream 作为管道的输入/输出流。管道输入流作为一个通信管道的接收端,管道输出流则作为发送端。管道流必须是输入/输出并用,即在使用管道前,两者必须进行连接。如图 12-3 所示。

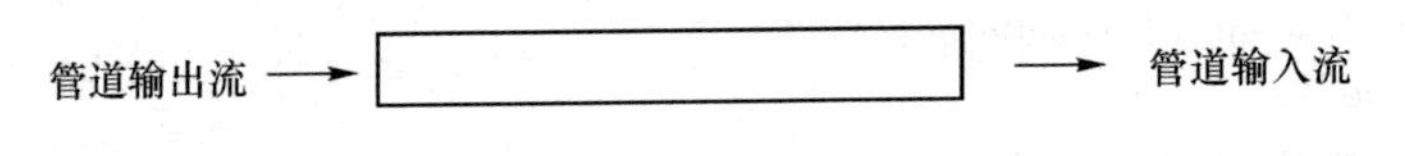

图 12-3 管道流

管道输入/输出流可以用两种方式进行连接。

(1) 在构造方法中进行连接

```
PipedInputStream(PipedOutputStream pos);
PipedOutputStream(PipedInputStream pis);
```

(2) 通过各自的 connect()方法连接

在类 PipedInputStream 中,调用方法

```
connect(PipedOutputStream pos);
```

在类 PipedOutputStream 中,调用方法

```
connect(PipedInputStream pis);
```

例 12-6 将数据从输出管道进,从输入管道出。

下面的程序将两个 byte 型数据从输出管道进去,从输入管道出来,并显示出来。

```
//pipedstream.java
import java.io.*;
class pipedstream
{ public static void main(String args[]) throws IOException
  {  byte aByteData1 = 123, aByteData2 = 111;
     PipedInputStream pis = new PipedInputStream();
     PipedOutputStream pos = new PipedOutputStream(pis);
     System.out.println("PipedInputStream");
```

```
        try
        {   pos.write(aByteData1);
            pos.write(aByteData2);
            System.out.println((byte)pis.read());
            System.out.println((byte)pis.read());
        }
        finally
        {  pis.close();pos.close();
        }
    }
}
```

12.3.8 内存读写流

java.io 中提供了类 ByteArrayInputStream、ByteArrayOutputStream 和 StringBufferInputStream 以支持内存的读写。

ByteArrayInputStream 从指定的字节数组中读取数据。

ByteArrayOutputStream 提供了缓冲区可以存放数据(缓冲区大小可以在构造方法中设定,缺省为 32 bit),可以用 write()方法向其中写入数据,然后用 toByteArray()方法将缓冲区中的有效字节写到字节数组中去。

StringBufferInputStream 与 ByteArrayInputStream 相类似,不同点在于它是从字符缓冲区 StringBuffer 中读取 16 bit 的 Unicode 数据,而不是 8 bit 的字节数据。

12.3.9 序列输入流

java.io 中提供了类 SequenceInputStream,使应用程序可以将几个输入流顺序地连接在一起,让程序员看起来就像是一个比较长的流一样。序列输入流提供了将多个不同的输入流统一为一个输入流的功能,这使得程序可能变得更加简洁。例如:

```
FileInputStream f1,f2;
String s;
f1 = new FileInputStream("file1.txt");
f2 = new FileInputStream("file2.txt");
SequenceInputStream fs = new SequenceInputStream(f1, f2);
                                //将文件输入流 f1 和 f2 连接在一起
DataInputStream ds = new DataInputStream(fs);
while( (s = ds.readLine()) != null )
  System.out.println(s);
```

12.4 基于字符的输入和输出类及应用实例

除了基于字节的流外,Java 还提供了用于字符流处理的类,它们是以 Reader 和 Writer 为基础派生的一系列类,它们是 Unicode 用两个字节表示的基于字符的流。类似于 InputStream、OutputStream,Reader、Writer 都是抽象类,这两个类提供了一组用于处理字符流的接口;接口中声明的方法也与 InputStream 和 OutputStream 类似,只不过其中的参数换成了字符或字符数组。例如:

```
String readLine()  ; // 其返回类型是字符串型
```

12.4.1 InputStreamReader 类和 OutputStreamWriter 类

InputStreamReader（Reader 的子类）和 OutputStreamWriter（Writer 的子类）是用于处理字符流的最基本的类，用来在字节流和字符流之间作为中介。在进行字符处理时，这两个类在构造函数中应指定一定的平台规范，以便把以字节方式表示的流转换为特定平台上的字符表示。

InputStreamReader 类和 OutputStreamWriter 类的构造函数为：

```
InputStreamReader(InputStream in);                //缺省规范
InputStreamReader(InputStream in, String enc);    //指定规范 encode
OutputStreamWriter(OutputStream out);             //缺省规范
OutputStreamWriter(OutputStream out, String enc); //指定规范 encode
```

如果读取的字符流不是来自本地（比如网上某处与本地编码方式不同的机器）时，那么在构造字符输入流时就不能简单地使用缺省编码规范，而应该指定一种统一的编码规范"ISO 8859_1"，这是一种映射到 ASCII 码的编码方式，能够在不同平台之间正确转换字符。例如：

```
InputStreamReader ir = new InputStreamReader( is,"8859_1" );
```

12.4.2 BufferedReader 类和 BufferedWriter 类

java.io 中提供的 BufferedReader 和 BufferedWriter 是基于字符缓冲流的类。其构造方法与 BufferedInputStream 和 BufferedOutputStream 相类似。另外，除了 read()和 write()方法外，还提供了整行字符处理方法：

```
public String readLine()
```

从输入流中读取一行字符。行结束标志为"\n"、"\r"，或两者一起。

例 12-7 应用字符缓冲流对文件进行读取。

```
import java.io.*;
class FileUnicodeRead
{ public static void main(String args[])
  { try
      {   FileInputStream fis = new FileInputStream("file1.txt");
          InputStreamReader dis = new InputStreamReader(fis);
          BufferedReader reader = new BufferedReader(dis);
          String s;
          while( (s = reader.readLine()) != null ) //为 null 时到文件尾
          {   System.out.println("read: " + s);
          }
          dis.close();
      }catch(IOException e)
      { System.out.println(e);
      }
  }
}
```

12.4.3 其他字符流

CharArrayReader 和 CharArrayWriter 是对字符数组进行读写的字符流。FileReader 和 FileWriter 是实现文件读写的字符流。StringReader 和 StringWriter 对字符串进行读写。PipedReader 和 PipedWriter 实现管道字符流。LineNumberReader 实现行处理字符流。

PrintWriter 是将字符写入流中。

12.5 小　结

在 Java 中有数据传输的地方都会用到输入/输出流(通常是文件、网络、内存和标准输入/输出等)。InputStream 和 OutputStream 是所有字节流的祖先(只有 RandomAccessFile 类是一个例外),read 和 write 是它们最基本的方法,读写单位是字节。Reader 和 Writer 是所有字符流的祖先,read 和 write 是它们最基本的方法,读写单位是字符。在众多的流对象中,并不是每一种都单独使用,其中一些流对象需要另外的流对象相关联。

File、FileInputStream、FileOutputStream、RandomAccessFile 是处理本地文件的类。

DataInputStream、DataOutputStream 是一个过滤流的子类,借此可以读写各种基本类型数据,在文件和网络中经常使用。如 readByte、writeBoolean 等。

BufferedInputStream、BufferedOutputStream 的作用是在数据送到目的地之前先缓存,达到一定数量时再送到目的地,以减少阻塞次数。

PipedInputStream、PipedOutputStream 适用于一个处理的输出作为另一个处理的输入的情况。

习　题

12.1　什么叫输入流?什么叫输出流?

12.2　字节流和字符流有何区别?

12.3　字节流的抽象的基类是什么类?字符流的抽象的基类是什么类?

12.4　RandomAccessFile 类用什么方法从指定流中读取一个整数、一行文本?

12.5　RandomAccessFile 类用什么方法将文件位置设置为文件中的某个位置以进行输入和输出?

12.6　写出输入/输出流的语句用以完成下面的任务,假设它们是同一程序的任务。

① 编写一条打开"oldmast. dat"文件以供输入的语句;使用 ObjectInputStream 对象的引用 inOldmaster 来包装一个 FileInputStream 对象。

② 编写一条打开"newmast. dat"文件以供输出的语句;使用 ObjectOutputStream 对象 outNewmaster 来包装一个 FileOutputStream 对象。

③ 编写一条打开"oldmast. dat"文件以供输入的语句,该记录是 AccountRecord 类的对象 record;使用 ObjectInputStream 对象 inOldmaster。

④ 编写一条向文件"newmast. dat"输出记录的语句,该记录是 AccountRecord 类的对象 record。使用 ObjectOutputStream 对象 outNewmaster 来包装一个 FileOutputStream 对象。

12.7　将键盘输入的内容逐行写入文件中,当输入"exit"时中止,文件名由运行时决定。

12.8　使用字符流 BufferedReader 类,统计文件 file. txt 包含的字符个数和行数。

12.9　在工资支付系统中,雇员的记录包括雇员号、姓名(String)、地址(String)和工资(Double)。编写一个程序,完成雇员记录的添加、修改、删除、查询功能。要求在系统退出时将系统所有的雇员记录写入文件中;在下次进入系统时,从该文件中恢复雇员信息。

第13章 网络技术和应用开发

13.1 Java网络技术概述

Java是一种流行的网络编程语言,它提供了两种功能强大的网络支持机制:访问网络资源的URL类和网络通信的Socket类,用来满足不同的要求。URL是用于访问Internet网上资源的应用;而基于TCP/IP协议中传输层接口Socket,是针对Client/Server模型的网络应用以及实现某些特殊协议的网络应用。

在网络应用的Client/Server模型中,由客户端提出某些操作请求,服务器执行该操作,并为客户做出响应。一种常见的Client/Server请求响应模型的实现,是在WWW浏览器和WWW服务器之间。当用户通过Web浏览器(客户应用程序)选择一个Web站点时,一个请求被发送到相应的Web服务器(服务器应用程序)。Web服务器在响应客户的请求时,通常会发送回一个合适的HTML网页。

Socket可以看成在两个程序进行通信连接中的一个端点,一个程序将一段信息写入Socket中,该Socket将这段信息发送给另外一个Socket,使这段信息能传送到其他程序中。基于Socket的通信,使得应用程序可将网络互联看成是文件I/O——程序读写套接字就像读写文件一样。

Java提供了两种网络通信的基本方式:流套接字(stream sockets)和数据报套接字(datagram sockets)。如果用流套接字,服务程序等待客户程序的连接,一旦客户程序与服务程序之间建立连接后,就可以通过流操作的方式来实现发送和接收数据的双向数据传递。因此流套接字提供了面向连接的网络服务,即服务主机与客户机是时时连接的,用于传送的协议是TCP(Transmission Control Protocol)协议。

如果用数据报套接字,则程序将要传递的数据打包,分成一个个小的数据包。每一个数据包都有它要传送到的计算机的地址;一旦数据包发送,就不能够保证它一定能够到达目的地址。同样,在数据的传递过程中,也不能够保证数据不被破坏或者发送方能够得到应答。因此,这种方式中服务主机跟客户机不是时时连接的,对于重要数据的传递是不太适用的。它用于传送的协议是UDP(User Datagram Protocol)协议。

java.net中包含网络通信所需要的类和接口。例如,InetAddress类指明数据要达到的目的地和服务方的地址;URL类封装了网络资源的访问;ServerSocket类和Socket类实现面向连接的网络通信;DatagramSocket类和DatagramPacket类实现数据报的收发通信。

本章将详细讨论网络数据传递的相关知识和上面提到的两种网络数据传递方式的实现,并通过实例介绍网络程序设计的方法。

13.2　URL 与网络应用

URL(Uniform Resource Locator,统一资源定位符)用来表示 Internet 上的网络资源,或者说确定数据在网络中的位置。利用 URL 对象中提供的方法,可直接读写网络中的数据。一个 URL 通常由 4 个部分组成,包括协议名、主机名、端口号、路径文件(含有文件路径及文件名)。例如 http://www. tsinghua. edu . cn: 80/home/homepage. htm 表示协议为 http,主机地址为 www. tsinghua. edu . cn,端口号为 80,路径文件为/home/homepage. htm。

13.2.1　URL 类

Java 将 URL 封装成 URL 类,URL 的组成部分为 Java 的 URL 类提供了不同的构造方法。

- URL(String spec):用指定的一个 String 来创建一个 URL 对象;
- URL(String protocol,String host,int port,String file):用指定的协议、主机名、端口号、路径文件来创建一个 URL 对象;
- URL(String protocol, String host, String file):用指定的协议、主机名、路径及文件名来创建 URL 对象;
- URL(URL context, String spec):用已存在的 URL 对象来创建 URL 对象。

利用 URL 类提供的方法,来获取 URL 对象的属性。常用的方法如下。

- String getProtocol():获取 URL 传输的协议;
- String getHost():获取 URL 的机器名称;
- int getPort():获取 URL 的端口号;
- String getFile():获取 URL 的文件名,包括路径和文件名;
- Obect getContent():获取 URL 的内容;
- InputStram openStream():打开与 URL 的连接,返回一个输入流,通过这个输入流读取数据;
- URLConnection openConnection():返回与 URL 进行连接的 URLConnection 对象。

例 13-1　URL 对象的创建及使用。

```
import java.net. * ;
public class Myurl{
    public static void main(String args[])
    {try {
        URL url = new URL("http://www.tsinghua.edu.cn:80/home/homepage.htm");
        System.out.println("the Protocol: " + url.getProtocol());
        System.out.println("the hostname: " + url.getHost());
        System.out.println("the port: " + url.getPort());
        System.out.println("the file: " + url.getFile());
        System. out. println(url.toString());
    }
    catch(MalformedURLException e)
    {System.out.println(e);
    }
   }
}
```

该程序的运行结果如图 13-1 所示。

```
the Protocol: http
the hostname: www.tsinghua.edu.cn
the port: 80
the file: /home/homepage.htm
http://www.tsinghua.edu.cn:80/home/homepage.htm
```

图 13-1 例 13-1 程序运行结果

13.2.2 用 applet 访问 URL 资源

利用 applet 的 AppletContext 接口中带有一个 URL 参数的方法 showDocument()，将使执行 applet 的浏览器显示指定的 URL 资源。

例 13-2 利用 applet 显示网络上其他 HTML 文档。

showdoc. java 的程序代码如下：

```
import javax.swing.*;
import java.awt.*;
import java.net.*;
public class showdoc extends JApplet
{  URL docur = null;
   public void paint(Graphics g) {
       try {
           docur = new URL("http://www.163.com");
       }
       catch (MalformedURLException e) {
          System.out.println("Can ope the URL ");
       }
       if (docur != null)
          getAppletContext().showDocument(docur,"_blank");
   }
}
```

例 13-2 程序中，ShowDocument 方法的格式为

```
showDocument(URL url,String target)
```

target 参数指定显示 URL 内容的窗体，这里 target 的值是"_bank"表示在新开的窗体显示 163 网站的内容。

showdoc. htm 的代码如下：

```
<html>
<applet code = "showdoc.class" width = "600" height = "500">
</applet>
</html>
```

在 IE 浏览器执行 showdoc. htm 文件后，将在新的 IE 窗口中显示 163 网站的内容。

13.2.3 Web 浏览器的设计

用 Java 的 application 实现一个简单的 Web 浏览器。应用程序使用 swing GUI 组件 JEditorPane 显示 Web 服务器上的文件内容。用户在窗体顶部的 JTextField 中输入 URL 后，将 URL 的 HTML 格式的文本在 JEditorPane 中显示。

JEditorPane 组件一般显示无格式的文本，也能显示带格式的 HTML 文本。JEditorPane

的方法 setPage(URL url)将下载由 URL 定位的文档并显示在 JEditorPane 组件中。

如果 JEditorPane 含有的 HTML 文档中含有超链接，当用户单击其中一个超链接时，则 JEditorPane 产生事件 HyperlinkEvent(javax. swing. event 包中)，并通知所有已注册的 HyperlinkListener 对象。由 HyperlinkListener 接口中的方法 hyperlinkUpdate()处理超链接事件 HyperlinkEvent，在 JEditorPane 中显示超链接的 Web 网页。HyperlinkEvent 类中包含一个嵌套内部类 EventType，此内部类声明的 3 个静态常量为：ACTIVED(表示用户单击一个超链接以改变 Web 网页)、ENTERED(表示用户把鼠标移到一个超链接上)和 EXITED(表示用户把鼠标移离一个超链接)。

例 13-3 用 Java 的应用程序实现一个简单的 Web 浏览器。

Browser. java 程序的代码如下：

```
// Use a to display the contents of a file on a Web server.
import java.awt. * ;
import java.awt.event. * ;
import java.net. * ;
import java.io. * ;
import javax.swing. * ;
import javax.swing.event. * ;
public class Browser extends JFrame {
   private JTextField enterField;
   private JEditorPane contentsArea;
   // set up GUI
   public Browser()
   { super( "Web 浏览器" );
     Container container = getContentPane();
     // create enterField and register its listener
     enterField = new JTextField( "http://www.google.com" );
     enterField.addActionListener(
       new ActionListener() {
          // get document specified by user
          public void actionPerformed( ActionEvent event )
          {
             getThePage( event.getActionCommand() );
          }
       } ); // end call to addActionListener
     container.add( enterField, BorderLayout.NORTH );
     // create contentsArea and register HyperlinkEvent listener
     contentsArea = new JEditorPane();
     contentsArea.setEditable( false );
     contentsArea.addHyperlinkListener(
       new HyperlinkListener() {
          // if user clicked hyperlink, go to specified page
          public void hyperlinkUpdate( HyperlinkEvent event )
          { if ( event.getEventType() ==
                 HyperlinkEvent.EventType.ACTIVATED )
              getThePage( event.getURL().toString() );
          }
       } ); // end call to addHyperlinkListener
     container.add( new JScrollPane( contentsArea ),
        BorderLayout.CENTER );
```

```
        setSize( 400, 300 ); setVisible( true );
    } // end constructor
    // load document
    private void getThePage( String location )
    { try {
            contentsArea.setPage( location );
            enterField.setText( location );
        }
        catch ( IOException ioException ) {
            System.out.println("Bad URL");
            }
      } // end method getThePage
      public static void main( String args[] )
      { Browser application = new Browser();
         application.setDefaultCloseOperation( JFrame.EXIT_ON_CLOSE );
      }
    } // end class Browser
```

程序执行后的输出结果如图 13-2 所示。假定用户在窗体顶部的 JTextField 中输入 Google 的 Web 地址后，将在 JEditorPane 中显示 Google 的 Web 页面〔如图 13-2(a)〕。用户单击其中一个超链接“高级搜索”，将显示超链接“高级搜索”的 Web 网页〔如图 13-2(b)〕。

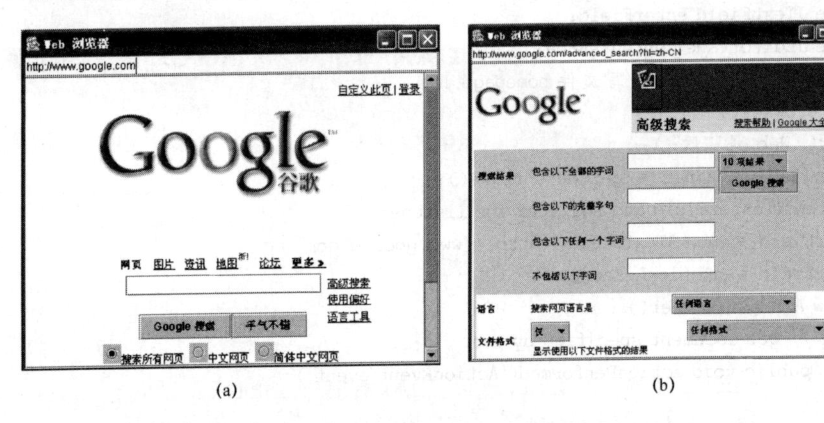

图 13-2　例 13-3 程序执行的部分输出结果

13.2.4 URLConnection 类

URLConnection 是一个抽象类，它提供了与 URL 资源的双向通信(读写操作)。URLConnection 对象的构造，可通过 URL 对象的 openConenection()来得到一个 URLConnection 对象，但这并未建立与指定的 URL 的连接，还必须调用 URLConnection 对象的 connect()方法建立连接。URLConnection 对象的常用方法如下。

(1) void connect()：打开与此 URL 引用资源的通信链接；

(2) int getContentLength()：获得文件的长度；

(3) String getContentType()：获得文件的类型；

(4) long getDate()：获得文件创建的时间；

(5) long getLastModified()：获得文件最后修改的时间；

(6) InputStream getInputStream()：获得输入流，以便读取文件的数据；

(7) OutputStream getOutputStream()：获得输出流，以便写文件。

例 13-4 从 Web 服务器上将 URL 文件下载到本机，并将文件的信息显示到屏幕。

urlConn. java 文件的代码如下：

```
import java.net. * ;
import java.io. * ;
class urlConn{
  public static void main(String args[])
  {InputStream is;
   OutputStream os;
   int b;
   try{
    URL url = new URL("http://www.tsinghua.edu.cn:80/home/homepage.htm");
    URLConnection uric =  url.openConnection(); //建立连接
    int fileLength = uric.getContentLength();
    if(fileLength> = 1 )
    { byte[] buffer = new byte[fileLength]; //定义字节数组用于读写文件
      is = uric.getInputStream(); //得到输入流
      is.read(buffer,0,fileLength); //从输入流读数据
      String fileName = url.getFile();//从 URL 得到路径文件
      fileName = fileName.substring(fileName.lastIndexOf('/') + 1);
      //得到文件名 homepage.htm
      os = new FileOutputStream(fileName);//在当前文件夹下建立文件输出流
      os.write(buffer); //写文件 homepage.htm
      System.out.write(buffer); //将文件 homepage.htm 内容显示在屏幕上
    }
    else
    {System.out.println("no content");
    }
  }
  catch(Exception e) {
     System.out.println(e);
  }
 }
}
```

该程序运行后，将文件 http://www.tsinghua.edu.cn:80/home/homepage.htm 下载到本机的当前文件夹下，并在屏幕上输出文件 homepage.htm 内容。

13.3 基于流套接字的客户/服务器通信

在基于流套接字的数据传递方式下，在两个网络应用程序(一个称为服务器应用，另一个称为客户机应用)之间发送和接收信息时，需要建立一个可靠的连接。Java 中与流套接字相关联的类有 3 个：InetAddress、ServerSocket 和 Socket。InetAddress 对象描绘了 32 位或 128 位 IP 地址，ServerSocket 对象用在服务器应用中，Socket 对象是建立网络连接时使用的。

13.3.1 InetAddress 类

Internet 上通过 IP 地址或域名标识主机，而 InetAddress 类用来表示与 Internet 地址相关的操作。InetAddress 类没有构造方法，要创建该类的实例对象，可以通过该类的静态方法

获得该对象。常用的一组静态方法如下。

(1) InetAddress getLocalHost():获得本地机的 InetAddress 对象。

(2) InetAddress getByName(Stung host):获得由 host 指定的 InetAddress 对象,host 是计算机的域名。

(3) InetAddress[]getAllmyName(String host):在 Web 中,可以用相同的名字代表一组计算机获得具有相同名字的一组 InetAddress 对象。

对于创建的 InetAddrss 对象,常用的一组操作方法如下。

(1) byte[] getAddress():得到 IP 地址;

(2) String getHostName():得到主机名字;

(3) String toString():得到主机名和 IP 地址的字符串。

例 13-5 InetAddress 类的应用。

InternetAddress.java 程序代码如下:

```
import java.net. * ;
class InternetAddress {
 public static void main(String args[])
 {   try{
     InetAddress iads;
     iads = InetAddress.getByName("www.163.com" );
     System.out.println("hostname =" + iads.getHostName());
     System.out.println(iads.toString() );
     }
     catch(Exception e)
     {System.out.println(e);
     }
   }
}
```

如果运行程序的机器已经与 Internet 连接好,则程序运行后的结果如图 13-3 所示。

```
hostname=www.163.com
www.163.com/202.108.9.34
```

图 13-3 例 13-5 程序运行结果

13.3.2 Socket 类

Socket 是建立网络连接时使用的,在连接成功时,客户和服务器应用程序两端都会产生一个 Socket 实例,操作这个实例,完成所需的会话。对于一个网络连接来说,套接字是平等的,并没有差别,不会因为在服务器端或在客户端而产生不同级别。

当客户程序需要与服务器程序通信时,客户程序在客户机要创建一个 Socket 对象。Socket 类有几个构造函数。两个常用的构造方法为:

(1) Socket(InetAddress addr, int port)

(2) Socket(String host, int port)

两个构造方法都创建了一个用于连接服务器主机和指定端口的客户端流套接字。

创建了一个 Socket 对象后,应用程序可通过调用 Socket 对象的 getInputStream()方法获得输入流,以读取对方传送来的信息;通过调用 Socket 对象的 getOutputStream()方法获得输出流,向对方发送消息。

在读写活动完成之后,应用程序调用 close()方法关闭流和流套接字。

下面的代码创建了一个服务器应用的主机地址为 198.163.227.6,端口号为 50000 的客户端 Socket 对象，然后从这个新创建的 Socket 对象中读取输入流和输出流，最后再关闭 Socket 对象和流。

```
Socket s = new Socket ("198.163.227.6",50000);
InputStream inputstream = s.getInputStream ();
OutputStream outputstream = s.getOnputStream ();
// Read or write from the inputstream/outputstream
inputstream.close ();
outputstream.close ();
connection.close ();
```

13.3.3 ServerSocket 类

服务器程序需要创建一个 ServerSocket 对象，用于在指定端口上监听客户端连接请求。ServerSocket 类有 2 个构造方法：

(1) ServerSocket(int port)：在指定端口上创建一个 ServerSocket 对象。服务器应用使用 ServerSocket 监听指定的端口，端口可以随意指定(由于 1024 以下的端口通常属于保留端口，在一些操作系统中不可以随意使用，所以建议使用大于 1024 的端口)。

(2) ServerSocket(int port，int queueLength)：Server 在指定端口上监听客户端连接请求，并指定允许连接的客户最大数目。

接下来服务程序进入无限循环之中，无限循环从调用 ServerSocket 的 accept()方法开始，在调用开始后，accept()方法将导致调用线程阻塞直到连接建立。在建立连接后，accept()返回一个最近创建的 Socket 对象，该 Socket 对象绑定了客户程序的 IP 地址或端口号。然后双方就可利用建立起来的流套接字进行通信。

13.3.4 基于流套接字的客户/服务器的通信过程

基于流套接字的客户/服务器模型的通信工作过程如图 13-4 所示。从图中可知，服务器应用由 5 个步骤组成。客户端应用由 4 个步骤组成。服务器应用使用 ServerSocket 对象监听指定的端口，等待客户连接请求。客户连接后，会话产生；在完成会话后，关闭连接。客户端应用使用 Socket 对网络上某一个服务器的某一个端口发出连接请求，一旦连接成功，打开会话；会话完成后，关闭 Socket。

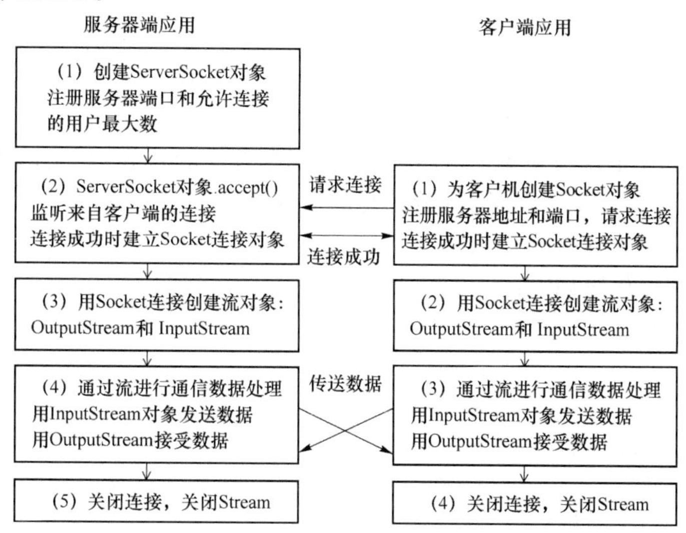

图 13-4　基于流套接字的客户/服务器的通信工作过程

下面通过例子说明如何建立服务器的应用和客户机的应用。

例 13-6 服务器端的应用。

Server.java 程序代码如下：

```
import java.net.*;
import java.io.*;
public class Server
{ private ServerSocket ss;
  private Socket socket;
  private BufferedReader instream;
  private PrintWriter outstream;
  public Server()
{ try
 { ss = new ServerSocket(10000);
   while (true) {
   socket = ss.accept(); //线程被阻塞直到连接建立,连接成功建立 Socket 对象
   System.out.println(" a client is connected");
   //从 Socket 对象创建输入流和输出流对象
   instream = newBufferedReader(new InputStreamReader
                    (socket.getInputStream()));
   outstream = new PrintWriter(socket.getOutputStream(),true);
   String line = instream.readLine(); //read a line from a client
   System.out.println("massage sended by a client is : " + line);
   //echo a line to the client
   outstream.println("echo to client :" + line);
   //close
   outstream.close(); instream.close(); socket.close();
 }
}
catch (IOException e)
 { e.printStackTrace();}
}
public static void main(String[] args)
{ new Server();
}
}
```

例 13-6 程序中,语句

```
outstream = new PrintWriter(socket.getOutputStream(),true);
```

是基于 Socket 建立了输出流对象 PrintWriter,第二个参数 true 表示要刷新输出缓冲区,会将头信息发送到客户端,使客户端做好接收信息的准备。

例 13-7 客户端的应用。

Client.Java 程序代码如下：

```
import java.io.*;
import java.net.*;
public class Client {
 Socket socket;
 BufferedReader instream;
 PrintWriter outstream;
 ChatServer = "127.0.0.1"
 public Client()
```

```
{  try
   {//create a socket to make connection
   socket = new Socket(InetAddress.getByName(chatServer), 10000);
   //get the input and output streams
   instream =
   new BufferedReader(new InputStreamReader(socket.getInputStream()));
   outstream = new PrintWriter(socket.getOutputStream(),true);
   //input a line from keyboad
   System.out.print("Please input a line string to be sended to a server: ");
   //建立标准设备的输入流,并从键盘读一行字符串
   BufferedReader line = new
      BufferedReader(new InputStreamReader(System.in));
   outstream.println(line.readLine()); //send a line string to server
   String s = instream.readLine(); //read a line from server
   //print the line on screen
   System.out.println("a line string echoed from server : " + s);
   line.close();
   outstream.close();
   instream.close();
   socket.close();
  }
  catch (IOException e)
  {e.printStackTrace();
  System.exit(1);}
 }
public static void main(String[] args)
{new Client();
}
}
```

例 13-7 程序中,语句

```
socket = new Socket("127.0.0.1", 10000);
```

创建了一个用于连接服务器主机(这里指本机)和端口为 10000 的客户端 Socket 对象,这里的服务器主机地址和端口号必须与服务器应用定义的一致。

上述两个程序 Server.java 和 Client.java 经编译后运行:先运行服务器应用 Server.class,等待客户连接请求,再运行客户端应用 Client.class。客户端应用 Client.class 运行并与服务器连接上后,服务器上显示"a client is connected";在客户端输入"Hello,frends"并发送,服务器接收到信息并显示出来,同时服务器将信息反馈到客户端,客户端收到此信息并显示。

客户端应用的运行结果如图 13-5 所示。

```
Please input a line string to be sended to a server: Hello ,frends
a line string echoed from server : echo :Hello ,frends
```

图 13-5 例 13-7 程序运行结果一

服务器应用 Server.class 运行结果如图 13-6 所示。

```
 a client is connected
massage sended by a client is : Hello ,frends
```

图 13-6 例 13-7 程序运行结果二

13.3.5 多线程实现多用户网上聊天

存在单个服务器程序与多个客户程序通信的可能。为了实现多用户网上能同时聊天,要求:

- 服务器程序响应客户程序不应该用很多时间,否则客户程序在得到服务前有可能用很多时间来等待通信的建立。
- 服务程序和客户程序的会话有可能是很长的(这与电话类似),因此,为加快对客户程序连接请求的响应,典型的方法是在服务器主机响应每一个客户连接请求时运行一个后台通信服务线程,这个后台线程处理和客户程序之间的通信。又由于在服务器主机上创建每一个后台通信服务线程,都需要花费一定的主机资源,因此对允许与服务器连接的客户数应有一定的限制。

例 13-8 服务器聊天程序的实现。

```
import java.net. * ;
import java.io. * ;
public class ChatServer
{  private int clientNum = 0; //存放当前连接的用户数
   private int maxClients = 10;//允许连接的用户最大数
   private ServerSocket ss;
   private CommunicationThread communications[];
   public ChatServer()
   { try {
       ss = new ServerSocket( 50000, maxClients );
     }
     // process problems creating ServerSocket
     catch( IOException ioException ) {
       ioException.printStackTrace();
       System.exit( 1 );
     }
     communications = new CommunicationThread [maxClients];
     for (int i = 0;i<maxClients;i ++ ) //循环等待用户连接
     { try {
       //等待用户连接,连接成功时创建 server 端通信线程并启动它
       communications[i] = new CommunicationThread(ss.accept(),i);
       communications[i].start();
       clientNum ++ ;
     }
     catch( IOException ioException ) {
       ioException.printStackTrace();
       System.exit( 1 );
     }
   }
}
private class CommunicationThread extends Thread {
    private Socket socket;
    private int clientID;
    private DataInputStream input;
    private DataOutputStream output;
    public CommunicationThread(Socket ss,int number) {
      socket = ss; //取连接对象 Socket
      clientID = number; //取分配给用户的 ID
```

```
        try {
        //从 Socket 得到输入/输出流
        input  = new DataInputStream( socket.getInputStream() );
        output  = new DataOutputStream( socket.getOutputStream() );
        }
      catch( IOException ioException ) {
        ioException.printStackTrace();
        System.exit( 1 );
        }
}
public void run() {
      try {
          //String s = null;
          //s = s + clientId;
          output.writeInt(clientID); //给用户发送 clientID
      }
      catch( IOException ioException ) {
       ioException.printStackTrace();
       System.exit( 1 );
       }
       //循环读一用户 clientID 发送来的信息,发送给聊天的各个用户
       while (true) {
            try {
               //读用户 clientID 发送来的信息
               String message = input.readUTF();
               for (int i = 0;i<clientNum;i++) { //发送给聊天的各个用户
                   communications[i].output.writeUTF
                        ("客户 " + clientID + ":" + message);
               }
               //用户 clientIDID 发送的信息是"Bye",则用户 ID 结束聊天
               if (message.equals("Stop chat!")) break;
            }
           catch( IOException ioException ) {
            ioException.printStackTrace();
             System.exit( 1 );
           }
       }
   try {
   output.close();
   input.close();
   socket.close();
   }
   catch( EOFException ioException ) {
      System.err.println( "Client terminated connection" );
   }
   catch( IOException ioException ) {
        ioException.printStackTrace();
        System.exit( 1 );
   }
  }
}
public static void main(String[] args)
```

```
  {  new ChatServer();
  }
}
```

ChatServer 文件声明了一个外部类 ChatServer 和一个内部类 CommunicationThread。在 ChatServer 类的 main()方法中创建了一个 ChatServer 对象，此时通过调用构造方法来创建 ServerSocket 对象，以监听端口 50000 上的连接请求，如果连接成功，ChatServer 进入一个无限循环中，交替调用 ServerSocket 的 accept() 方法来等待连接请求，同时启动后台线程处理连接(accept()返回的请求)。线程由 CommunicationThread 继承的 start()方法开始，并执行 CommunicationThread 的 run()方法中的代码。

一旦 run()方法运行，线程将向当前连接的客户应用发送用户号 clientID，接着循环读此用户发送来的聊天信息，并转发给当前参与聊天的各个用户，直到此用户发出信息"stop chat"为止。

一旦编译了 ChatServer.java 的源代码，在开始运行 ChatServer.class 后，就可以运行一个或多个 ChatClient 程序。

例 13-9 客户端聊天程序的实现。

客户端聊天程序是一个 Applet，嵌入在 HTML 文档中。Applet 的功能为：①提供客户聊天界面，用于输入信息和显示信息；②接收通过服务器转发的其他客户聊天信息并显示。

小应用程序 ChatClient.java 的代码如下：

```
import java.awt.*;
import java.awt.event.*;
import java.net.*;
import java.io.*;
import javax.swing.*;
public class ChatClient extends JApplet implements Runnable {
private JLabel clientLabel;
private int clientID;
private JTextField inputField;
private JTextArea displayArea;
private Socket connection;
private DataInputStream input;
private DataOutputStream output;
// Set up user-interface
public void init()
{Container container = getContentPane();
JPanel panel1,panel2;
panel1 = new JPanel();
clientLabel = new JLabel("Client ");
inputField = new JTextField(20); //inputField 用于输入用户聊天信息
inputField.addActionListener( new ActionListener() {
  public void actionPerformed(ActionEvent e) {
      try {//发送用户输入的聊天信息到 server
          output.writeUTF(inputField.getText());
      }
      catch ( IOException ioException ) {
        ioException.printStackTrace();
      }
  } });
  panel1.setLayout(new FlowLayout());
```

```
    panel1.add(clientLabel );
    panel1.add( inputField );
    container.add(panel1,BorderLayout.NORTH);
    panel2 = new JPanel();
    displayArea = new JTextArea( 10, 30 );
    displayArea.setEditable( false );
    panel2.add( new JScrollPane( displayArea ), BorderLayout.CENTER );
    container.add(panel2,BorderLayout.CENTER );
} // end method init
// Make connection to server and get associated streams.
// Start separate thread to allow this applet to
// continually update its output in textarea display.
public void start()
{   // connect to server, get streams and start commThread
   try {              // make connection
       connection = new Socket( getCodeBase().getHost(), 50000 );
      // get streams
      input = new DataInputStream( connection.getInputStream() );
      output = new DataOutputStream( connection.getOutputStream() );
   }
   // catch problems setting up connection and streams
   catch ( IOException ioException ) {
      ioException.printStackTrace();
    }
    // create and start commThread
    Thread commThread = new Thread( this );
    commThread.start();
 } // end method start
 // control thread that allows continuous update of displayArea
 public void run()
 {   // get client identifier
    try {// 读 server 发送来的用户号 clientID 并显示
       clientID = input.readInt();
    // display clientID in event-dispatch thread
    SwingUtilities.invokeLater(
       new Runnable() {
          public void run()
          {
               clientLabel.setText( "客户 " + clientID );
          }
       } );
    while ( true ) { //循环读聊天的各个用户发送来的信息并显示
        String message = input.readUTF();
        displayMessage( message );
    }
} // end try
catch( EOFException ioException ) {
       displayMessage( "Server terminated connection" );
}
 catch ( IOException ioException ) {
     ioException.printStackTrace();
 }
```

```
} // end method run
private void displayMessage( final String messageToDisplay )
{ // display message from event-dispatch thread of execution
  SwingUtilities.invokeLater(
    new Runnable() { // inner class to ensure GUI updates properly
      public void run() // updates displayArea
      {  displayArea.append( messageToDisplay + "\n");
         displayArea.setCaretPosition(
            displayArea.getText().length() );
      }
    } ); // end call to SwingUtilities.invokeLater
  }
} // end class
```

例 13-9 程序中，由于小应用程序的线程要处理 GUI 界面的组件响应事件，所以又创建了一个线程 commThread 完成通信处理。但 Applet 和 commThread 两个线程需要同时更新 GUI 界面上的组件，会带来不正确性的结果。每次应只有一个线程执行与 GUI 组件的所有交互。通常，该线程称为 Java 系统中的事件调度线程(event-dispatch thread)。SwingUtilities 类(javax.swing 包)提供静态方法 invokeLater 帮助实现这个线程。InvokeLater 方法指定 GUI 的处理语句在事件调度线程中执行。InvokeLater 方法以一个实现 Runnable 接口的对象作为参数，将 GUI 处理语句写在 Runnable 接口的 run()方法中。例如：

```
SwingUtilities.invokeLater(
   new Runnable() { // inner class to ensure GUI updates properly
      public void run()// updates displayArea
      {
        displayArea.append( messageToDisplay + "\n");
        displayArea.setCaretPosition(
           displayArea.getText().length() );
      } } // end inner class
); // end call to SwingUtilities.invokeLater
```

在程序调用 invokeLater 方法时，该 GUI 组件的更新操作将在事件调度线程中排队以等待执行，然后作为事件调度线程的一部分，将调用 run 方法以执行输出操作。这样以线程安全的形式更新 GUI 组件。

HTML 文档文件 ChatClient.htm 的内容如下：

```
<html>
<applet code = "ChatClient.class" width = 500 height = 225>
</applet>
</html>
```

每个要聊天的用户，在 Web 浏览器中打开 ChatClient.htm 文件，将得到用户聊天的界面。如图 13-7 所示是两个用户在网上进行聊天的聊天界面。

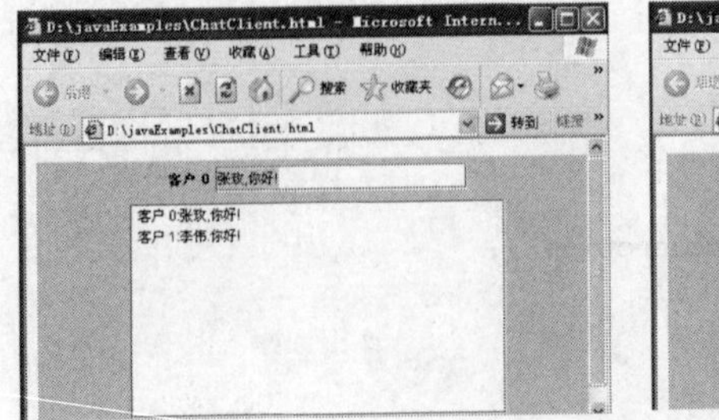

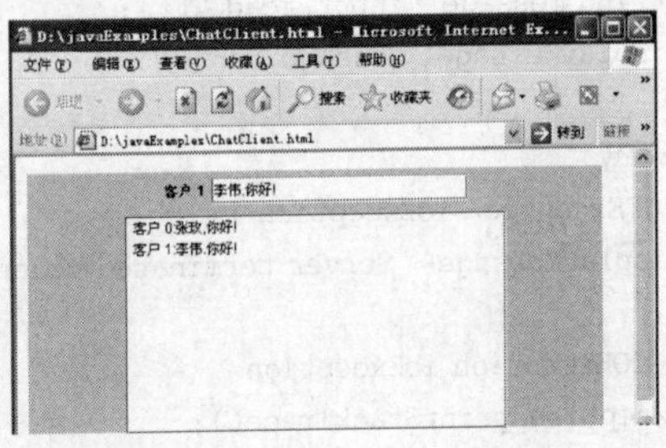

图 13-7　两个用户的聊天界面

13.4 基于数据报套接字方式的客户/服务器通信

数据报套接字是一种无连接的数据传递方式。它的传输速度快，但不能保证所有数据都能到达目的地，所以一般用于非重要的数据传输。与数据报套接字相关的有3个类：DatagramPacket、DatagramSocket和MulticastSocket。DatagramPacket类用来创建数据包；DatagramSocket用于发送和接收数据包；MulticastSocket描绘了能进行多点传送的套接字，这3个类均位于java.net包。

13.4.1 DatagramPacket类

使用数据报方式首先将数据打包，DatagramPacket类用来创建数据包。数据包分为两种：一种是要发送数据的数据包，该数据包有要到达的目的地址；另一种是用来接收数据的数据包。

（1）要创建接收数据包，使用DatagramPacket类的构造方法：

```
DatagramPacket(byte ibuft[],int ilength)
```

ibuf[]为接收数据包的存储数据的缓冲区，ilength为从传递过来的数据包中读取的字节数。

（2）要创建发送数据包，使用DatagramPacket类的构造方法：

```
DatagramPacket(byte ibuf[],int ilength,InetAddrss iaddr,int port)
```

ibuf为要发送数据的缓冲区；ilength为发送数据的字节数；iaddr为数据包要传递到的目标地址；iport为目标地址的程序接收数据包的端口号。

数据包也是对象，有如下操作方法用来获取数据包的信息：

① inetAddrss getAddrss()　　//获得数据包要发送的目标地址
② byte[]getData()　　//获得数据包中的数据
③ public int getLength()　　//获得数据包中数据的长度
④ int getPort( )　　//获得数据包中的目标地址的主机端口号

13.4.2 DatagramSocket类

发送和接收数据包，需要发送和接收数据包的DatagramSocket对象，用DatagramSocket类构造此对象。DatagramSocket类的构造方法如下。

（1）Datagramsocket()：用本地机上任何一个可用的端口创建一个发送/接收数据包的套接字。

（2）DatagramSocket(int port)：用一个指定的端口创建一个发送/接收数据包的套接字。

DatagramSocket对象相应的方法如下。

（1）void receive(Datagrampacket p)：接收数据包。

（2）void send(DatagramPacket p)：发送数据包。

（3）int getLocalPort( )：得到本地机的端口。

13.4.3 基于数据报套接字的客户/服务器的通信应用实例

在基于数据报套接字的方式下，服务器应用程序和客户机应用程序之间发送和接收信息。

完成的功能是:在客户端应用程序中,用户在文本框中输入一行信息,按下回车键后,会将此行信息发送到服务器上。服务器接收到此信息后,将此信息在服务器的 GUI 界面上显示,并发回到客户机上。客户机收到此信息后,即将此信息显示。

例 13-10 服务器应用程序的实现。

服务器应用程序类 Server.java 中,完成的任务如下。

(1) 构造方法用于建立 GUI 界面显示接收到的数据包,建立服务器套接字 DatagramSocket 对象,并将服务器套接字绑定在端口 5000 上。

(2) 声明方法 waitForPackets(),用无限循环等待客户发送过来的数据分组。首先建立一个接收数据包 DatagramPacket 对象,用 Socket 的方法 receive 等待数据包到达服务器,此方法的调用一直阻塞,直到分组到来,则将分组存储到 receivePacket 中。接着调用方法 displayMessage,将分组信息显示。最后调用 sendPacketToClient()将服务器接收到的分组发回到客户机。

(3)声明方法 sendPacketToClient(),带有一个参数的接收数据包对象。首先创建了发送数据包 DatagramPacket 对象 sendPacket,然后调用 Socket 的方法 send 将数据包发送出去。

(4)在 main 方法中,先创建服务器应用程序类 Server 的对象,然后调用方法 waitForPacket(),等待客户发送过来的数据分组。

Server.java 的代码如下:

```
import java.io.*;
import java.net.*;
import java.awt.*;
import java.awt.event.*;
import javax.swing.*;
public class Server extends JFrame {
  private JTextArea displayArea;
  private DatagramSocket socket;
  // set up GUI and DatagramSocket
  public Server()
  { super("服务器应用");
    displayArea = new JTextArea();
    getContentPane().add( new JScrollPane( displayArea ),
       BorderLayout.CENTER );
    setSize( 400, 300 );
    setVisible( true );
    // create DatagramSocket for sending and receiving packets
    try {
       socket = new DatagramSocket( 5000 );
    }
    catch( SocketException socketException ) {
       socketException.printStackTrace();
       System.exit( 1 );
    }
  }
 // wait for packets to arrive, display data and echo packet to client
 private void waitForPackets()
 { while ( true ) { // loop forever
     // receive packet, display contents, return copy to client
     try {
```

```
            // set up packet
            byte data[] = new byte[ 100 ];
            DatagramPacket receivePacket =
              new DatagramPacket( data, data.length );
            socket.receive( receivePacket ); // wait for packet
            // 将接收的数据包显示
            displayMessage( "\n接收的数据包内容：" + new String
              ( receivePacket.getData(),0, receivePacket.getLength()) +
              "\n长度：" + receivePacket.getLength() +
              "\n\n对方主机地址：" + receivePacket.getAddress() +
              "\n端口号：" + receivePacket.getPort());
            sendPacketToClient( receivePacket ); // send packet to client
         }
         catch( IOException ioException ) {
            displayMessage( ioException.toString() + "\n" );
            ioException.printStackTrace();
         }
      } // end while
   }
   // echo packet to client
   private void sendPacketToClient( DatagramPacket receivePacket )
      throws IOException
   { // create packet to send
     DatagramPacket sendPacket = new DatagramPacket(
        receivePacket.getData(), receivePacket.getLength(),
        receivePacket.getAddress(), receivePacket.getPort() );
     socket.send( sendPacket ); // send packet
   }
   private void displayMessage( final String messageToDisplay )
   { // display message from event-dispatch thread of execution
     SwingUtilities.invokeLater(
       new Runnable() { // inner class to ensure GUI updates properly
         public void run() // updates displayArea
         { displayArea.append( messageToDisplay );
           displayArea.setCaretPosition(
              displayArea.getText().length() );
         }
       } // end inner class
     ); // end call to SwingUtilities.invokeLater
   }
   public static void main( String args[] )
   {
      Server application = new Server();
      application.setDefaultCloseOperation( JFrame.EXIT_ON_CLOSE );
      application.waitForPackets();
   }
}
```

服务器应用启动执行后，等待客户端发送过来的数据分组。

例 13-11 客户机应用程序的实现。

客户应用程序类 Client.java 中，完成的任务如下。

(1) 构造方法用于建立了 GUI 界面供输入信息和显示接收到的数据包。建立套接字

DatagramSocket 对象。在输入文本字段的事件处理代码 actionPerformed 方法中,取用户输入的信息,创建要发送给服务器的数据包,在此构造数据包对象时,将服务器的主机地址和端口号捆绑。接着用 DatagramSocket 对象的方法 send()发送数据包。

(2) 声明方法 waitForPackets(),用无限循环等待服务器发送过来的数据分组。首先建立一个接收数据包 DatagramPacket 对象,用 socket 的方法 receive 等待数据包到达服务器,此方法的调用一直阻塞,直到分组到来,则将分组存储到 receivePacket 中。接着调用方法 displayMessage,显示分组信息。

(3) 在 main 方法中,先创建客户应用程序类 Client 的对象,然后调用方法 waitForPackets(),等待服务器发送过来的数据分组。

Client.java 的代码如下:

```
import java.io.*;
import java.net.*;
import java.awt.*;
import java.awt.event.*;
import javax.swing.*;
public class Client extends JFrame {
   private JTextField enterField;
   private JTextArea displayArea;
   private DatagramSocket socket;
   // set up GUI and DatagramSocket
   public Client()
   { super("客户应用");
     Container container = getContentPane();
     enterField = new JTextField(20 );
     enterField.addActionListener(
       new ActionListener() {
          public void actionPerformed( ActionEvent event )
          { // create and send packet
            try {
               // get message from textfield and convert to byte array
               String message = event.getActionCommand();
               byte data[] = message.getBytes();
               // create sendPacket
               DatagramPacket sendPacket = new DatagramPacket( data,
                 data.length, InetAddress.getLocalHost(), 5000 );
               socket.send( sendPacket ); // send packet
            }
            catch ( IOException ioException ) {
               ioException.printStackTrace();
            }
          } // end actionPerformed
       } // end inner class
     ); // end call to addActionListener
     JPanel jpanel1 = new JPanel();
     jpanel1.add(new JLabel("输入要发送的信息:"));
     jpanel1.add(enterField);
     container.add( jpanel1, BorderLayout.NORTH );
     displayArea = new JTextArea();
     container.add( new JScrollPane( displayArea ),
     BorderLayout.CENTER );
     setSize( 400, 300 );setVisible( true );
     // create DatagramSocket for sending and receiving packets
```

```
        try {
            socket = new DatagramSocket();
        }
        // catch problems creating DatagramSocket
        catch( SocketException socketException ) {
            socketException.printStackTrace();
            System.exit( 1 );
        }
    }
    // wait for packets to arrive from Server, display packet contents
    private void waitForPackets()
    {
        while ( true ) { // loop forever
            // receive packet and display contents
            try {
                // set up packet
                byte data[] = new byte[ 100 ];
                DatagramPacket receivePacket = new DatagramPacket(
                    data, data.length );
                socket.receive( receivePacket ); // wait for packet
                // display packet contents
                displayMessage( "\n被发回的数据包内容:" +
                        new String( receivePacket.getData(),
                        0, receivePacket.getLength() ) +
                    "\n长度为:" + receivePacket.getLength() +
                    "\n\n对方主机:" + receivePacket.getAddress() +
                    "\n端口号:" + receivePacket.getPort() );
            }
            // process problems receiving or displaying packet
            catch( IOException exception ) {
                displayMessage( exception.toString() + "\n" );
                exception.printStackTrace();
            }
        } // end while
    }
    // displayArea in the event-dispatch thread
    private void displayMessage( final String messageToDisplay )
    { // display message from event-dispatch thread of execution
        SwingUtilities.invokeLater(
            new Runnable() { // inner class to ensure GUI updates properly
                public void run() // updates displayArea
                { displayArea.append( messageToDisplay );
                }
            } // end inner class
        );
    }
    public static void main( String args[] )
    {
        Client application = new Client();
        application.setDefaultCloseOperation( JFrame.EXIT_ON_CLOSE );
        application.waitForPackets();
    }
}
```

先运行服务器应用 Server.class,再运行客户应用 Client.class。运行结果如图13-8所示。

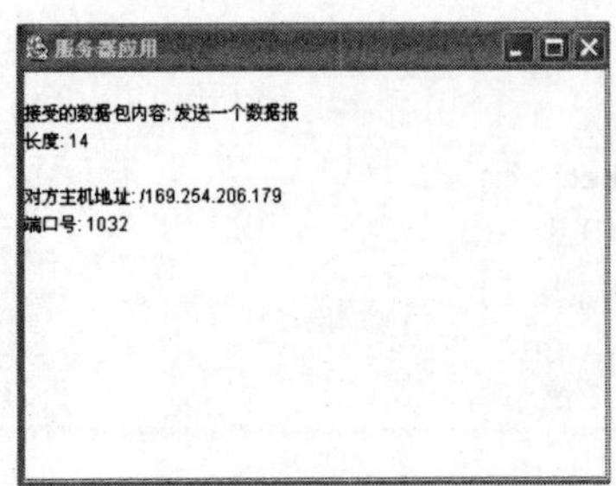

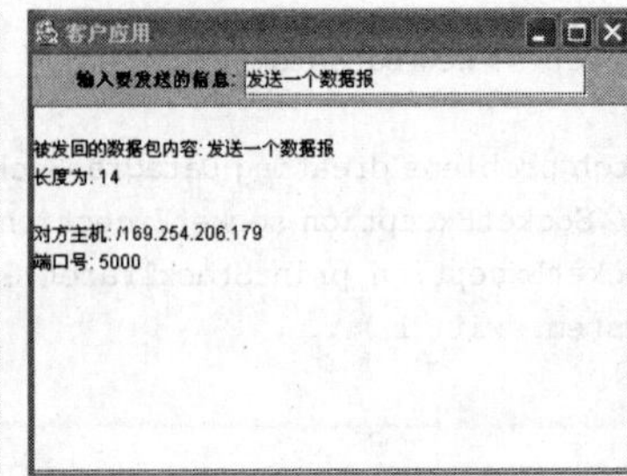

图 13-8　基于数据报套接字的客户/服务器的通信界面

13.5　小　结

Java 提供的网络功能按层次使用分为 URL、流套接字和数据报套接字。

URL 类是为访问网络资源提供的 API 接口，利用 URL 对象中的方法可以直接读取网络中的资源；而 URLConnection 类可双向访问(即可读可写)网络上的资源。

流套接字是一种面向连接的网络服务，即服务主机与客户机是时时连接的，用于传送的协议是 TCP 协议。

数据报套接字是一种面向无连接的网络服务，程序将要传递的数据打包，用于传送的协议是 UDP 协议。

Java 语言中的网络编程技术，为构建基于 Internet 的 Web 应用、支持客户/服务器应用模型提供了极大的方便。

习　题

13.1　概念题。

(1) Internet 上的主机地址如何表示？

(2) URL 包含哪些部分？URL 与 URLConnection 的区别是什么？

(3) 简述基于流套接字的网络编程工作原理，基于数据报套接字网络编程的工作原理。两者有何区别？

(4) 简述 Applet 的 AppletContect 接口中的方法 showDocument 的作用。

(5) 简述 Socket 类和 ServerSocket 类的作用。

(6) 简述 DatagramPacket 类和 DatagramSocket 类的作用。

13.2　编制程序：要求利用 URL 和 InputStream 对 URL 文件读取并显示。

13.3　设计一个类似于 Internet Explorer 的浏览器。

13.4　用数据报形式改写多用户聊天程序。

13.5　对例 13-8、例 13-9 的网上多用户聊天程序进行修改：使用户能注册用户身份，输入用户身份后才能进入聊天室聊天。

13.6　利用流套接字设计一个网上五子棋游戏。

第14章 JDBC技术和数据库应用开发

Java 语言使用 JDBC(Java Database Connectivity,Java 数据库连接)API 与数据库进行通信,并用它操纵数据库中的数据。这种 API 与特定数据库的驱动程序相分离,使开发人员能够改变底层的数据库,而不必修改访问数据库的 Java 代码。本章首先介绍 JDBC 技术,包括 JDBC 的体系结构、JDBC 的主要类和接口;应用 JDBC 访问数据库的过程;MySQL 数据库服务器的安装、配置和使用,学生信息数据库 study 的创建过程;最后给出了如何应用 JDBC 技术、开发基于 Client/Server 模式的数据库应用实例,详细叙述了实体层、数据库访问层和视图层的程序设计过程。

14.1 JDBC 技术

JDBC 为 Java 定义了一个"调用级"(call-level)的 SQL 接口,通过该接口可执行 SQL 语句并且得到访问结果。通过使用 JDBC,开发人员可以很方便地将 SQL 语句传送给几乎任何一种数据库。

14.1.1 JDBC 的体系结构

JDBC 的体系结构如图 14-1 所示。由图中可以看出,JDBC API 的作用就是屏蔽不同的数据库驱动程序之间的差别,使得程序设计人员有一个标准的、纯 Java 的数据库程序设计接口,为 Java 访问任意类型的数据库提供技术支持。驱动程序管理器(Driver Manager)为应用程序装载数据库驱动程序。数据库驱动程序是与具体的数据库相关,用于向数据库提交 SQL 查询请求。

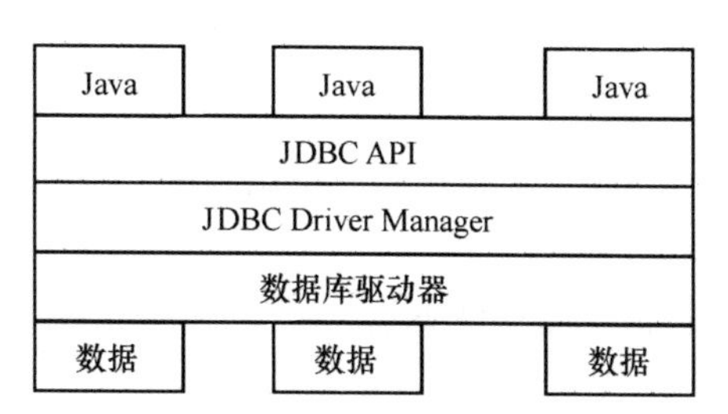

图 14-1 JDBC 的体系结构

14.1.2 JDBC 驱动程序类型

使用 JDBC API 存取特定数据库系统前,必须拥有适当的 JDBC 数据库驱动程序,读者可以在网站 http://www.oracle.com/technetwork/java/javase/jdbc/index.html 查询。

JDBC 支持如下 4 种类型的驱动。

1. JDBC-ODBC bridge plus ODBC driver(类型 1)

JDBC-ODBC 桥接驱动程序。其底层通过 ODBC(Open databas Connectivity)驱动程序来连接数据库。Java 软件开发工具包中包含 JDBC 对 ODBC 驱动程序(sun.jdbc.odbc.jdbcOdbcDriver)。该驱动程序利用 JDBC-ODBC 桥接驱动程序时,客户端必须先安装适当的 ODBC 驱动程序,同时还要配置 ODBC 数据源。Java 应用程序先从 JDBC Driver Manager 驱动 JDBC 驱动程序,再调用 ODBC Driver Manager 去驱动 ODBC 驱动程序连接至数据库。

2. Native-API partly-Java driver(类型 2)

本地 API 部分用 Java 来编写的驱动程序。此种方式先将 JDBC 函数调用转换成数据库客户端函数库的 API(位于客户端计算机),然后与数据库相连。Oracle、Sybase、DB2 等可使用此种驱动程序,要求必须在客户端操作系统内安装特定软件。

3. JDBC-Net pure Java driver(类型 3)

JDBC 网络纯 Java 驱动程序。首先 JDBC 驱动程序会将 JDBC 函数调用解释成与数据库无关的网络通信协议,经过中介服务器的第二次解析,最后才转换成相对应的数据库通信协议,后台数据库发生变化时,只需要更换中介层与数据库之间的 JDBC 驱动程序。

4. Native-protocol pure Java driver(类型 4)

本地协议纯 Java 驱动程序。这种驱动程序将 JDBC 调用直接转换为 DBMS 所使用的网络协议。这将允许从客户机器上直接调用 DBMS 服务器,是 Intranet 访问的一个很实用的解决方法。

14.1.3 JDBC API 的主要类和接口简介

JDBC API 由一系列与数据库访问有关的类和接口组成,它们放在 java.sql 包中。其中主要的类和接口如下所示。

① DriverManager 类: 管理各种数据库驱动程序加载,建立数据库连接。以便将 Java 应用程序对应至正确的 JDBC 驱动程序,DriverManager 允许在内存中同时加载多个 JDBC 驱动程序,分别指向不同数据库。

② Connection 接口 :用来建立数据库连接。通常需传入表示连接方式的 URL 字符串,然后返回一个 Connection 对象;

③ Statement、PreparedStatement 与 CallableStatement 接口:用于处理连接中的 SQL 语句;

④ ResultSet:用于处理数据库操作结果集。

14.2 创建 MySQL 的数据库

MySQL 是最受欢迎的开源关系数据库管理系统,由瑞典 MySQL AB 公司开发,目前属于 Oracle 公司。MySQL 的最新版本可从 http://www.mysql.com/downloads/免费下载,

MySQL 被广泛地应用在 Internet 上的中小型网站中。由于其体积小、速度快、总体拥有

成本低，尤其是开放源码这一特点，使得许多中小型企业的管理信息系统纷纷选择了 MySQL 作为其数据库。

MySQL 使用结构化查询语言 SQL 进行数据库管理和操作。

用户数据库 study 是我们本章要用的数据库，存放着学生的基本信息、课程情况和学生学习各门课程的学习成绩。study 数据库中有学生情况表 student、课程情况表 course 和学习成绩表 sc。student 由学号、姓名、性别、年龄和所在系各字段组成；course 由课程号、课程名和学分各字段组成；sc 由学号、课程号和学习成绩各字段组成。

14.2.1 MySQL 的安装和配置

1. 下载 Windows 版的安装软件包

(1) MySQL 数据库管理软件：mysql-essential-5.1.68-win32.msi(必备)。

(2) MySQL 的 JDBC 驱动程序：mysql-connector-java.jar(必备)。

(3) MySQL 的图形化管理界面软件：mysql-gui-tools-5.0-r17-win32.msi (可选)。

2. 安装和配置 MySQL 软件

双击程序 mysql-essential-5.1.68-win32.msi，将进入安装界面，不断单击[Next]，单击[Finish]后，开始配置服务器，在配置过程中，选择[Standard Configuration]选项；[Character Set]选择字符集为[utf8]；[Modify Secuity Setings]中，设置 root 账号的用户口令，如为 root；其他选项选择默认值则可，直到最后单击[Execute]完成服务器配置。

启动或停止 MySQL 服务器的过程：安装正确完成后，在 Windows 的"管理—>服务"中，会观察到有"MySQL"服务自动启动成功的一栏。可以通过此界面"启动或停止"MySQL 服务器。

为了避免通过 MySQL 命令行窗口或者通过 JSP 访问 MySQL 数据库中的汉字信息时出现乱码，还需配置 MySQL 安装目录下的文件 my.ini，将此文件内容中两处的字符集(character-set)改为 utf8 或者 gb2312。修改完 my.ini 后要重新启动 MySQL 服务器才生效。

MySQL 安装完成后，在 Windows 桌面出现程序项"MySQL—>MySql Server5.1—>MySQL Command Line Client"。双击程序项"MySQL Command Line Client"，打开 MySQL 的 SQL 命令行处理窗口，提示输入先前设置的 Password，如 root，将进入如图 14-2 所示的 DOS 界面，这是输入执行 MySQL 各种命令的窗口。

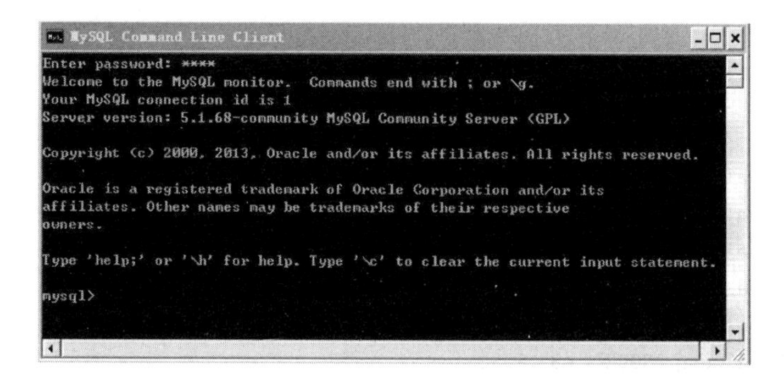

图 14-2　的 DOS 界面(命令行窗口)

mysql-essential-5.1.68-win32.msi 安装完成后，只能以 DOS 命令行方式访问 MySQL 命令。接着可以选择安装 mysql-gui-tools-5.0-r17-win32.msi，安装完成后将出现 MySQL 图形化管理界面。

mysql-connector-java.jar 是 Java 访问 MySQL 的 JDBC 驱动程序，其存放的路径位置应在 classPath 环境变量中进行设置，或在 Eclipse 程序项目的库文件[libraries]中进行设置。

14.2.2 创建数据库 study

先将 study.sql 文件复制到 C 盘根目录下，打开 MySQL 命令行窗口（MySQL Command Line Client），通过运行下面的脚本命令：

```
source c:\study.sql
```

完成创建数据库 study 和三张表 student、course 和 sc。脚本文件见课件中的例子文件 study.sql，其文件内容如下：

```
create database study;
use study;
CREATE TABLE course (
    cno char (2),
    cname char (16),
    credit decimal(2, 1));
insert into  course values('c1','数据库',4);
insert into  course values('c2','数学',2.5);
insert into  course values('c3','信息系统',4);
insert into  course values('c4','英语',3);
CREATE TABLE student (
    sno char (5),
    sname char (10),
    ssex char (2),
    sage int,
    sdept char( 20 ));
insert into student values('20001','张小明','女',23,'信息系');
insert into student values('20002','李强','男',21,'计算机系');
insert into student values('20003','王方','女',28,'信息系');
insert into student values('20004','刘晨','男',18,'计算机系');
CREATE TABLE SC(
  Sno char(5),
  Cno char(2),
  Grade decimal(3,0)
  );
insert into SC values('20001','c1',91);
insert into SC values('20001','c2',92);
insert into SC values('20001','c3',93);
insert into SC values('20001','C4',94);
insert into SC values('20002','c2',81);
insert into SC values('20002','c3',80);
insert into SC values('20004','C1',70);
insert into SC values('20004','C2',71);
insert into SC values('20004','C3',72);
```

14.3 Java 应用程序通过 JDBC 存取数据库的过程

本节首先介绍通过 JDBC 存取数据库的步骤，然后通过一个简单实例演示整个过程。

14.3.1 应用 JDBC 存取数据库的步骤

Java 应用程序通过 JDBC 存取数据库时应该遵循以下 4 个步骤：

- 加载适当的 JDBC 驱动程序，建立数据库连接；
- 建立与执行 SQL 语句；
- 处理结果集；
- 关闭数据库连接。

1. 加载 JDBC 驱动程序，建立数据库连接

使用 Class.forName()加载 JDBC 驱动程序，其语句的语法格式如下：

```
Class.forName(String jdbc_driver);
```

例如，加载 MySQL 驱动程序的语句：

```
Class.forName("com.mysql.jdbc.Driver");
```

使用 DriverManger.getConnection()，建立一个新的数据库连接。其方法格式如下：

```
Connnetion getConnection(String connURl,String loginName,String password);
```

返回连接到特定数据库的 Connnetion 对象。其中 connURl 用来定义连接数据库的参数。

例 14-1 连接 MySQL 的数据库 study，用户名为"root"，口令为"root"，则建立数据库连接的代码如下：

```
try {
    Class.forName("com.mysql.jdbc.Driver ");
    String url = "jdbc:mysql://localhost:3306/study";
    String login = "root", password = = "root";
    Connection conn = DriverManager.getConnection
        (url,login,password);
  }catch (ClassNotFoundException e) {//输出捕获到的异常信息
        e.printStackTrace();
  }catch ( SQLException sqlException ) {
        e.printStackTrace();    //输出捕获到的异常信息
}
```

2. 建立 SQL 语句对象，执行查询

建立数据库连接以后，必须先建立一个 Statement 对象才能执行 SQL 语句。在 Java 中，定义了三种类型的 Statememt：Statement、PreparedStatement 和 CallableStatement。三种类型均包含用于进行数据库操作的 SQL 语句。

(1) 使用 Statement

示例如下：

```
Statement Stmt = conn.createStatement();
String sqlQuery = "select * from Employee";
ResultSet rs = Stmt.executeQuery(sqlQuery);
```

(2) 使用 PreparedStatement

用于处理预编译的 SQL 语句，可重复执行。如果某个 SQL 语句必须重复执行，建议使用

PreparedStatement。示例如下：

```
PreparedStatement prepStmt = conn.prepareStatement
("SELECT * FROM Employee ");
ResultSet   rset = prepStmt.executeQuery();
```

用于带参数的SQL语句的执行，参数用问号(?)表示。下面的示例中，将执行带有两个输入参数的UPDATE语句，其语句如下：

```
PreparedStatement prepStmt = conn.prepareStatement
   ("UPDATE emp SET sal = ? WHERE ename = ? ");
prepStmt.setInt(1,100000);       //设置要传入的第一个 int 型参数值为 100000
prepStmt.setString(2,"Rich");    //设置要传入的第二个字符串参数值为"Rich"
prepStmt.executeUpdate();
```

(3) 使用CallableStatement

使用CallableStatement调用数据库存储过程。语法格式为：

```
CallableStatement 变量名 = conn.prepareCall("call 存储过程名称");
```

下面的示例，将调用带有两个输入参数的存储过程update_salary，代码如下：

```
CallableStatement callStmt = conn.prepareCall("call update_salary(?,?)");
callStmt.setInt(1,7788);  //设置要传入的第一个 int 型输入参数值是 7788
callStmt.setInt(2,10000); //设置要传入的第二个 int 型输入参数值是 10000
callStmt.execute();
```

上面建立Statement对象之后，有3种方法执行SQL语句，分别是：

```
executeQuery();    //执行 select 的 SQL 查询语句
executeUpdate();   //执行 insert、delete、update 更新语句，以及 DDL 语句
execute();         //执行 SQL 查询语句
```

3. 处理结果集

可用while循环打印出ResultSet记录集内的所有记录。例如：

```
while(rs.next()){
    System.out.println(rs.getInt(1));  //1、2 表示各字段相对位置
    System.out.println(rs.getString(2));
  }
```

4. 关闭数据库连接

最后一个操作是关闭Connection、Statement、ResultSet等对象。

```
rs.close();
stmt.close();
con.close();
```

14.3.2 创建MyEclipse的项目实例：完成JDBC访问MySQL数据库study

例14-2 通过JDBC存取MySQL的数据库study的表course的程序。

1. 程序代码如下：

```
//JdbcExample.java
import java.sql.*;
class JdbcExample1 {
    public static void main(String args[]) {
        String driverClassName = "com.mysql.jdbc.Driver";
        String url =
        "jdbc:mysql://localhost:3306/study? useUnicode = true&amp;characterEncoding = UTF-8";
        String username = "root";
```

```
      String password = "root";
    Connection conn = null;
    try {
        Class.forName(driverClassName); //加载 JDBC 驱动程序
        //通过数据源连接
        conn = DriverManager.getConnection(url,username,password);
        //建立 SQL 语句对象
        Statement statement = conn.createStatement();
        String sqlQuery = "select * from course";
        //执行查询
        ResultSet rs = statement.executeQuery(sqlQuery);
        while(rs.next()){ //处理结果集
               System.out.print(rs.getString(1) + ",");
               System.out.print(rs.getString(2) + ",");
               System.out.println(rs.getInt(3));
      }
      //关闭数据库连接
      rs.close();
      statement.close();
        conn.close();
      }
      //一定要捕获异常
      catch ( SQLException e ) {
          e.printStackTrace();  //输出捕获到的异常信息
      }
      catch(Exception e){
          e.printStackTrace();
      }
  } //end of main
}
```

2. 使用 MyEclipse 环境,完成 JDBC 存取 MySQL 的数据库 study,整个项目 ch14 的创建过程如下。

(1) 在 MyEclipse 环境下,创建 Java Project,起名为 ch14。

(2) 为项目 ch14 的库文件[libraries]添加 MySQL 的 JDBC 驱动程序包 mysql-connector-java.jar 的过程:

单击项目 ch14,选择 MyEclipse 主菜单上的"project—>Properties—>Java Build Path—>Libraries",如图所示 14-3,单击"Add External JARs"按钮,出现选择添加外部的 JARs 的界面,选中已经下载好的 MySQL 的 JDBC 驱动程序包,如 mysql-connector-java.jar,然后单击"打开",单击"OK"即可。完成了给项目添加外部包后,检查一下项目 ch14 的 Refrenced Libraries 栏目下是否出现了刚刚加入的 JDBC 驱动程序包,如图 14-4 所示。

(3) 在项目 ch14 中,创建类文件 JdbcExample.java,输入正确的文件内容如前面所示。

(4) 运行项目的 Java 应用程序。选择"Run As—>Java application",将运行 Java 应用程序(即运行主类 JdbcExample.class)。运行结果如下:

```
c1 , 数据库,     4
c2 , 数学,       2.5
c3 , 信息系统,   4
c4 , 英语,       3
```

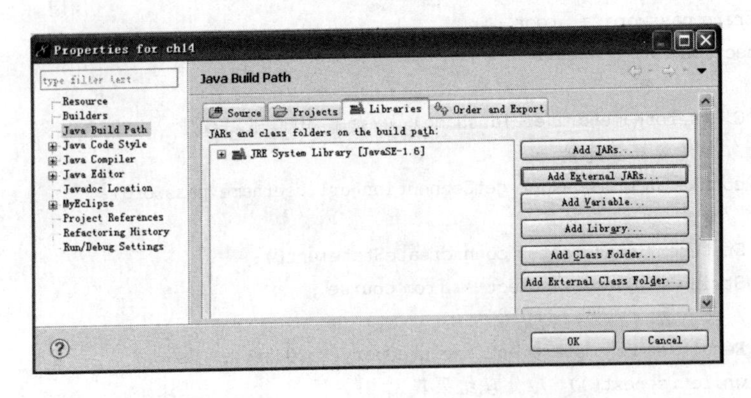

图 14-3 给项目添加 Java 外部 JARs 包界面

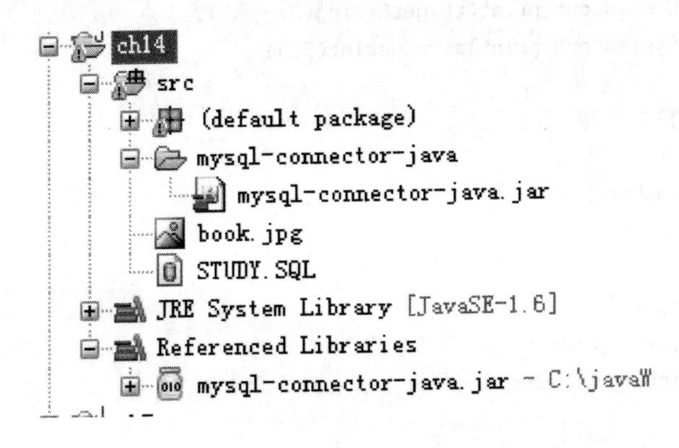

图 14-4 项目 ch14 完成添加了 JDBC 驱动的 Package View

14.4 JDBC 中的主要接口和类

Java 应用程序访问数据库主要是通过 JDBC API 实现的，本节将介绍下面的几个重要接口和类，它们是 DriverManager 类、Connection 接口；Statement 接口、PreparedStatement 接口、CallableStatement 接口；ResultSet 接口、ResultSetMetaData 接口、DatabaseMetaData 接口。并介绍 Java 的基本数据类型和 SQL 中支持的数据类型的对应关系。

14.4.1 DriverManager 类

DriverManager 类是 Java 中用于数据库驱动程序管理的类，作用于用户和驱动程序之间。它跟踪可用的驱动程序，并在数据库和相应驱动程序之间建立连接，也处理诸如驱动程序登录时间限制及登录、跟踪消息的显示等事务。DriverManager 类直接继承自 java.lang.object，其主要成员方法有：

① static Connection getConnection(String url)：通过指定的 URL 创建数据库的连接；

② static Connection getConnection(String url，String user，String password)：通过指定的 URL 及用户名、密码创建数据库的连接。

14.4.2 连接 SQL Server、Oracle 和 Access 数据库的程序举例

例 14-3 通过创建好的数据源名 studyDSN，连接 Microsoft SQL Server2008 或 SQL Server 2008 express 的数据库。

首先创建 ODBC 数据源：用 Windows 控制面板中数据源“ODBC”管理器，创建一个连接到 SQL Server 的数据源，名称为 studyDSN，连接的数据库是 study，用户为 sa，口令为 sa。此数据源名 studyDSN 在程序中使用。则建立数据库连接的代码如下：

```
try {
    String url ="jdbc:odbc:studyDSN";
    String login = "sa"; String password ="sa";
    Class.forName("sun.jdbc.odbc.JdbcOdbcDriver");
    Connection conn = DriverManager.getConnection(url,login,password);
}catch ( SQLException sqlException ) {
        e.printStackTrace();//输出捕获到的异常信息
} catch (ClassNotFoundException e) {
        e.printStackTrace();
}
```

例 14-4 通过 JDBC-ODBC 驱动程序连接 SQL Server 2008 express 版本。但不必事先建立 ODBC 的数据源名。省略了 try-catch 的主要代码如下：

```
Class.forName("sun.jdbc.odbc.JdbcOdbcDriver");
String url ="jdbc:odbc:Driver = {SQL Server};Server = machineName" +"\\" +
    "SQLEXPRESS;Database = study;uid = sa;pwd = sa ";
Connection conn = DriverManager.getConnection(connURL);
```

上述代码中的 machineName 是机器名，也可将 machineName 改为 IP 地址，如 127.0.0.1。

例 14-5 从网站下载 SQL Server 2005 或 SQL Server 2008 的 sqljdbc_2.0 驱动程序包 sqljdbc4.jar，并将此包放在用户应用能访问的类路径或项目的库路径中。机器名为 localhost（本地机器），端口号为 1030，数据库为 study。省略了 try-catch 的主要代码如下：

```
String url = "jdbc:sqlserver://localhost:1030;DatabaseName = study";
String user = "sa",pwd = "sa";
Class.forName("com.microsoft.sqlserver.jdbc.SQLServerDriver");
Connection conn = DriverManager.getConnection(url,user,password);
```

例 14-6 连接 Oracle 数据库。

从网站下载 Oracle 驱动程序包 classes12.zip，并将此包的路径放在 classpath 环境变量中或放在项目的库路径中。连接数据库 ORCL，机器的 ip 为 192.168.0.20，端口号为 1521。省略了 try-catch 的主要代码如下：

```
String url = "jdbc:oracle:thin:@192.168.0.20:1521: ORCL";
Class.forName("oracle.jdbc.driver.OracleDriver");
Connection conn = DriverManager.getConnection(url,user,password);
```

例 14-7 通过 JDBC-ODBC 连接 access 的数据库 study.mdb。省略了 try-catch 的主要代码如下：

```
Class.forName("sun.jdbc.odbc.JdbcOdbcDriver");
String url ="jdbc:odbc:driver = {Microsoft Access Driver ( * .mdb)};DBQ = study.mdb"; Connection
conn = DriverManager.getConnection(url);
```

14.4.3 Connection 接口

Connection 接口是用来表示数据库连接的对象，对数据库的一切操作都是在这个连接的

基础上进行的。Connection 接口的主要成员方法包括以下几种。

① void clearWarnings()：清除连接的所有警告信息；

② Statement createStatement()：创建一个 Statement 对象；

③ Statement createStatement(int resultSetType, int resultSetConcurrency)：创建一个 Statement 对象，它将生成具有特定类型和并发性的结果集。

其中，resultSetType 有以下几种值：

- TYPE_FORWARD_ONLY 结果集不可滚动；
- TYPE_SCROLL_INSENSITIVE 结果集可滚动，不反映数据库的变化；
- TYPE_SCROLL_SENSITIVE 结果集可滚动，反映数据库的变化；

resultSetConcurrency 有以下几种值：

- CONCUR_READ_ONLY 不能用结果集更新数据；
- CONCUR_UPDATABLE 能用结果集更新数据。

④ boolean getAutoCommit()：返回 Connection 类对象的 AutoCommit 状态；

⑤ void setAutoCommit(boolean autoCommit)：设定 Connection 类对象的 AutoCommit 状态，为 true 时，SQL 语句组被提交作为一个单独的事务；否则 SQL 语句组的事务分组由 commit 或 roolback 语句确定；

⑥ void commit()：提交对数据库的改动并释放当前连接持有的数据库的锁；

⑦ void rollback()：回滚当前事务中的所有改动并释放当前连接持有的数据库的锁；

⑧ 10、void setReadOnly()：设置连接的只读模式；

⑨ void close()：立即释放连接对象的数据库和 JDBC 资源；

⑩ boolean isClosed()：测试是否已经关闭 Connection 类对象对数据库的连接；

⑪ DatabaseMetaData getMetaData()：建立 DatabaseMetaData 类对象；

⑫ PreparedStatement prepareStatement(String sql)：建立 PreparedStatement 类对象。

14.4.4 Statement 接口

Statement 接口用于在已经建立连接的基础上向数据库发送 SQL 语句的对象。有 3 种类型的 Statement 对象：Statement、PreparedStatement（继承自 Statement）和 CallableStatement（继承自 PreparedStatement）。它们都作为在给定连接上执行 SQL 语句的容器，每个都专用于发送特定类型的 SQL 语句：Statement 对象用于执行不带参数的简单 SQL 语句；PreparedStatement 对象用于执行带或不带 IN 参数的预编译 SQL 语句；CallableStatement 对象用于执行对数据库中存储过程的调用。

创建 statement 对象的方法如下：

```
Statement stmt = conn.createStatement();
```

Statement 接口中包括的主要方法如下。

① ResultSet executeQuery(String sql)：使用 SELECT 命令对数据库进行查询，返回结果集。

② int executeUpdate(String sql)：使用 INSERT、DELETE、UPDATE 对数据库进行新增、删除和修改记录操作。

③ boolean execute(String sql)：执行 SQL 语句。

④ void close()：结束 Statement 类对象对数据库的联机。

⑤ void addBatch(String sql) ：在 Statement 语句中增加用于数据库操作的 SQL 批处理语句。

⑥ Connection getConnection()：获取对数据库的连接。

值得注意的是，Statement 接口提供了 3 种执行 SQL 语句的方法：executeQuery、executeUpdate 和 execute。使用哪一个方法由 SQL 语句所产生的内容决定。executeQuery 方法用于产生单个结果集的 SQL 语句，如 SELECT 语句。executeUpdate 方法用于执行 INSERT、UPDATE、DELETE 及 DDL（数据定义语言）语句，例如 CREATE TABLE 和 DROP TABLE。executeUpdate 的返回值是一个整数，表示它执行的 SQL 语句所影响的数据库中的表的行数（更新计数）。execute 方法用于执行返回多个结果集或多个更新计数的语句。

14.4.5 PreparedStatement 接口

PreparedStatement 接口和 Statement 类的不同之处：PreparedStatement 类对象会将传入的 SQL 命令事先编译等待使用，如果有单一的 SQL 指令需多次执行时，用 PreparedStatement 会比 Statement 效率更高。

PreparedStatement 接口的主要方法如下。

① ResultSet executeQuery()：使用 SELECT 命令对数据库进行查询。

② int executeUpdate() ：使用 INSERT\DELETE\UPDATE 对数据库进行新增、删除和修改操作。

③ ResultSetMetaData getMetaData()：取得 ResultSet 类对象有关字段的相关信息。

④ void setXXX (int parameterIndex，XXX x) ：给 PreparedStatement 类对象的 IN 参数设定为 XXX 类型的值。这里 XXX 可为 int、float、String 、Date 和 Time。

⑤ void setArray(int index，Array x)：设置为数组类型。

⑥ void setAsciiStream(int index，InputStream stream，int length)：设置为 ASCII 输入流。

⑦ void setBinaryStream(int index，InputStream stream，int length)：设置为二进制输入流。

⑧ void setCharacterStream(int index，InputStream stream，int length)：设置为字符输入流。

PreparedStatement 构造的 SQL 语句往往不是完整的语句，需要在程序中进行动态设置。这增强了程序设计的灵活性。

PreparedStatement 对象的创建方法如下：

```
PreparedStatement pstmt = con.prepareStatement("update tbl_User set reward = ? where userId = ?");
```

在该语句中，包括两个可以进行动态设置的字段：reward 和 userId。

如果我们想给第一个注册的用户 5000 点奖励，则可以用下面的方法设置两字段的内容：

```
pstmt.setInt(1, 5000);
pstmt.setInt (2, 1);
```

如果我们想给前 50 个注册的用户每人 5000 点奖励，则可以用循环语句对两字段进行设置：

```
pstmt.setInt(1, 5000);
for (int i = 0; i < 50; i ++ )
```

```
{
  pstmt.setInt(2,i);
  int rowCount = pstmt.executeUpdate();
}
```

如果传递的数据量很大，则可以通过将 IN 参数设置为 Java 输入流来完成。当语句执行时，JDBC 驱动程序将重复调用该输入流，读取其内容并将它们当作实际参数数据传输。JDBC 提供了 3 种将 IN 参数设置为输入流的方法：setBinaryStream 用于含有字节的流；setAsciiStream 用于含有 ASCII 字符的流；setUnicodeStream 用于含有 Unicode 字符的流。这些方法比其他的 setXXX 方法要多一个用于指定流的总长度的参数，因为一些数据库在发送数据之前需要知道它传送的数据的大小。

下面是一个使用流作为 IN 参数发送文件内容的例子：

```
java.io.File file = new java.io.File("/tmp/data");
int fileLength = file.length();
java.io.InputStream fin = new java.io.FileInputStream(file);
java.sql.PreparedStatement pstmt
    = con.prepareStatement("update table set stuff = ? where index = 4");
pstmt.setBinaryStream (1, fin, fileLength);
pstmt.executeUpdate();
```

当语句执行时，将反复调用输入流 fin 以传递其数据。

14.4.6 CallableStatement 接口与 Java 执行 MySQL 存储过程的程序举例

CallableStatement 接口用于执行对数据库中存储过程的调用。在 CallableStatement 对象中，有一个通用的成员方法 call，这个方法以名称的方式调用数据库中的存储过程。在数据库调用过程中，可以通过设置 IN 参数向调用的存储过程提供执行所需的参数。另外，在存储过程的调用中，通过 OUT 参数获取存储过程的执行结果。

创建 CallableStatement 对象的方法如下：

```
CallableStatement cstmt =
    con.prepareCall("call procedure_name (?,?,?,…)");
```

其中(?,?,?,…)定义了存储过程的输入参数和(或)输出参数。向存储过程传递输入参数(IN)值的方法是通过 setXXX 语句完成的。而存储过程的输出参数(output)，则需要调用 registerOutParameter 方法设置存储过程的 OUT(输出)参数，然后调用 getXXX 方法来获取输出参数的值结果。

例 14-8 利用 JDBC 接口调用 Mysql 的存储过程 Query_Student 和 Query_Study1。

(1) 打开 MySQl Command Client Line 窗口，分别执行下面的脚本内容，将在数据库 study 中创建存储过程 Query_Student 和 Query_Study1。

创建存储过程 Query_Study1 的脚本内容如下：

```
use study;
DELIMITER  //
CREATE PROCEDURE Query_Study1 ()
BEGIN
SELECT sname,sdept,sage
FROM student;
END
//
```

创建存储过程 Query_Student 的脚本内容如下：

```
DELIMITER  //
CREATE PROCEDURE Query_Student(
IN p_no char(6),
OUT p_name char(20) ,
OUT p_dept char(10) )
BEGIN
  SELECT sname ,sdept into p_name, p_dept
  FROM student
  WHERE sno = p_no;
END
//
```

(2) 编写 JdbcCallProcedure. java 的代码如下：

```
import java.sql. * ;
import javax.swing. * ;
class jdbcCallProcedure {
     public static void main(String args[]) {
     String driverClassName = "com.mysql.jdbc.Driver";
     String url =
    "jdbc:mysql://localhost:3306/study? useUnicode = true&amp;characterEncoding = UTF-8";
     String username  =  "root";      String password  =  "root";
     try {
           Class.forName(driverClassName).newInstance();
           Connection conn  =  DriverManager.getConnection
(url + "&user = " + username + "&password = " + password);
           //带有入口参数和出口参数的存储过程的调用,select 的结果是单值
           CallableStatement callStmt =
             conn.prepareCall("{call Query_Student(?,?,?)}");
           //设置要传入的第一个输入参数(?)值是 20001
           callStmt.setString(1,"20001");
           //注册输出参数
           //callStmt.registerOutParameter(1,java.sql.Types.FLOAT);
           callStmt.registerOutParameter(2,java.sql.Types.VARCHAR);
           callStmt.registerOutParameter(3,java.sql.Types.VARCHAR);
           callStmt.execute();
           String   b1 = callStmt.getString(2);
           String   b2 = callStmt.getString(3);
           System.out.println("调用存储过程 Query_Student 的结果:");
           System.out.println(b1 + "," + b2);
           //存储过程的调用,select 的结果是结果集合
           System.out.println("调用存储过程 Query_Study1 的结果:");
           callStmt =
             conn.prepareCall("{call Query_Study1}");
           callStmt.execute();
           ResultSet rs  =  callStmt.getResultSet();
           while (rs.next()) {
           System.out.println(rs.getString(1)  +  "\t"  +  rs.getString(2)  +  "\t"  +  rs.getInt(3));
           }
           callStmt.close();
           conn.close();
    }         //一定要捕获异常
```

```
    catch ( SQLException e ) {
        e.printStackTrace();   //输出捕获到的异常信息
    }
   catch(Exception e){
        e.printStackTrace();
    }
   } //end of main
}
```

程序运行的输出结果如下：

```
调用存储过程Query_Student的结果:
张小明,信息系
调用存储过程Query_Study1的结果:
张小明    信息系   23
李强      计算机系 21
王方      信息系   28
刘晨      计算机系 18
```

14.4.7 Java 数据类型和 SQL 中支持的数据类型的对应关系

从上面的程序可以看出，Java 的基本数据类型和 SQL 中支持的数据类型有一定的对应关系。这种对应关系如表 14-1 所示。

表 14-1 SQL 的数据类型与 Java 数据类型的对应关系

SQL 数据类型	Java 数据类型	SQL 数据类型	Java 数据类型
CHAR	String	REAL	float
VARCHAR	String	FLOAT	float
LONGVARCHAR	String	DOUBLE	double
NUMERIC	java. math. BigDecimal	BINARY	byte[]
DECIMAL	java. math. BigDecimal	VARBINARY	byte[]
BIT	boolean	LONGVARBINARY	byte[]
TINYINT	byte	DATE	java. sql. Date
SMALLINT	short	TIME	java. sql. Time
INTEGER	int	TIMESTAMP	java. sql. Timestamp
BIGINT	long		

14.4.8 ResultSet 接口

结果集(ResultSet)是存储 SQL 查询结果的对象，通过维护记录指针(Cursor)的移动，能操作结果集中的每一行记录，并提供一系列的方法对数据库进行新增、删除和修改操作。

ResultSet 接口主要方法有以下几种。

① boolean next()：移动记录指针到结果集的下一条记录，若移动成功返回 true。

② 类型 get 类型(int columnIndex)或者类型 get 类型(String columnName)：读取结果集中当前行的指定字段的值。

③ boolean absolute(int row)：移动记录指针到结果集的指定的记录。

④ void beforeFirst()：移动记录指针到结果集的第一条记录之前。

⑤ void afterLast():移动记录指针到结果集的最后一条记录之后。

⑥ boolean first():移动记录指针到结果集的第一条记录。

⑦ boolean last():移动记录指针到结果集的最后一条记录。

⑧ boolean previous():移动记录指针到结果集的上一条记录。

⑨ ResultSetMetaData getMetaData():取得 ResultSetMetaData 类对象。

14.4.9 ResultSetMetaData 接口

ResultSetMetaData 接口的对象保存了所有 ResultSet 类对象中关于字段的元信息,并提供许多方法来取得这些信息。

ResultSetMetaData 接口中的主要方法如下。

① int getColumnCount(): 取得 ResultSet 对象的字段个数。

② int getColumnDisplaySize():取得 ResultSet 对象的字段长度。

③ String getColumnName(int column):取得 ResultSet 对象的字段名称。

④ String getColumnTypeName(int column):取得 ResultSet 对象的字段类型名称。

⑤ String getTableName(int column) :取得 ResultSet 对象的字段所属数据表的名称。

14.4.10 DatabaseMetaData 接口

DatabaseMetaData 接口保存了数据库的所有特性,并且提供许多方法来取得这些信息。

DatabaseMetaData 类的主要方法如下。

① String getDatabaseProductName() :取得数据库名称。

② String getURL() :取得连接数据库的 JDBC URL。

③ String getUserName():取得登录数据库的使用者账号。

14.5 基于 C/S 模式的学生信息数据库管理系统的开发

例 14-9 利用前面建立的 MySQL 的数据库 study,对数据库中表 student 进行插入、修改、删除和浏览。

开发学生信息管理系统时,程序设计时的考虑原则包括以下两个方面。

(1) 数据库应用程序架构上,采用 C/S(Clieng/Server)的多层设计模式,使程序结构清楚,各层功能相互独立,各个层次只能对其下层进行调用,使整个系统易于维护和修改。

下面以例 14-9 的程序设计为例,说明如何应用 C/S 架构的过程。

整个学生信息管理系统分为三层结构(从上层到下层)。

第 1 层,视图层 (用户界面层):用于管理信息的显示,是与用户交互的图形界面。

由类 MainFrame. java、类 AddStudentPanel. java、类 UpdateStudentPanel、类 DeleteStudentPanel. java、类 ListStudentsPanel. java 和类 StudentUI. java 组成。

第 2 层,数据库访问层:完成与数据库的所有操作。

由类 StudentManager 和类 DBConnection 组成,它们封装了 student 表的所有操作。其中,StudentManager 类定义了插入、修改、删除和查询表 student 的所有操作。而 DBConnection. java 定义了访问 MySql 数据库的连接操作,如果要连接其他类型的数据库,只要修改此类中的少量代码即可。

第 3 层,实体层:对数据库的每个对象基表,都定义了 Java 中对应的实体类。

定义了实体类 Student. java,对应着数据库中的表 student。

(2) 通用性原则,应用通用性的程序设计技术。其好处是:可以利用本节示范的程序设计例子很容易地编写 Java 访问其他数据库的应用程序。

14.5.1 创建实体层 Bean

针对 study 数据库中的表 student,该表包含的字段有:学号 Sno、姓名 Sname、性别 Ssex、年龄 Sage 和所在系 Sdept。编写 student 对应的实体类代码如下。

```
//Student.java
public class Student {
    private String sno;
    private String sname;
    private String ssex;
    private  int sage;
    private String Sdept;
    public String getSno() {
        return sno;
    }
    public void setSno(String sno) {
        this.sno = sno;
    }
    public String getSname() {
        return sname;
    }
    public void setSname(String sname) {
        this.sname = sname;
    }
    public String getSsex() {
        return ssex;
    }
    public void setSsex(String ssex) {
        this.ssex = ssex;
    }
    public int getSage() {
        return sage;
    }
    public void setSage(int sage) {
        this.sage = sage;
    }
    public String getSdept() {
        return Sdept;
    }
    public void setSdept(String sdept) {
        Sdept = sdept;
    }
}
```

14.5.2 创建数据库访问层:插入、修改、删除和浏览

访问数据库层由 StudentManager. java 和 DBConnection. Java 组成。

1. DBConnection.java 的程序设计

DBConnection 类定义了访问 MySQL 数据库的打开连接和关闭操作，如果要连接其他类型的数据库，只要修改此类中的变量 driverClassName 和 url 的值就可以了，而不影响整个系统的其他代码。DBConnection 类的代码编写如下。

```
/DBConnection.java
//Note 加载 Mysql JdbcDrive
import java.io.IOException;
import java.io.InputStream;
import java.sql.*;
import javax.swing.JOptionPane;
public class DBConnection{
    static  Connection conn = null;
    private static String username ="root";
    private static String password ="root" ;
    private static String driverClassName ="com.mysql.jdbc.Driver";
    private static String url =
"jdbc:mysql://localhost:3306/study? useUnicode=true&amp;characterEncoding=UTF-8";
    public  Connection getConnection()  {
    try {
            Class.forName(driverClassName);
            //或者 Class.forName(driverClassName).newInstance();
            conn = DriverManager.getConnection(url,username,password);
      }
      catch(Exception e){
        e.printStackTrace();
      }
      return conn;
  }
  public void closeConnection() throws SQLException {
        if (conn!= null) {
            try {
                conn.close();
            } catch (SQLException e) {
                e.printStackTrace();
            }
        }
    }
}
```

2. StudentManager.java 的程序设计

StudentManager 类封装了访问 student 表的插入、修改、删除和查询操作。此类的设计思路如下。

(1) 继承父类 DBConnection，即子类的连接和关闭操作来自于父类。

(2) 在构造方法中调用父类的连接方法，即在创建 StudentManager 对象时，完成连接数据库操作。类的域变量 Connection conn 在各方法中共享。

(3) 方法 int execUpdate(String sql)：定义了对 student 表进行的更新操作，包括插入、修改、删除。入口参数是 SQL 语句的字符串，返回更新操作的结果，返回结果大于 0 则更新成功，为 0 则更新不成功，并及时关闭数据库连接。

(4) 方法 List<Student> Query(String sqlString)：定义了对 student 表查询的结果，以

List 结构返回结果，并及时关闭数据库连接。

StudentManager 类的使用例子：对 student 插入一条记录的操作代码。

```
String sqlString = "INSERT INTO student VALUES
    ('2001','张红','女',22,'计算机系')";
StudentManager studentmanager = new  StudentManager();
int result = studentmanager.execUpdate(sqlString);
```

StudentManager.java 代码如下：

```
import java.sql.*;
import java.util.ArrayList;
import java.util.List;
public class StudentManager extends DBConnection {
    private  Connection conn = null;
    //构造方法中完成打开数据库连接
    public StudentManager() {
       conn = getConnection();
    }
//对 student 表进行插入、修改、删除操作
public int execUpdate(String sql)throws SQLException {
        int result = 0;
        Statement statement = null;
        try {
                  statement  =  conn.createStatement();
                  result  =  statement.executeUpdate(sql);
          }
          catch (SQLException e)
          { System.out.println(e); }
          finally {
          if (statement!= null)
              statement.close();
          closeConnection();
        }
        return result;
}
//对 student 表行查询，并以 List 结构返回结果，并及时关闭数据库连接
public List<Student> Query(String sqlString) throws SQLException
{       List<Student> list  =  new ArrayList<Student>();
        Statement statement = null;
        ResultSet rs = null;
          try {
              statement  =  conn.createStatement();
               rs =  statement.executeQuery(sqlString);
              while(rs.next()) {
                     Student m = new Student();
                     m.setSno(rs.getString("sno"));
                     m.setSname(rs.getString("sname"));
                     m.setSsex(rs.getString("ssex"));
                     m.setSage(rs.getInt("sage"));
                     m.setSdept(rs.getString("Sdept"));
                     list.add(m);
              }
        }
```

```
        catch (SQLException e)
        { System.out.println(e); }
        finally {
          if (statement != null)
            statement .close();
          if (rs!= null)
            rs.close();
          closeConnection();
        }
      return list;
    }
  }
```

14.5.3 创建用户图形界面层:主窗口、主菜单、插入、修改、删除和浏览

用户图形界面层 GUI,由主类 Main.java、主窗口主菜单 MainFrame.java、插入记录界面 AddStudentPanel.java、修改记录界面 UpdateStudentPanel、删除记录界面 DeleteStudentPanel.java 和浏览学生信息界面 ListStudentsPanel.java 和公用界面 StudentUI.java 组成。其中 StudentUI.java 定义了一个学生信息的显示界面,可作为图形组件被组合在插入、修改和删除的图形界面中。项目的完整代码见课件例子。下面分别介绍它们的程序设计特点。

1. 主类 Main.java 和主窗口主菜单 MainFrame.java 的程序设计

主类 Main.java 是包含 main 方法,是整个项目执行的入口点。Main 主类中,主要完成创建学生管理系统的主窗口对象,并使主窗口在屏幕居中显示。Main 主类代码如下。

```
import java.awt. * ;
//创建系统主类
class Main  {
    public static void main(String[] args)
    {   // 创建主界面,并使主窗口在屏幕居中
        MainFrame frame = new MainFrame();
        // 获取屏幕尺寸
        Dimension screenSize = Toolkit.getDefaultToolkit().getScreenSize();
        Dimension frameSize = frame.getSize();// 获取主界面的窗体尺寸
        // 令主界面窗体居中
        if (frameSize.height > screenSize.height)
           frameSize.height = screenSize.height;
        if (frameSize.width > screenSize.width)
           frameSize.width = screenSize.width;
           frame. setLocation ((screenSize. width - frameSize. width)/2, (screenSize. height - frameSize.height) / 2);
           //frame.setResizable(false);
           frame.setVisible(true);// 令主界面显示
  }
}
```

MainFrame.java 定义了主窗口,在主窗口中创建显示主菜单,并给各菜单项注册事件监听处理对象。此程序设计界面的运行结果如图 14-5 所示。MainFrame 主窗口的设计思路如下。

(1) 从 JFrame 继承,实现 ActionListener 接口,它是所有菜单项的事件处理接口。

(2) MainFrame 内容面板的布局设计为 BorderLayout 方位布局,且加入的组件都显示在

布局对象的中心位置 BorderLayout. CENTER。

(a) 主窗口开始显示时,"book.jpg"图像标签组件显示在内容面板上,且放在方位布局对象的中心位置 BorderLayout. CENTER。

(b) 当用户单击某个菜单项时,则内容面板上显示对应的 JPanel 组件。例如,当用户单击"学生管理"菜单中的菜单项"添加学生"时,则在内容面板上被设置为添加学生界面 AddStudentPanel 对象,且显示在方位布局的中心位置。实现的代码如下。

```
AddStudentPanel addPanel = new AddStudentPanel();   //创建添加学生面板对象
this.remove(this.getContentPane());                 //移除主界面上原有的内容
this.setContentPane(addPanel);                      //设置内容面板为 addPanel 对象
this.setVisible(true);                              // 设置界面可见
```

(c) 由于给各菜单项注册的是同一个事件监听处理对象,所以在 actionPerformed 事件处理方法中必须判断用户当前单击的哪个菜单项(事件源),如 if (actionEvent. getSource() == jMenuItem1),则转向"退出"菜单项的处理。

图 14-5 学生信息管理系统的主界面

MainFrame 主窗口类的源代码如下:

```
import java.awt. * ;
import java.awt.event. * ;
import javax.swing. * ;
//创建主界面类
public class MainFrame extends JFrame implements ActionListener
    {    // 初始化主菜单
         JMenuBar jMenuBar1 = new JMenuBar();
         JMenu jMenuFile = new JMenu("文件");
         JMenuItem jMenuFileExit = new JMenuItem("退出");
         JMenu jMenu1 = new JMenu("学生管理");
         JMenuItem jMenuItem1 = new JMenuItem("添加学生");
```

```
JMenuItem jMenuItem2 = new JMenuItem("删除学生");
JMenuItem jMenuItem3  = new JMenuItem("修改学生信息");
JMenu jMenu2 = new JMenu("学生查询");
JMenuItem jMenuItem4 = new JMenuItem("浏览全部学生");
// 构造方法,界面初始化
public MainFrame() {
    // 创建内容面板,设置其布局
    JPanel contentPane = (JPanel) getContentPane();
    contentPane.setLayout( new BorderLayout() );
    // 框架的大小和其标题
    setSize(new Dimension(450, 450));
    setTitle("学生信息管理系统");
    // 添加菜单条
    setJMenuBar(jMenuBar1);
    // 添加菜单组件到菜单条
    jMenuBar1.add(jMenuFile);
    jMenuBar1.add(jMenu1);
    jMenuBar1.add(jMenu2);
    jMenuBar1.add(jMenuFileExit);
    // 添加菜单项组件到菜单组件
    jMenuFile.add(jMenuFileExit);
    jMenu1.add(jMenuItem1);
    jMenu1.add(jMenuItem2);
    jMenu1.add(jMenuItem3);
    jMenu2.add(jMenuItem4);
    // 给菜单添加事件监听器
    jMenuFileExit.addActionListener(this);
    jMenuItem1.addActionListener(this);
    jMenuItem2.addActionListener(this);
    jMenuItem3.addActionListener(this);
    jMenuItem4.addActionListener(this);
    //jMenuItem5.addActionListener(this);
    // 添加图像标签到内容面板
     Icon bug = new ImageIcon(getClass().getResource("book.jpg" ));
     JLabel label = new JLabel( bug, SwingConstants.CENTER );
     contentPane.add(label,BorderLayout.CENTER);
    // 关闭框架窗口时的默认事件方法
    setDefaultCloseOperation(EXIT_ON_CLOSE);
}
// 菜单事件的处理方法
public void actionPerformed(ActionEvent actionEvent)
    {  // 单击“文件”菜单下的“退出”菜单项
       if (actionEvent.getSource() = = jMenuFileExit)  {
                 System.exit(0);
       }
       // 单击“学生管理”菜单下的“添加学生”菜单项
       if (actionEvent.getSource() = = jMenuItem1)
       {  // 创建添加学生面板对象
             AddStudentPanel add = new AddStudentPanel();
             // 移除主界面上原有的内容
             this.remove(this.getContentPane());
             this.setContentPane(add);
```

```
            this.setVisible(true); // 令界面可见
        }
        // 删除学生
        if (actionEvent.getSource() == jMenuItem2){
            DeleteStudentPanel delete = new DeleteStudentPanel();
            this.remove(this.getContentPane());
            this.setContentPane(delete);
            this.setVisible(true);
        }
        // 修改学生
        if (actionEvent.getSource() == jMenuItem3 )  {
            UpdateStudentPanel mod = new UpdateStudentPanel();
            this.remove(this.getContentPane());
            this.setContentPane(mod);
            this.setVisible(true);
        }
        // 学生信息全部浏览
        if (actionEvent.getSource() == jMenuItem4)  {
            ListStudentsPanel list = new ListStudentsPanel();
            this.remove(this.getContentPane());
            this.setContentPane(list);
            this.setVisible(true);
          }
    }
}
```

2. 编辑学生信息界面的通用类 StudentUI 的程序设计

类 StudentUI 是一个定制的 JPanel，完成学生信息的编辑界面的设计，它是输入记录、修改记录和显示记录的公共的界面部分，被组合在输入界面类 AddStudentPanel、修改界面类 UpdateStudentPanel 和删除界面类 DeleteStudentPanel 中。类 StudentUI 完成的界面设计的效果如图 14-6 所示。通用性体现在：画面记录的列标题可变，下面两个按钮的文本提示也可变，且两个按钮的事件处理也可变。

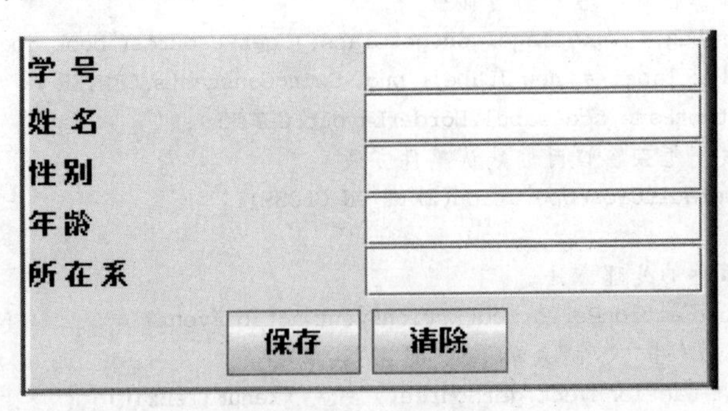

图 14-6　StudentUI 类的界面设计效果

类 StudentUI 的设计思路如下。

(1) 类 StudentUI 是一个定制的 JPanel，布局管理器为 BoderLayout 方位：在 Center 方位加入 innerPanelCenter 学生编辑界面组件，在 SOUTH 方位加入 innerPanelSouth 两个按钮界面。

(2) 调用者利用 StudentUI 类构造方法的参数(字符型数组 String arrayString[])，传递编辑学生记录界面上的一组列的标题。在构造方法中，从数组 arrayString 的属性 length 中可得到列的个数，并取出列的标题放在 labels 数组中。

(3) 定义一组文本字段 fields[]用于编辑一张表的记录各字段，并且调用者通过方法 getFields()和 setFields()可读取和设置 fields 值。

(4) 类 StudentUI 界面上提供两个通用 Button 按钮 doTask1、doTask2，按钮的标签内容和按钮的事件处理过程，可通过调用者来设置。

类 StudentUI 的程序代码如下。

```
//StudentUI.java
import java.awt.*;
import javax.swing.*;
import javax.swing.border.*;
public class StudentUI extends JPanel {
  // GUI components; protected for future subclasses access
  protected JLabel labels[];
  protected JTextField fields[];
  protected JButton doTask1, doTask2;
  protected JPanel innerPanelCenter, innerPanelSouth;
  protected int size; // number of text fields in GUI
  // constants representing text fields in GUI
  public static final int SNO = 0, SNAME = 1, SSEX = 2,
    SAGE = 3, SDEPT = 4;
  //  构造方法的参数是字符型数组参数,其值为 GUI 的各字段列标题
  public StudentUI( String arrayString[] )   {
     size = arrayString.length;
     labels = new JLabel[ size ];
     fields = new JTextField[ size ];
     // create labels
     for ( int count = 0; count < labels.length; count ++ )
        labels[ count ] = new JLabel( arrayString[ count ] );
     // create text fields
     for ( int count = 0; count < fields.length; count ++ )
        fields[ count ] = new JTextField();
     // create panel to lay out labels and fields
     innerPanelCenter = new JPanel();
     innerPanelCenter.setLayout( new GridLayout( size, 2 ) );
     // attach labels and fields to innerPanelCenter
     for ( int count = 0; count < size; count ++ ) {
        innerPanelCenter.add( labels[ count ] );
        innerPanelCenter.add( fields[ count ] );
     }
     // create generic buttons; no labels or event handlers
     doTask1 = new JButton();  doTask2 = new JButton();
     // create panel to lay out buttons and attach buttons
     innerPanelSouth = new JPanel();
     innerPanelSouth.add( doTask1 );
     innerPanelSouth.add( doTask2 );
     // set layout of this container and attach panels to it
     setLayout( new BorderLayout() );
     add( innerPanelCenter, BorderLayout.CENTER );
     add( innerPanelSouth, BorderLayout.SOUTH );
```

```
    setBorder(BorderFactory.createBevelBorder(BevelBorder.RAISED));
    validate(); // validate layout
  } // end constructor
  // return reference to generic task button doTask1
  public JButton getDoTask1Button()    {
     return doTask1;
  }
  // return reference to generic task button doTask2
  public JButton getDoTask2Button()    {
     return doTask2;
  }
  // return reference to fields array of JTextFields
  public JTextField[] getFields()    {
     return fields;
  }
  // clear content of text fields
  public void clearFields()    {
    for ( int count = 0; count < size; count ++ )
        fields[ count ].setText( "" );
  }
  // set text field values; throw IllegalArgumentException if
  // incorrect number of Strings in argument
  public void setFieldValues( String strings[] )
    throws IllegalArgumentException
  {  if ( strings.length != size )
        throw new IllegalArgumentException( "There must be " +
           size + " Strings in the array" );
    for ( int count = 0; count < size; count ++ )
        fields[ count ].setText( strings[ count ] );
  }
  // get array of Strings with current text field contents
  public String[] getFieldValues()    {
    String values[] = new String[ size ];
    for ( int count = 0; count < size; count ++ )
        values[ count ] = fields[ count ].getText();
    return values;
  }
} // end class BankUI
```

3. 添加学生记录界面的程序设计

类 AddStudentPanel 是一个定制的 JPanel，且设置为 BoderLayout 布局。在此类中创建 StudentUI 对象并将它添加到方位布局的北方位置。在创建 StudentUI 对象时，调用构造方法，将要显示的记录的一组列标题传递给 StudentUI 对象；并从 StudentUI 对象中得到用户输入的记录的各字段值，两个通用按钮的内容和事件处理代码在此类中得到设置。

当用户单击“添加学生”菜单项时，程序执行后的显示结果如图 14-7 所示。左边图是显示输入数据的界面，用户可输入记录的各字段值；单击“保存”按钮后，将记录成功添加到数据库中，并出现右图成功添加的提示信息。若添加记录不成功，将显示各种原因提示信息。

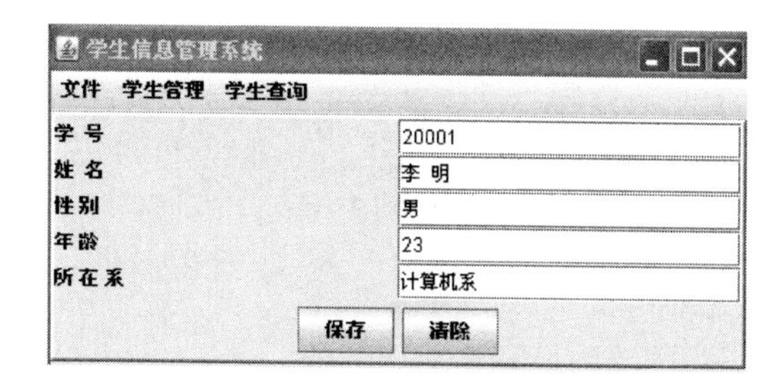

图 14-7 增加记录的程序运行结果

类 AddStudentPanel 的代码如下。

```
import java.awt.*;
import java.awt.event.*;
import java.sql.*;
import java.util.*;
import javax.swing.*;
public class AddStudentPanel extends JPanel {
    private JButton clearButton, writeButton;
    private StudentUI userInterface;
    String names1[] = {"学  号","姓  名","性 别","年 龄","所 在 系"};
  // set up GUI
  public AddStudentPanel(){
     setLayout( new BorderLayout() );
     userInterface = new StudentUI( names1 );
     this.add( userInterface, BorderLayout.NORTH );
     // configure button doTask1 for use in this program
     writeButton = userInterface.getDoTask1Button();
     writeButton.setText( "保存" );
     // register listener to call addRecord when button pressed
     writeButton.addActionListener(
        // anonymous inner class to handle writeButton event
        new ActionListener() {
           public void actionPerformed( ActionEvent event )
           {  String fieldValues[] = userInterface.getFieldValues();
              if ( ! fieldValues[ StudentUI.SNO ].equals( "" ) ) {
              // output values to student
              try {
                 int numberAge = Integer.parseInt(
                 fieldValues[ StudentUI.SAGE ] );
                 //define string for sql insert  statement
                 String sqlString = "INSERT INTO student " +
                  "VALUES ('" +   fieldValues[0] + "', '" +
                  fieldValues[1]  + "', '" + fieldValues[2] + "', "
                  + numberAge + ",'" + fieldValues[4] + "')";
                 StudentManager studentmanager = new  StudentManager();
                 int result = studentmanager.execUpdate(sqlString);
                 if (result!= 0) {
                    userInterface.clearFields();
                    JOptionPane.showMessageDialog( AddStudentPanel.this,
                    "Inserted sucess!", "Insert Result",
                     JOptionPane.INFORMATION_MESSAGE );
```

```
                }
            } // end try
            // process invalid age number
            catch ( NumberFormatException formatException ) {
                JOptionPane.showMessageDialog( AddStudentPanel.this,
                    "Bad age number ", "Invalid Number Format",
                    JOptionPane.ERROR_MESSAGE );
            }
            // process exceptions from file output
            catch (SQLException ee)
            { System.out.println(ee);  }
        }  //end of if sno field value is not empty
        else  //if sno field value is  empty
            JOptionPane.showMessageDialog( AddStudentPanel.this,
            "Bad sno number ", "student number is  empty!", JOptionPane.ERROR_MESSAGE );
            }
        }
    ); // end call to addActionListener
    // configure button doTask2 for use in this program
    clearButton = userInterface.getDoTask2Button();
    clearButton.setText( "清除" );
    // register listener to call userInterface clearFields()
    clearButton.addActionListener(
    // anonymous inner class to handle clearButton event
        new ActionListener() {
            // call userInterface clearFields() when button pressed
            public void actionPerformed( ActionEvent event )
            {   userInterface.clearFields(); }
        }
    ); // end call to addActionListener
  }
}
```

4. 修改学生记录界面的程序设计

UpdateStudentPanel是一个定制的JPanel,完成修改学生信息界面的工作。当用户单击菜单栏上"修改学生信息"菜单项时,将出现修改学生信息的图形界面,如图14-8所示。根据用户输入的要修改的学号值,程序定位并显示要修改的student表中的记录。用户修改后单击"确认修改"按钮,则修改结果保存到数据库中。若数据库中对应的记录号的修改记录不存在,则显示无此记录的提示。

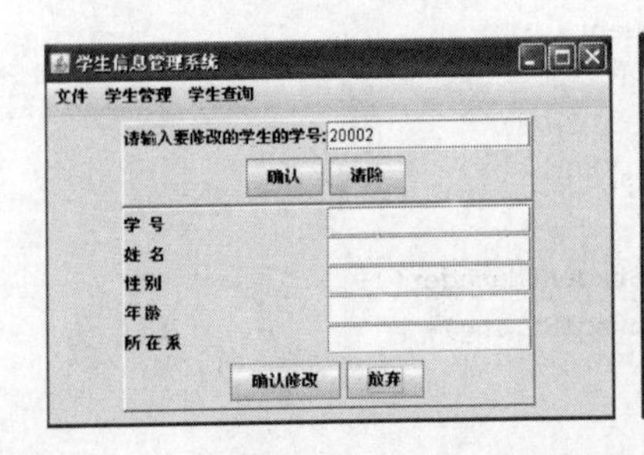

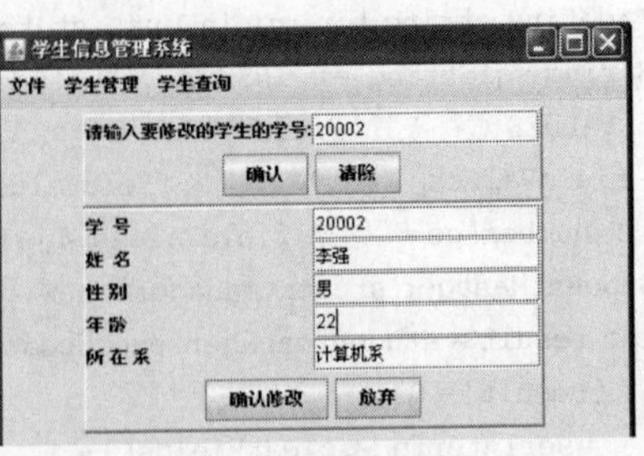

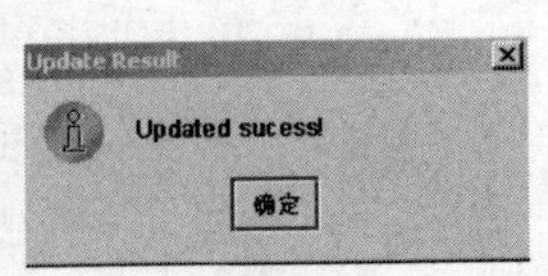

图14-8 修改记录的程序运行结果

UpdateStudentPanel类的代码如下。

```
import java.awt. * ;
import java.awt.event. * ;
import java.sql. * ;
import java.util. * ;
import java.util.List;
import javax.swing. * ;
import javax.swing.border. * ;
class UpdateStudentPanel extends JPanel{
  private StudentUI userInterface1,userInterface2;
  private JButton firstButton1,secondButton1, firstButton2,secondButton2;
  String snoUpdate;
  String sqlString ;
   // set up GUI
   public UpdateStudentPanel( )
   {
      //修改条件
      String names1[] = { "请输入要修改的学生的学号:"};
      userInterface1 = new StudentUI( names1 );  // four textfields
      //设置显示要修改的记录画面
      String names2[] = { "学  号","姓  名","性 别","年 龄","所 在 系"};
      userInterface2 = new StudentUI(names2 );
      this.setLayout(new FlowLayout());
      //Container c = getContentPane();
      Box box = Box.createVerticalBox();
      box.add(userInterface1  );
      box.add(userInterface2 );
      this.add(box);
      // configure button doTask1 for userInterface1 in this program
      firstButton1 = userInterface1.getDoTask1Button();
      firstButton1.setText( "确认" );
      // register listener to call firstButton1 when button pressed
      firstButton1.addActionListener(
         // anonymous inner class to handle firstButton1 event
         new ActionListener() {
            // call DisplayRecord() when firstButton1 pressed
            public void actionPerformed( ActionEvent event )
            {     DisplayRecord();
            }
         } // end anonymous inner class
      ); // end call to addActionListener
      // configure button doTask2 for userInterface1 in this program
      secondButton1 = userInterface1.getDoTask2Button();
      secondButton1.setText( "清除" );
    /* register listener to call serInterface1.clearFields() when secondButton1 pressed *、
      secondButton1.addActionListener(
         // anonymous inner class to handle secondButton1 event
         new ActionListener() {
            public void actionPerformed( ActionEvent event )
            {          userInterface1.clearFields();
            }
         } // end anonymous inner class
      );
```

```
    // configure button doTask1 for userInterface2 in this program
    firstButton2 = userInterface2.getDoTask1Button();
    firstButton2.setText("确认修改");
    // register listener to call UpdateRecord() when button pressed
    firstButton2.addActionListener(
                new ActionListener() {
        // call UpdateRecord(); when button pressed
        public void actionPerformed( ActionEvent event )
        {    UpdateRecord();
        }
      } // end anonymous inner class
    ); // end call to addActionListener
    // configure button doTask2 for userInterface2 in this program
    secondButton2 = userInterface2.getDoTask2Button();
    secondButton2.setText("放弃");
    // register listener to call userInterface2.clearFields() when button pressed
    secondButton2.addActionListener(
      // anonymous inner class to handle secondButton2 event
      new ActionListener() {
        // call addRecord when button pressed
        public void actionPerformed( ActionEvent event )
        {     userInterface2.clearFields();
        }
      } // end anonymous inner class
    ); // end call to addActionListener
    setSize( 400, 260 );
    setVisible( true );
  } // end  constructor
  public void DisplayRecord()   {
   String fieldValues1[] = userInterface1.getFieldValues();
   String fieldValues2 [] = new String[5] ;
   if ( ! fieldValues1[ StudentUI.SNO ].equals( "" ) ) {
      snoUpdate = fieldValues1[0];
      try {
        String sqlString = "select * from student " +
                          " where sno = '" + fieldValues1[0] + "'";
        StudentManager studentmanager = new StudentManager();
            List<Student> list = studentmanager.Query(sqlString);
        if ( list.size()!= 0 ) {
          fieldValues2[0] = list.get(0).getSno();
          fieldValues2[1] = list.get(0).getSname();
          fieldValues2[2] = list.get(0).getSsex();
          fieldValues2[3] = String.valueOf(list.get(0).getSage()) ;
          fieldValues2[4] = list.get(0).getSdept();
          userInterface2.setFieldValues(fieldValues2);
        }
        else
        {  userInterface2.clearFields();
           JOptionPane.showMessageDialog( UpdateStudentPanel.this,
             "Not fund this record!", "Find Result",
                JOptionPane.INFORMATION_MESSAGE );
        }
```

```
            } // end try
          // process exceptions from sql
          catch (Exception ee)
      {       System.out.println(ee);   }
    }  //end of if sno field value is not empty
    else //if sno field value is  empty
          JOptionPane.showMessageDialog( this,
              "Bad sno number ", "Invalid Number Format",
              JOptionPane.ERROR_MESSAGE );
  }
  // Update record of student
  public void UpdateRecord()   {
      String fieldValues[] = userInterface2.getFieldValues();
      // if sno field value is not empty
      if ( ! fieldValues[ StudentUI.SNO ].equals( "" ) ) {
         // update values from student
         try {
             int numberAge = Integer.parseInt(
             fieldValues[ StudentUI.SAGE ] );
             //the string sql statement
             String sqlString = "Update student set " +
                               "sno = '" + fieldValues[0] + "'," +
                               "sname = '" + fieldValues[1] + "'," +
                               "ssex = '" + fieldValues[2]  + "', " +
                               "sage = " + numberAge + ", " +
                               "sdept = '" + fieldValues[4] +
                               "' where sno = '" + snoUpdate + "'";
             System.out.println(sqlString );
             StudentManager studentmanager = new  StudentManager();
             int result = studentmanager.execUpdate(sqlString);
             if (result!= 0) {
                 JOptionPane.showMessageDialog( this,
                 "Updated sucess!", "Update Result",
                  JOptionPane.INFORMATION_MESSAGE );
             }
         } // end try
         // process invalid age number format
         catch ( NumberFormatException formatException ) {
             JOptionPane.showMessageDialog( this,
                "Bad age number ", "Invalid Number Format",
                JOptionPane.ERROR_MESSAGE );
         }
         // process exceptions from sql
         catch (SQLException ee)
                { System.out.println(ee);   }
  }  //end of if sno field value is not empty
  else // if sno field value is  empty
             JOptionPane.showMessageDialog( this,
                "Bad sno number ", "Invalid Number Format",
                JOptionPane.ERROR_MESSAGE );
  } // end method updateRecord
} // end class
```

5. 删除学生记录的界面程序设计

DeleteStudentPanel 是一个定制的 JPanel,完成删除学生信息界面的工作。当用户单击菜单栏上的"删除学生"菜单项时,将出现删除学生信息的图形界面,如图 14-9 所示。根据用户输入的要删除的学号值,程序定位并显示对应学号的学生记录。用户单击"确认删除"按钮后,则删除 student 表中的对应记录。若数据库中对应的删除记录不存在,则显示无此记录的提示。DeleteStudentPanel 类的程序代码如下。

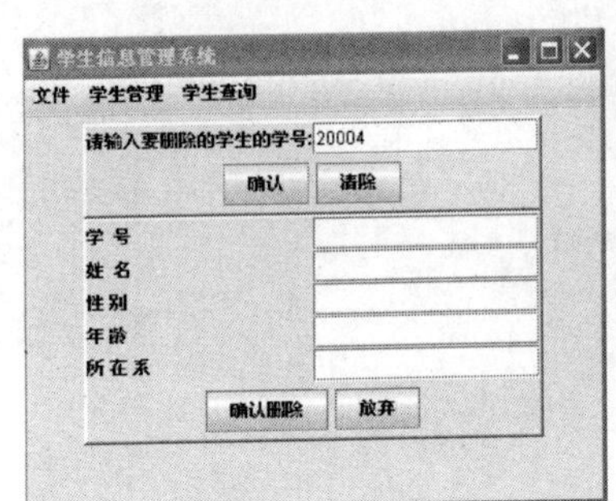

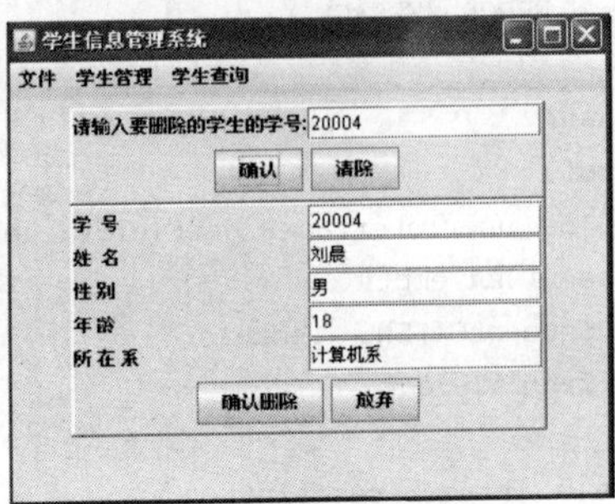

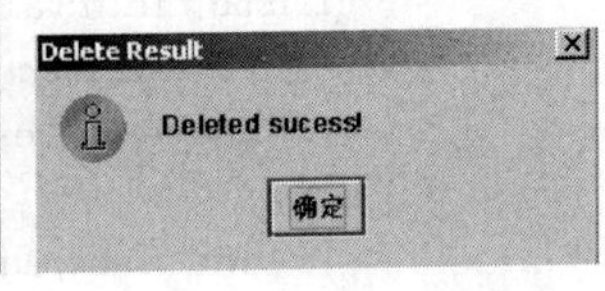

图 14-9　删除记录的运行结果

```
//DeleteStuden JPanel.java
import java.awt. * ;
import java.awt.event. * ;
import java.sql. * ;
import java.util. * ;
import java.util.List;
import javax.swing. * ;
import javax.swing.border. * ;
class DeleteStudentPanel extends JPanel {
    private StudentUI userInterface1,userInterface2;
    private JButton   firstButton1,secondButton1, firstButton2,secondButton2;
    String snoUpdate;
    String sqlString ;
    // set up GUI
    public DeleteStudentPanel()  {
        String names1[] = { "请输入要删除的学生的学号:"};
        userInterface1 = new StudentUI( names1 );
        //set up column names
        String names2[] = { "学  号","姓  名","性 别","年 龄","所 在 系"};
        userInterface2 = new StudentUI(names2 );
        this.setLayout(new FlowLayout());
        Box box = Box.createVerticalBox();    box.add(userInterface1  );
        box.add(userInterface2 );             this.add(box);
        firstButton1 = userInterface1.getDoTask1Button();
        firstButton1.setText( "确 认" );
        // register listener to call openFile when firstButton1 pressed
        firstButton1.addActionListener(
            // anonymous inner class to handle firstButton2 event
            new ActionListener() {
                public void actionPerformed( ActionEvent event )
                {  DisplayRecord();   // call DisplayRecord() when button pressed
                }
```

```
        } // end anonymous inner class
    ); // end call to addActionListener
    // configure button doTask2 for userInterface1 in this program
    secondButton1 = userInterface1.getDoTask2Button();
    secondButton1.setText( "清除" );
    // register listener to call clearFields when button pressed
    secondButton1.addActionListener(
        new ActionListener() {
            public void actionPerformed( ActionEvent event )
            {               userInterface1.clearFields();
            }
        } // end anonymous inner class
    );
    // configure button doTask1 for userInterface2 in this program
    firstButton2 = userInterface2.getDoTask1Button();
    firstButton2.setText( "确认删除" );
    // register listener to call firstButton2 when button pressed
    firstButton2.addActionListener(
      new ActionListener() {
            ublic void actionPerformed( ActionEvent event )
            { deleteRecord();           }
        } // end anonymous inner class
    ); // end call to addActionListener
    // configure button doTask2 for serInterface2 in this program
    secondButton2 = userInterface2.getDoTask2Button();
    secondButton2.setText( "放弃" );
    // register listener to call clearFields when button pressed
    secondButton2.addActionListener(
        // anonymous inner class to handle secondButton2 event
        new ActionListener() {
            // call addRecord when button pressed
            public void actionPerformed( ActionEvent event )
            {         userInterface2.clearFields();
            }
        } // end anonymous inner class
    ); // end call to addActionListener
    setSize( 400, 260 );
    setVisible( true );
} // end  constructor
public void DisplayRecord()
{ String fieldValues1[] = userInterface1.getFieldValues();
  String fieldValues2 [] = new String[5] ;
  if ( ! fieldValues1[ StudentUI.SNO ].equals( "" ) ) {
      snoUpdate = fieldValues1[0];
      try {
          String sqlString = "select * from student " +
                             " where sno ='" + fieldValues1[0] + "'";
          StudentManager studentmanager = new  StudentManager();
          List<Student> list = studentmanager.Query(sqlString);
          if ( list.size()!= 0 ) {
            fieldValues2[0] = list.get(0).getSno();
              fieldValues2[1] = list.get(0).getSname();
```

```
                fieldValues2[2] = list.get(0).getSsex();
                fieldValues2[3] = String.valueOf(list.get(0).getSage()) ;
                fieldValues2[4] = list.get(0).getSdept();
                userInterface2.setFieldValues(fieldValues2);
            }
            else
            {   userInterface2.clearFields();
                JOptionPane.showMessageDialog( DeleteStudentPanel.this,
                  "Not fund this record!", "Find Result",
                    JOptionPane.INFORMATION_MESSAGE );
            }
          }
        catch (Exception ee)
        {   System.out.println(ee);  }
    }  //end of if sno field value is not empty
    else //if sno field value is  empty
         JOptionPane.showMessageDialog( DeleteStudentPanel.this,
             "Bad sno number ", "Invalid Number Format",
             JOptionPane.ERROR_MESSAGE );
  }
  // Delete record of student
  public void deleteRecord()    {
     String fieldValues[] = userInterface2.getFieldValues();
     // if sno field value is not empty
     if ( ! fieldValues[ StudentUI.SNO ].equals( "" ) ) {
        // delete record from student
        try {
            String sqlString = "delete from student  " +
                               "where sno = '" + snoUpdate + "'";
            StudentManager  studentmanager = new  StudentManager();
                int result = studentmanager.execUpdate(sqlString);
            if (result!= 0) {
                JOptionPane.showMessageDialog( this,
                "Deleted sucess!", "Delete Result",
                 JOptionPane.INFORMATION_MESSAGE );
            }
        } // end try
        catch (SQLException ee)
        { System.out.println(ee);  }
        }  //end of if sno field value is not empty
        else
            JOptionPane.showMessageDialog( this,
              "Bad sno number ", "Invalid Number Format",
              JOptionPane.ERROR_MESSAGE );
    } // end method updateRecord
} // end class
```

6. 浏览学生所有记录的界面程序设计

ListStudentsPanel 类是一个定制的 JPanel。当用户单击菜单栏上"浏览全部学生"菜单项时，将显示所有学生信息的图形界面，如图 14-10 所示。ListStudentsPanel 类中，利用 JTextArea 组件可多行文本编辑的特性，显示从数据库查询得到的 student 表的记录集合。类 ListStudentsPanel 的代码如下。

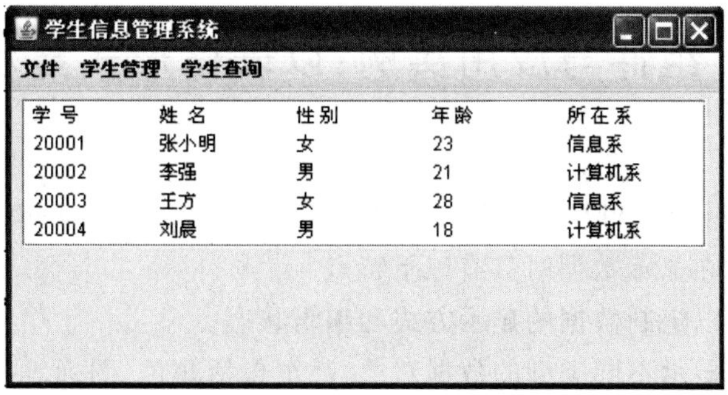

学生信息管理系统

文件　学生管理　学生查询

学 号	姓 名	性 别	年 龄	所 在 系
20001	张小明	女	23	信息系
20002	李强	男	21	计算机系
20003	王方	女	28	信息系
20004	刘晨	男	18	计算机系

图 14-10　浏览全部学生信息

```
// ListStudentsPanel.java
import java.awt.*;
import java.sql.*;
import java.util.*;
import javax.swing.*;
import java.util.List;
public class ListStudentsPanel extends JPanel {
   public ListStudentsPanel() {
       try {
          String names2[] = {"学　号","姓　名","性 别","年 龄","所 在 系"};
          String sqlString = new String("select sno ,sname ,ssex ,sage ,sdept from student");
          StudentManager  studentmanager = new  StudentManager();
          List<Student> list = studentmanager.Query(sqlString);
          StringBuffer results = new StringBuffer();
          results.append( "  " );
          for (int i = 0;i<names2.length;i ++ )
              results.append( names2[ i ] + "\t" );
         results.append( "\n" );
         for (int i = 0;i<list.size();i ++ ){
              results.append( "  " );
              results.append(list.get(i).getSno() + "\t");
              results.append(list.get(i).getSname() + "\t");
              results.append(list.get(i).getSsex() + "\t");
              results.append(String.valueOf(list.get(i).getSage() + "\t") );
              results.append(list.get(i).getSdept() + "\t");
              if (i!= list.size()-1)
                   results.append( "\n" );
         }
         JTextArea textArea  =  new JTextArea( results.toString() );
         this.add( new JScrollPane( textArea ) );
         setSize( 300, 100 );   // set window size
       }
       catch ( SQLException sqlException ) {
         JOptionPane.showMessageDialog( null, sqlException.getMessage(),
              "Database Error", JOptionPane.ERROR_MESSAGE );
       }
   }
}
```

14.6 JTable 组件与应用实例:以表格形式显示数据库内容

1. 类 JTable

JTable 组件是 Swing 组件中比较复杂的组件,属于 javax.swing 包,它能以二维表的形式显示数据。类 JTable 在显示数据时具有以下特点。

(1) 可定制性:可以定制数据的显示方式和编辑状态。

(2) 异构性:可以显示不同类型的数据对象,甚至包括颜色、图标等复杂对象。

(3) 简便性:可以以缺省方式轻松地建立起一个二维表。

使用类 JTable 显示数据之前,必须根据情况事先生成定制的表格模型、单元绘制器或单元编辑器。类 AbstractListModel 用来定制用户自己的表格模型,这个类将在下面介绍。TableCellRenderer接口用来定制单元绘制器,TableCellEditor 接口用来定制单元编辑器,这两个接口主要用于颜色对象的处理上。

创建 JTable 的表格对象时,将捆绑定制的表格模型。例如:

```
JTable table = new JTable(dataModel); //dataModel 是定制的表格模型对象
JScrollPane scrollpane = new JScrollPane(table); //将表格添加到可滚动的面板
```

2. 类 AbstractTableModel

类 AbstractTableModel,提供了 TableModel 接口中绝大多数方法的缺省实现。类 AbstractTableModel隶属于 javax.swing.table。该类是一个抽象类,没有完全实现,不能实例化,使用时必须在程序中实现方法。

要想生成一个具体的 TableModel 作为 AbstractTableMode 的子类,至少必须实现以下三个方法:

① public int getRowCount():得到表格的行数。

② public int getColumnCount():得到表格的行数。

③ public Object getValueAt(int row, int column):得到表格的第 row 行、第 column 列的单元值。

例如,我们可以建立一个简单二维表(10×10),实现方法如下:

```
TableModel dataModel = new AbstractTableModel() { //定制自己的表格模型
        public int getColumnCount() { return 10; }
        public int getRowCount() { return 10;}
        public Object getValueAt(int row, int col)
        { return new Integer(row * col); }
};
JTable table = new JTable(dataModel); //捆绑定制的表格模型
JScrollPane scrollpane = new JScrollPane(table); //将表格添加到可滚动的面板
```

例 14-10 以表格格式显示 study 数据库的多表查询结果集。

此程序运行时,用户能自定义要查询的 SQL 语句,包括单表查询、多表查询,有条件的查询、无条件的查询,以及对表的统计查询。运行的结果以二维表格的形式表现。

程序设计思路如下。

为了在用户界面上以二维表格形式表现查询结果,利用 JTable 表格类和 AbstractTableModel 表格模型类。用 JTable 类以二维表的形式显示,数据表格中的数据从表格模型 AbstractTableModel 类的对象中获取。

为了实现在表格中显示数据库的查询结果，用一个 TableModel 对象把 ResultSet 数据提供给 JTable，在 JTable 中显示查询结果。

ResultSetTableModel 类执行与数据库的连接和生成 ResultSet 对象。DisplayQueryResults 类创建 GUI，建立 JTable 的对象，并从 ResultSetTableModel 的对象中获得数据。

ResultSetTableModel 类是 AbstractTableModel 类的子类，在此类中重写了 TableModel 的方法 getColumnCount()、getRowCount()、getValueAt 和 getColumnName(得到列名)、getColumnClass(得到列名的类)、getValueAt(得到表格中单元格的值)，而 TableModel 的方法 isCellEditable 和 setValueAt()的默认实现没有被重写，因而本例不支持 JTable 单元的编辑。

ResultSetTableModel 类的构造方法接收三个参数：数据库驱动程序列的名称、数据库的 URL 和要执行的查询 SQL 语句的字符串。类中使用的 createStatement 方法如下：

```
statement = connection.createStatement(
    ResultSet.TYPE_SCROLL_INSENSITIVE,
    ResultSet.CONCUR_READ_ONLY );
```

其版本带有两个参数：结果集类型和结果集的并发性。结果集类型规定 ResultSet 的游标移动的方向以及 Result 是否对变化敏感。结果集的并发性规定是否能用 ResultSet 的更新方法更新 ResultSet。上面语句规定了一个游标可沿两个方向滚动的、对变化敏感的和只读的 ResultSet。

ResultSetTableModel 类的方法 setQuery 执行查询 SQL 语句，并调用方法 fireTableStructureChanged(从 AbstractTableModel 类继承)，通知任何以对象 ResultStTableModel 为表格模型的 JTable，模型的结构已经发生变化，通知 JTable 用新的 ResultSet 数据重新填充 JTable 的行和列。SetQuery 方法中抛出的任何异常被传递给调用者。

ResultStTableModel 类的定义代码如下。

```
import java.sql.*;
import java.util.*;
import javax.swing.table.*;
// ResultSet rows and columns are counted from 1 and JTable
// rows and columns are counted from 0.
public class ResultSetTableModel extends AbstractTableModel {
   private Connection connection;
   private Statement statement;
   private ResultSet resultSet;
   private ResultSetMetaData metaData;
   private int numberOfRows;
   private boolean connectedToDatabase = false;
   // initialize resultSet and obtain its meta data object;
   // determine number of rows
   public ResultSetTableModel( String driver, String url,
      String query ) throws SQLException, ClassNotFoundException
   {     // connect to database
         Class.forName( driver );
      connection = DriverManager.getConnection( url );
      // create Statement to query database
      statement = connection.createStatement(
         ResultSet.TYPE_SCROLL_INSENSITIVE,
         ResultSet.CONCUR_READ_ONLY );
      // update database connection status
```

```
    connectedToDatabase = true;
    // set query and execute it
    setQuery( query );
}
// get class that represents column type
public Class getColumnClass( int column ) throws IllegalStateException
{   // ensure database connection is available
    if ( ! connectedToDatabase )
        throw new IllegalStateException( "Not Connected to Database" );
    // determine Java class of column
    try {
        String className = metaData.getColumnClassName( column + 1 );
        System.out.println("@@@@   "+ className);
        // return Class object that represents className
        return Class.forName( className );
    }
    // catch SQLExceptions and ClassNotFoundExceptions
    catch ( Exception exception ) {
        exception.printStackTrace();
    }
    // if problems occur above, assume type Object
    return Object.class;
}
//继承 AbstractTableModel 必须实现的方法
// get number of columns in ResultSet
public int getColumnCount() throws IllegalStateException
{   // ensure database connection is available
    if ( ! connectedToDatabase )
        throw new IllegalStateException( "Not Connected to Database" );
    // determine number of columns
    try {
        return metaData.getColumnCount();
    }
    // catch SQLExceptions and print error message
    catch ( SQLException sqlException ) {
        sqlException.printStackTrace();
    }
    // if problems occur above, return 0 for number of columns
    return 0;
}
//继承 AbstractTableModel 必须实现的方法
//得到列名 in ResultSet
public String getColumnName( int column ) throws IllegalStateException
{    // ensure database connection is available
    if ( ! connectedToDatabase )
        throw new IllegalStateException( "Not Connected to Database" );
    try {
        System.out.println("kkkk   "+ metaData.getColumnName( column + 1 ));
        return metaData.getColumnName( column + 1 );
    }
    catch ( SQLException sqlException ) {
        sqlException.printStackTrace();
```

```
    }
    // if problems, return empty string for column name
    return "";
}
//继承 AbstractTableModel 必须实现的方法
// return number of rows in ResultSet
public int getRowCount() throws IllegalStateException
{   // ensure database connection is available
    if ( ! connectedToDatabase )
       throw new IllegalStateException( "Not Connected to Database" );
    return numberOfRows;
}
//继承 AbstractTableModel 必须实现的方法,以获得编辑数据。
// obtain value in particular row and column
public Object getValueAt( int row, int column )
    throws IllegalStateException
{   // ensure database connection is available
    if ( ! connectedToDatabase )
       throw new IllegalStateException( "Not Connected to Database" );
    // obtain a value at specified ResultSet row and column
    try {
       resultSet.absolute( row + 1 );
       return resultSet.getObject( column + 1 );
    }
    catch ( SQLException sqlException ) {
       sqlException.printStackTrace();
    }
    // if problems, return empty string object
    return "";
}
// set new database query string
public void setQuery( String query )
    throws SQLException, IllegalStateException
{
    // ensure database connection is available
    if ( ! connectedToDatabase )
       throw new IllegalStateException( "Not Connected to Database" );
    // specify query and execute it
    resultSet = statement.executeQuery( query );
    // obtain meta data for ResultSet
    metaData = resultSet.getMetaData();
    // determine number of rows in ResultSet
    resultSet.last();                    // move to last row
    numberOfRows = resultSet.getRow();   // get row number
    // notify JTable that model has changed
    fireTableStructureChanged();
}
// close Statement and Connection
public void disconnectFromDatabase()
{   try {
       statement.close();
       connection.close();
```

```
        }
      catch ( SQLException sqlException ) {
         sqlException.printStackTrace();
        }
        // update database connection status
        finally {
         connectedToDatabase = false;
        }
    }
}   // end class ResultSetTableModel
```

DispalyQueryResults 类负责创建 GUI;建立数据库驱动程序列的名称、数据库的 URL 和要执行的查询 SQL 语句的字符串,并将它们传递给创建的 ResultStTableModel 对象;利用 ResultStTableModel 对象再创建 JTable 的对象,并将 JTable 对象添加到可滚动的面板中,此滚动的面板加入到 Frame 中。类 DisplayQueryResults.java 的代码如下。

```
//  DisplayQueryResults.java
//Note 加载 Mysql JdbcDrive
import java.awt.*;
import java.awt.event.*;
import java.sql.*;
import java.util.*;
import javax.swing.*;
import javax.swing.table.*;
public class DisplayQueryResults extends JFrame {
   // JDBC driver and database URL
   static String JDBC_DRIVER = "com.mysql.jdbc.Driver";
   static String url = "jdbc:mysql://localhost:3306/study";
static String username = "root", password = "root";
   String sex = "男";
   // 定义 SQL statement
   String DEFAULT_QUERY = "select student.sno as 学号,sname as 姓名," +
   "ssex as 性别,sage as 年龄,sdept as 所在系 ,cname as 课程名,grade as 成绩 from student,sc,
course "
    + "where student.sno = sc.sno and sc.cno = course.cno";
   private ResultSetTableModel tableModel;
   private JTextArea queryArea;
   // create ResultSetTableModel and GUI
   public DisplayQueryResults()     {
      super( "以表格形式显示通用查询结果" );
      // create ResultSetTableModel and display database table
      try {
         url = url + "? &user = " + username + "&password = " + password;
         // create TableModel for results of query SELECT * FROM authors
        tableModel = new ResultSetTableModel( JDBC_DRIVER, url,
           DEFAULT_QUERY );
        // set up JTextArea in which user types queries
        queryArea = new JTextArea( DEFAULT_QUERY, 3, 100 );
        queryArea.setWrapStyleWord( true );
        queryArea.setLineWrap( true );
        JScrollPane scrollPane = new JScrollPane( queryArea,
           ScrollPaneConstants.VERTICAL_SCROLLBAR_AS_NEEDED,
           ScrollPaneConstants.HORIZONTAL_SCROLLBAR_NEVER );
```

```
        // set up JButton for submitting queries
        JButton submitButton = new JButton( "Submit Query" );
        // create Box to manage placement of queryArea and
        // submitButton in GUI
        Box box = Box.createHorizontalBox();
        box.add( scrollPane );
        box.add( submitButton );

        // create JTable delegate for tableModel
        JTable resultTable = new JTable( tableModel );
        //将表格添加到可滚动的面板
        JScrollPane tableScrollPane = new JScrollPane(resultTable);
        // place GUI components on content pane
        Container c = getContentPane();
        c.add( box, BorderLayout.NORTH );
        c.add( new JScrollPane( tableScrollPane ), BorderLayout.CENTER );
        // create event listener for submitButton
        submitButton.addActionListener(
           new ActionListener() {
              // pass query to table model
              public void actionPerformed( ActionEvent event )
              { // perform a new query
                 try {
                    tableModel.setQuery( queryArea.getText() );
               }
                 catch ( SQLException sqlException ) {
                    JOptionPane.showMessageDialog( null,
                       sqlException.getMessage(), "Database error",
                       JOptionPane.ERROR_MESSAGE );
                    // try to recover from invalid user query
                    // by executing default query
                    try {
                       tableModel.setQuery( DEFAULT_QUERY );
                       queryArea.setText( DEFAULT_QUERY );
                    }
                    // catch SQLException when performing default query
                    catch ( SQLException sqlException2 ) {
                       JOptionPane.showMessageDialog( null,
                          sqlException2.getMessage(), "Database error",
                          JOptionPane.ERROR_MESSAGE );
                       // ensure database connection is closed
                       tableModel.disconnectFromDatabase();
                       System.exit( 1 );    // terminate application
                    }  // end inner catch
                 } // end outer catch
              }  // end actionPerformed
           }  // end ActionListener inner class
        ); // end call to addActionListener
        // set window size and display window
        setSize( 500, 250 );
        setVisible( true );
    }  // end try
```

```
        catch ( ClassNotFoundException classNotFound ) {
            JOptionPane.showMessageDialog( null,
                "Cloudscape driver not found", "Driver not found",
                JOptionPane.ERROR_MESSAGE );
            System.exit( 1 );    // terminate application
        }
        catch ( SQLException sqlException ) {
            JOptionPane.showMessageDialog( null, sqlException.getMessage(),
                "Database error", JOptionPane.ERROR_MESSAGE );
            // ensure database connection is closed
            tableModel.disconnectFromDatabase();

            System.exit( 1 );    // terminate application
        }
        setDefaultCloseOperation( DISPOSE_ON_CLOSE );
        // ensure database connection is closed when user quits application
        addWindowListener(
            new WindowAdapter() {
                // disconnect from database and exit when window has closed
                public void windowClosed( WindowEvent event )
                {
                    tableModel.disconnectFromDatabase();
                    System.exit( 0 );
                }
            }
        );
    } // end DisplayQueryResults constructor
    public static void main( String args[] )
    {      new DisplayQueryResults();
    }
} // end class DisplayQueryResults
```

程序执行结果如图 14-11 所示。

Displaying Query Results

select student.sno as 学号,sname as 姓名,ssex as 性别,sage as 年龄,sdept as 所在系,cname as 课程名,grade as 成绩 from student,sc,course where

Submit Query

学号	姓名	性别	年龄	所在系	课程名	成绩
20001	张小明	女	23	信息系 ...	数据库 ...	
20001	张小明	女	23	信息系 ...	数学	
20001	张小明	女	23	信息系 ...	信息系统 ...	
20001	张小明	女	23	信息系 ...	英语	
20002	李强	男	21	计算机系 ...	数学	
20002	李强	男	21	计算机系 ...	信息系统 ...	
20004	刘晨	男	18	计算机系 ...	数据库 ...	
20004	刘晨	男	18	计算机系 ...	数学	
20004	刘晨	男	18	计算机系 ...	信息系统 ...	

图 14-11 显示可定义的多表查询结果集

14.7 小 结

Java 程序使用 JDBC API 与数据库进行通信,并用它操纵数据库中的数据。JDBC 驱动器实现与某个数据库的接口。Java.sql包提供访问数据库的类和接口。SQL是所有关系数据库的查询通用语言。本章详细讨论了JDBC API中的主要类和接口,包括DriverManeger、Connection、Statement、PreparedStatement、CallableStatement、ResultSet、ResultSetMetaData、DatabaseMetaData,以及GUI组件JTable和AbstractTableModel类以二维表格形式显示查询结果,并结合实例讨论了如何利用这些类对数据库进行插入、修改、删除和查询操作的示范性的程序设计。

习　　题

14.1　填空题

1）在Java中,SQL的查询结果作为________对象进行处理。

2）________包提供了在Java中处理关系数据库的类和接口。

3）接口________帮助管理Java和数据库之间的连接。

4）________对象用来把一个查询提交给数据库。

5）________接口的对象保存了所有ResultSet类对象中关于字段的元信息,并提供许多方法来取得这些信息。

14.2　试说明Statement接口、PreparedStatement接口和CallableStatement接口的作用和区别?

14.3　说说Java应用程序通过JDBC存取数据库的过程。

14.4　编写程序:对study数据库中的表course完成插入、修改、删除和查询操作。

14.5　对study数据库,编写程序完成:

1）显示每个学生所学的全部课程。包括学号、姓名、课程号、课程名、成绩,以二维表格形式显示,并按学号排序。

2）统计各门课的平均成绩,包括课程号、课程名、平均成绩,并以二维表格形式显示。

第15章　JSP、Struts 2.x技术与Web应用开发

JSP (Java Server Pages)技术为创建动态生成内容的Web页面提供了一个简捷而快速的方法，是servlet技术的一种扩展。JSP技术的设计目的是使得Web应用程序员能够创建动态内容，重用预定义的组件JavaBeans，并使服务器脚本与组件进行交互，将应用程序逻辑和页面显示分离。

Struts 2.x技术使Web应用能灵活地应用MVC设计模型，开发者通过使用Struts提供的一个好的控制器和一套定制的标签库，提高了程序的开发效率。具备MVC架构的程序，使整个软件结构层次分明、可重用性强，增加了程序的健壮性和可伸缩性，便于开发与设计分工，提供集中统一的权限控制、校验、国际化等。

本章首先对JSP技术进行了全面概述，介绍了JSP运行环境的安装和配置过程，详细讨论了JSP的关键元素：指令、脚本、动作和标记库，结合例子给出了这些JSP元素的使用规则。讨论了JavaBeans在JSP中的使用格式和综合应用实例。

接着讨论Struts 2.x的程序设计基础知识：包括Struts 2.x的架构、Action的定义、标签库的使用、配置文件的书写，以及MyEclipse环境下开发一个Struts2应用的具体过程。最后结合实例给出了如何应用JSP技术和Struts 2.x技术开发具备MVC模式的Web应用实例。

15.1　JSP网站开发的基础知识

一个稍微复杂的网站都会包含许多HTML、CSS和JavaScript，本节通过举例介绍这些基本概念，使读者理解得直观明了，并介绍JSP概念和JSP应用的执行过程。

15.1.1　HTML、CSS与JavaScript介绍

HTML超文本标记语言(Hypertext Markup Language)是用于描述网页文档的标记(Tag)语言，它通过各种标记描述页面不同的内容，说明段落、标题、图像、字体等在浏览器中的显示效果。浏览器打开HTML文件时，将根据HTML标记来显示内容。标记的一般定义语法格式为：

```
<tag attributes> document content </tag>
```

或

```
<tag attributes />
```

其中，tag是标记名，attributes是该标记的一个或多个属性，document content是要显示的内容，显示的格式效果往往由tag描述。标记定义由<tag…>开始，结束于</tag>。而<tag

attributes />则表示开始和结束合为一体的特殊情况。标记作为 HTML 的基本元素，它们还可以互相嵌套、组合使用。

例如，<h1 align="center"> 学习 Dreamweaver </h1> ，是 h1 标记，即以 h1 为标题且居中的格式，显示内容为"学习 Dreamweaver"。

例如，<p/>是一个段落标记，但显示的内容为空，用于分隔段落。

例 15-1 一个简单的 HTML 页面文件，文件名 h1. html。

```
<html>
<head><title> 学习网页</title>
<meta http-equiv = "Content-Type" content = "text/html; charset = utf-8" />
</head>
<body>
<h2 align = "center">学习 HTML</h2>
<p> Homework will be given daily</p>
Test and quizzes will be used to check your understanding of concepts
</body>
</html>
```

从上例中可以看出，一个 HTML 文档的结构是：最外层是标签<html>…</html>，里面包含文档头<head>…</head>和文档体<body>…</body>，标记可以嵌套组合，但必须写完整。请读者在浏览器中打开此文件，并观察页面显示效果。

CSS(Cascading Style Sheet，级联样式表)，定义如何显示 HTML 元素，用于控制 Web 页面的外观，并允许将样式信息与网页内容分离的一种标记性语言。

JavaScript(简称 JS 脚本)是一种基于对象和事件驱动的在客户端(浏览器)执行的脚本语言，它常用来给 HTML 网页添加动态功能，比如验证输入数据的合法性，响应用户的各种事件。

例 15-2 一个含有 CSS 和 JavaScripts 脚本的 HTML 页面文件，文件名 h2. htm。

```
<html>
<meta http-equiv = "Content-Type" content = "text/html; charset = utf-8">
<style type = "text/css">
<!--
.c1 {
font-size: 14px;
color: #F00;
}
-->
</style>
<head>
<script language = "javascript">
function show(){
  var name = document.getElementById("username").value;
  var password = document.getElementById("pass").value;
  if (name == ""||password == "") {
       alert("输入内容不能为空!");
  return false;
  }
  else {
  alert("欢迎学习网页制作!");
  return true;
```

```
    }
    return true;
  }
</script>
</head>
<body>
<form  action = "h2.html" method = "post">
    <div class = "c1">姓名:<input type = "input" name = "username" id = "username" value = "" />
</div>
    <div class = "c1">口令:
     <input name = "pass" id = "pass" type = "password">
     <input type = "submit" name = "mybutton" value = "点击" onClick = "return show()"/>
     </div>
</form>
</body>
</html>
```

上面 HTML 页面代码中，<style type="text/css">…</ style>代码段是 CSS 内容，它定义了要显示的“姓名”和“口令”的样式由标记 span 的属性 c1 控制，c1 被称为样式名，c1 在上面 CSS 代码段定义中描述为“.c1 {font-size: 14px;color: #F00;”即字体尺寸 14，红色，所以浏览器会以字体尺寸 14、红色的效果显示“姓名”和“口令”内容。

CSS 内容可直接放在 html 文件中，也可单独存放在另一文件(样式文件)中，然后在要使用该样式的 html 文件中，用<link>标签引入样式文件。例如下面语句：

```
<link rel = "stylesheet" href = "./oa.css" type = "text/css"></link>
```

表示当前的 html 文件引入文件名“oa.css”的样式文件。CSS 单独存放在一个文件的好处是它可以被多个 HTML 页面文件共享引用，这样可使得同一网站的多个页面的表现风格效果统一起来。

上面代码中<script language="javascript">…</ script>之间的内容，是 JS 脚本内容，这里用于弹出提示框信息。同样 JS 脚本内容既可存放在 html 页面文件中，也可单独存放在另一个文件中，然后在要使用该脚本的 html 文件中，用< script >标签引入脚本文件。例如下面语句：

```
<script language = "JavaScript" src = "crossbrowser.js" type = "text/javascript">
</script>
```

表示当前的 html 文件引入文件名“crossbrowser.js”的脚本文件。脚本单独存放在一个独立文件中的好处是可以被多个 HTML 页面文件共享使用。

在浏览器中打开上例的页面文件 h2.html，运行结果如图 15-1 所示。当用户在输入姓名和口令中都输入了内容后，单击“点击”按钮时，将执行用 JS 脚本写的函数 show()，将弹出“欢迎学习网页制作”提示信息框；当用户在姓名或口令字段中至少有一个字段没输入内容而单击“点击”按钮时，执行函数 show()将弹出“输入内容不能为空”的提示信息对话框。

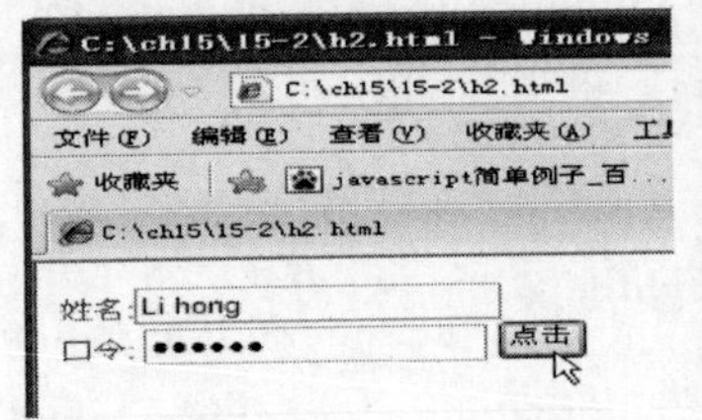

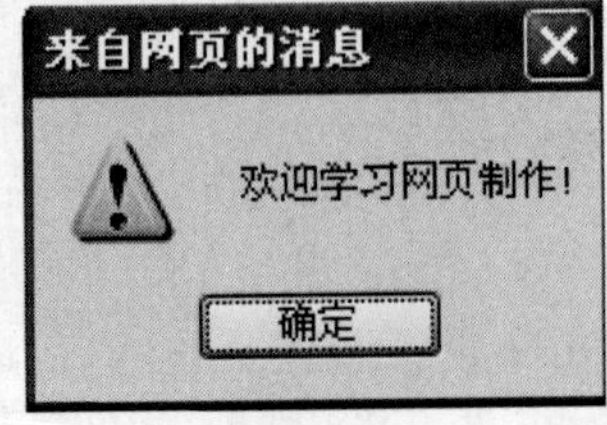

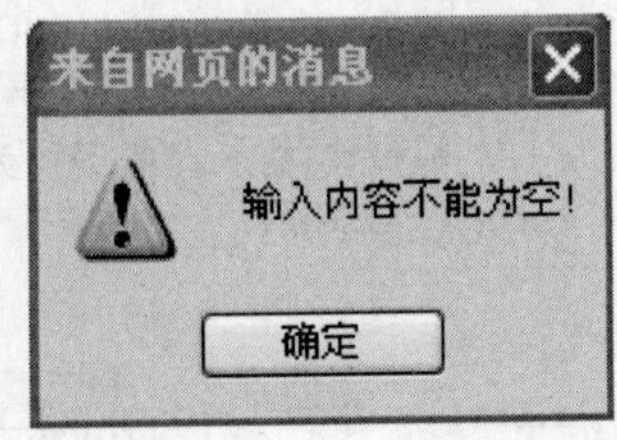

图 15-1　例 15-2 在浏览器中运行效果

从上例可知，页面中使用CSS使得网页显示内容和表现风格相分离，还可以使同一个样式作用于多个显示内容，这样使同一网站的表现风格一致。而JS是在IE浏览器中执行的脚本语言。

为了提高HTML页面文件的生成效率，开发者通常使用可视化的制作软件编写页面文件，如Dreamweaver CS4，它是Macromedia公司与Adobe公司合并后新推出的一款功能强大的网页制作软件，它将可视布局工具、应用程序开发功能和代码编辑支持组合为一个功能强大的工具系统，使每个级别的开发人员和设计人员都可利用它快速地创建网页界面。感兴趣的读者请从网络下载学习。

15.1.2 JSP概述

JSP是一种实现静态HTML和动态HTML混合编码的技术。JSP元素通常放入特殊标记之内，通常以标记“＜%”开始，以“%＞”结束。这些JSP元素将根据不同的请求产生网页的动态部分(即内容可变)。JSP页面文件通常以.jsp为扩展名。下面是一个简单的JSP页面。此页面在IE执行后显示两行“你好”。

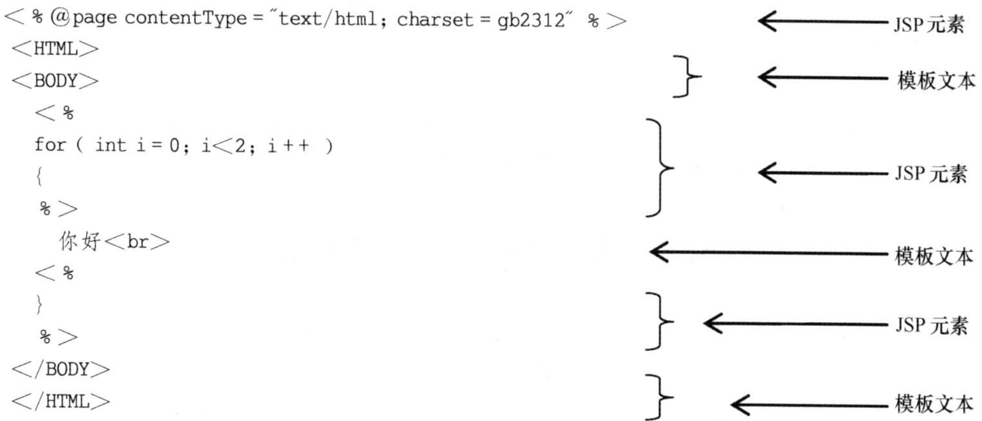

```
<%@page contentType="text/html; charset=gb2312" %>
<HTML>
<BODY>
  <%
  for ( int i=0; i<2; i++ )
  {
  %>
    你好<br>
  <%
  }
  %>
</BODY>
</HTML>
```


页面中任何不是JSP元素的内容被叫作固定模板文本(template text)。固定模板文本的大小，通常帮助程序员决定是使用servlet还是使用JSP。如果发送给客户的内容大部分是模板文本，而仅有一小部分的内容是由Java代码动态生成，则使用JSP，否则就使用servlet。实际上，有些servlet并不产生任何内容，而是代理客户执行一项任务，然后调用其他的servlet或JSP来做响应。注意，在大多数情况下，servlet和JSP技术是可以互换的。与servlet一样，JSP通常作为Web服务器的一部分来执行。执行JSP的Web服务器被称为JSP容器(Container)。

在处理一个JSP页面请求时，模板文本将与由JSP元素产生的动态内容组合起来，组合的结果将作为应答发回给浏览器。一个JSP文件的执行过程如图15-2所示。

(1) 客户端发出Request请求。

(2) JSP Container将JSP转换成servlet的源代码(其中的静态HTML被直接输出到与servlet service方法关联的输出流)。

(3) 将产生的servlet的源代码经过编译后，并加载到内存执行。

(4) 把结果Response(响应)至客户端。

一个JSP页面除了HTML静态文本外，还含有JSP本身的元素：指令(directive)、脚本(scriptlet)、动作(Action)和标记库(tag library)。JSP指令是那些发送给JSP容器的消息，它使程序员能够指定页面设置、包含其他资源中的内容和指定 JSP 中使用的标记库。脚本，即脚本元素(scripting element)，用来嵌入 Java 代码，这些 Java 代码被转换成为 servlet 的一部分。为了简化脚本元素，JSP 定义了一组可以直接使用的隐含(预定义好的)对象，如 request、out 等。JSP 动作将功能封装在预定义的标记中，然后程序员可以将它们(预定义的标记)嵌入 JSP 中，通常根据发送给服务器的特定客户请求中的信息来执行动作。它们还可以创建在 JSP scriptlet 中使用的 Java 对象。标记库是标记扩展机制的一部分，它使程序员能够创建定制的标记。这些 JSP 元素的详细内容将在本章讨论。

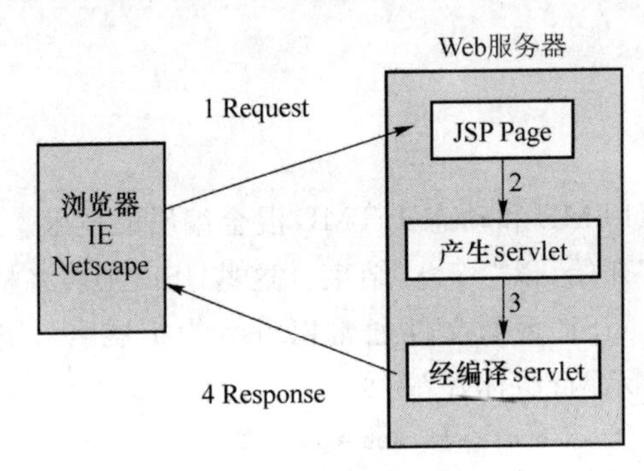

图 15-2　JSP 的执行过程

学习 JSP 和 servlet 开发，首先准备一个符合 Java servlet 和 Java server Pages 规范的开发和运行环境。下面首先讨论 JSP 运行环境的安装和配置。

15.2　JSP 运行环境的安装

Tomcat 是由 JavaSoft 和 Apache 开发团队共同提出合作计划(Apache Jakarta Project)下的产品，是 JSP 和 servlet 标准的完全功能实现。它包含一个 Web 服务器，因此它可以用作 JSP 和 servlet 的独立测试器。由于 Tomcat 技术先进、性能稳定，而且免费，因而深受 Java 爱好者的喜爱并得到了部分软件开发商的认可，成为目前比较流行的 Web 应用服务器。目前最新版本是 7.0。

15.2.1　Tomcat 6 的安装和配置

从 http://tomcat.apache.org/download-60.cgi 网站自行免费下载最新发布的 Tomcat。本书中下载的 Tomcat 的软件压缩包 apache-tomcat-6.0.20.zip，Tomcat 6 能支持 Servlet 2.4 和 JSP 2.0 以上版本。

Tomcat 必须在 JVM 的环境下工作。所以在安装 Tomcat 之前，必须先安装好 J2SDK。假设 J2SDK 的安装目录为 C:\Program Files\Java\jdk1.7.0_65。

安装 Tomcat 的过程：首先将 apache-tomcat-6.0.20.zip 解压缩到磁盘某一目录下，这里假设目录为 g:\apache-tomcat-6.0.20。

解压过程结束后，为了使 Tomcat 能正确工作，还必须设置系统环境变量 JAVA_HOME。其值指向 Java 的安装目录(C:\Program Files\Java\ jdk1.7.0_65)。过程是通过进入“我的电脑—>属性—>高级—>环境变量”，选择—>“系统变量”来设置。设置过程如下(具体参见 running.txt 中说明信息)：单击新建，在变量名中输入：JAVA_HOME，在变量值中输入：C:\

Program Files\Java\ jdk1. 7. 0_65。

在设置好环境变量后，就可以启动 Tomcat 服务器。将目录切换到 g:\apache-tomcat-6. 0. 20\bin。在该子目录下，startup. bat、shutdown. bat 是在 Windows 上启动或停止 Tomcat 服务器的批处理文件。要启动 Tomcat 服务器，执行命令 startup. bat，Tomcat 将运行于 TCP 端口 8080(缺省端口)上，不会与在 TCP 端口 80 上运行的标准服务器发生冲突。Tomcat 启动后会出现 DOS 的命令提示窗口。

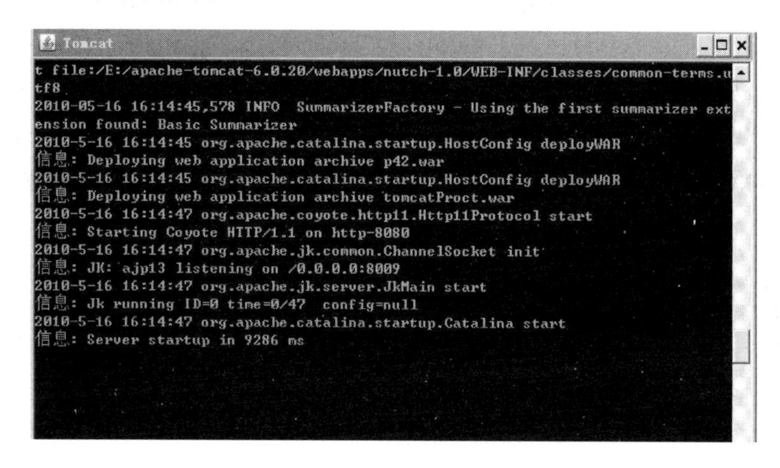

图 15-3　Tomcat 启动后的 DOS 窗口

为了验证 Tomcat 服务器正在运行，打开浏览器，如 IE，输入命令：

```
http://localhost:8080/
```

假若 Tomcat 安装成功，这时会看到如图 15-3 所示的 Tomcat 文档的主页画面。如果没有显示 Tomcat 文档的主页，则试着输入 URL：

```
http://127.0.0.1:8080/
```

本地主机名 localhost 转换成了 IP 地址 127. 0. 0. 1。

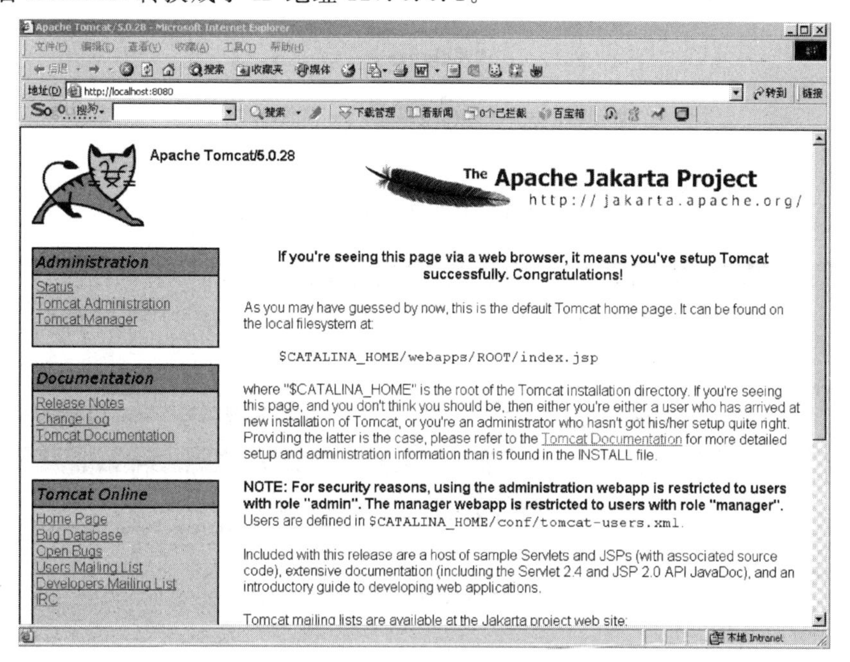

图 15-4　Tomcat 文档的主页

要关闭 Tomcat 服务器，在子目录 bin 下，运行批处理命令文件 shutdown. bat；或者关闭 Tomcat 启动后的 DOS 窗口。

15.2.2 在 MyEclipse 8.5 上配置 Tomcat 服务器

1. 在 MyEclipse 8.5 上配置 Tomcat 服务器

打开 MyEclipse 主窗口，在主菜单上单击 “Windows—>perference”，弹出 perference 窗口，单击“MyEclipse—>Server—>Tomcat—>Tomcat6. x”如图 15-5 的左边图所示；单击“Browse”，填入 Tomcat 的安装目录，单击上方的“Enable”，配置结果如图 15-5 的右图所示。单击下方的“Apply”，最后单击“OK”，就完成了配置 Tomcat 的过程。

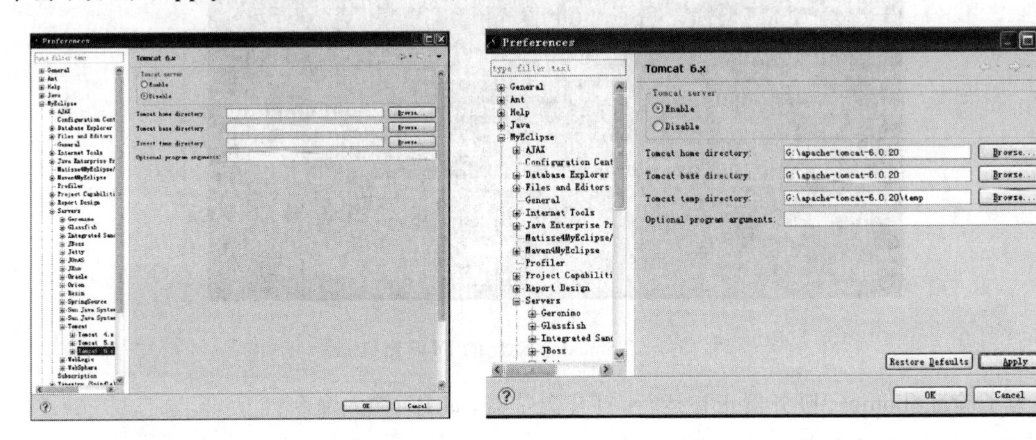

图 15-5　MyEclipse 上配置 Tomcat 6 界面

2. 在 MyEclipse 8.5 上启动和关闭 Tomcat 服务器

在 MyEclipse 主窗口，单击工具条上如图 15-6 所示的服务器图标，单击下拉式菜单如图 15-7 所示中“Start”，则启动 Tomat 运行，单击“Stop Server”则中断 Tomcat 运行。

图 15-6　Web 应用发布图标和服务器图标

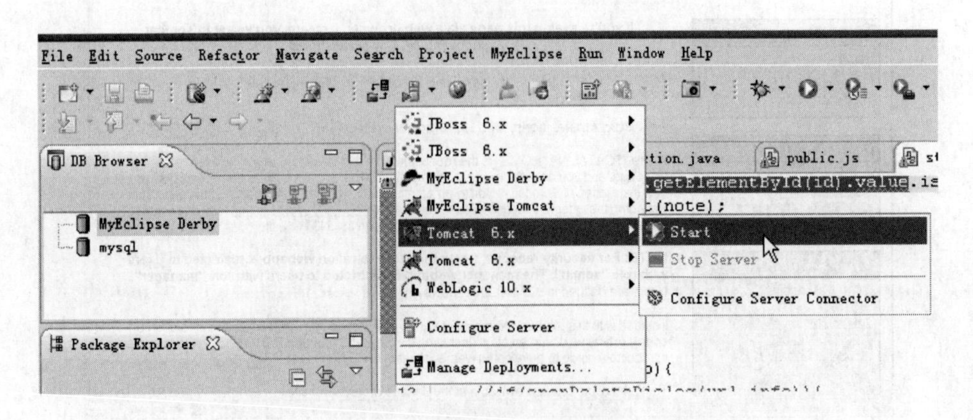

图 15-7　MyEclipse 上启动和停止 Tomcat 菜单

15.2.3 如何在 MyEclipse 上创建、发布和运行 Web 应用程序

在 Tomcat 服务器上成功执行一个 Web 应用程序，包括如下过程。

1. 新建一个 Web 项目

在 MyEclipse 主窗口的菜单中，单击“File->New->Web Project”后，出现如图 15-8 所示的创建新项目的界面，在“Project Name”输入栏中输入“ch15”，单击“Finish”即可。这时在主窗口的 Package Explore 视图面板上，可看到新创建的 Web 项目 ch15 的名字。

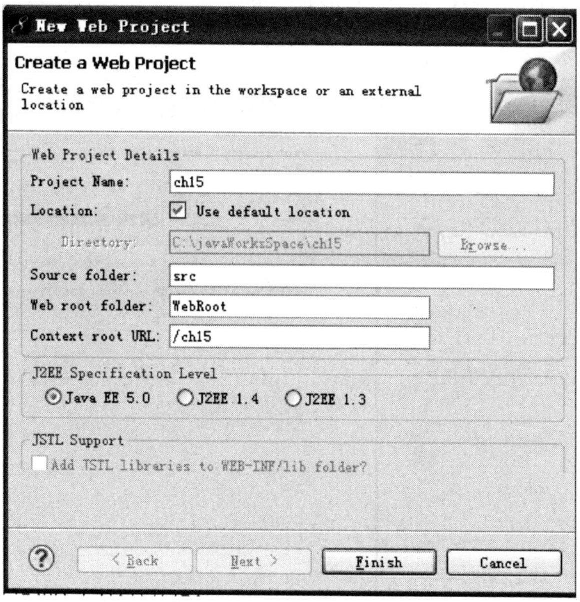

图 15-8 创建新项目界面

2. 在 Web 项目中创建或修改 JSP 文件

单击 ch15 项目下的“WebRoot”文件夹，可看到 MyEclipse 系统自动创建一个文件名 index.jsp 的 JSP 文件。单击 index.jsp 文件，右边出现 index.jsp 的编辑修改画面，修改此文件内容如下。

例 15-3 修改 index.jsp 文件，实现将系统当前日期和时间插入在 Web 页中。

```
<%@ page language="java" import="java.util.*"  import="java.text.*"
  pageEncoding="utf-8"%>
  <html>
    <head>
      <meta http-equiv="Content-Type" content="text/html; charset=utf-8">
      <title>My JSP 'index.jsp' starting page</title>
  </head>
  <body>
          这是一个名字为 index.jsp 的文件。<br/>
    <% = new SimpleDateFormat("yyyy-MM-dd HH:mm:ss").format(new Date()) %>
  </body>
</html>
```

可以在项目 ch15 的 WebRoot 文件下添加类型为.jsp 或.html 或.htm 的文件。也可以添加与任何页面相关的其他类型的文件。

3. 将 Web 项目 ch15 发布到 Tomcat 服务器

单击项目 ch15，单击如图 15-6 所示的工具条的发布图标，弹出“Project Deployment”窗口，如图 15-9 所示，单击“Add”按钮，弹出“New Project”，选择 Web 服务器窗口，在“Server”下拉式菜单上选择“Tomcat 6. x”，如图 15-10 所示，单击“Finish”后，返回前一界面，如图 15-11 所示，这时如果发布成功，会在此界面上出现“successfully deployed”的提示信息，单击“OK”，就成功完成了 Web 项目 ch15 的发布。

同样在图 15-11 界面还有其他按钮，它们的作用如下：“Remove”按钮是将已经发布到 Tomcat 上的项目去除，“Redepoy”是把当前项目重新发布到 Tomcat，是适用于项目被修改后的情况。

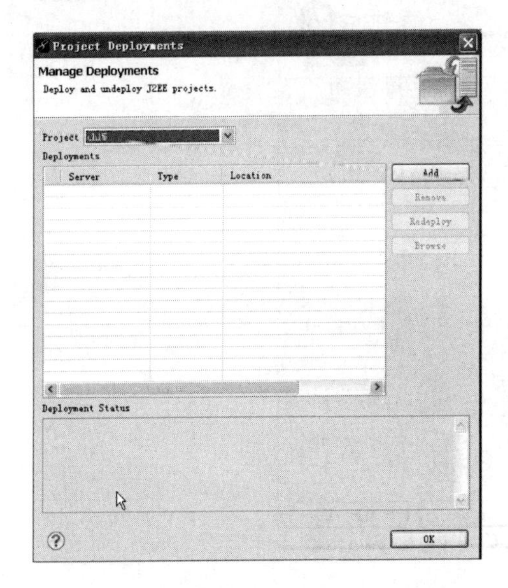

图 15-9　Project Deployment 窗口界面 1

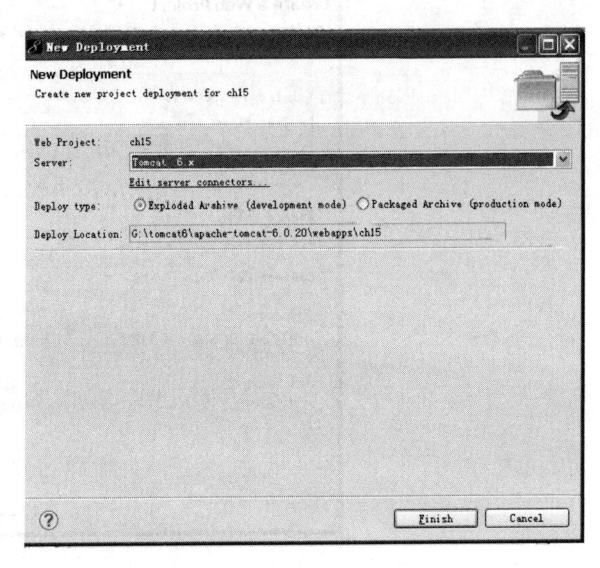

图 15-10　项目选择 Web 服务器窗口界面

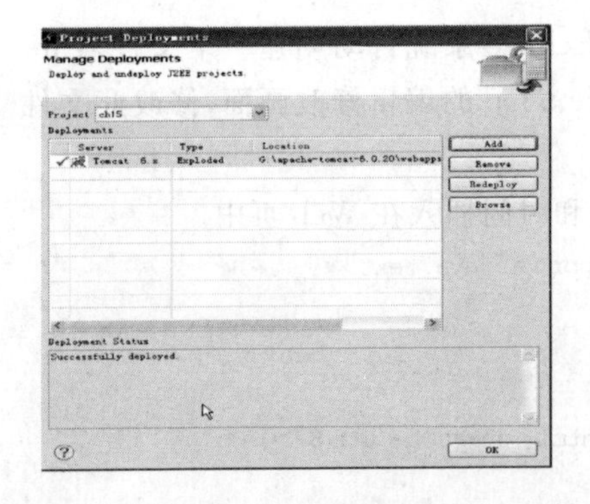

图 15-11　项目在 Tomact 发布成功界面

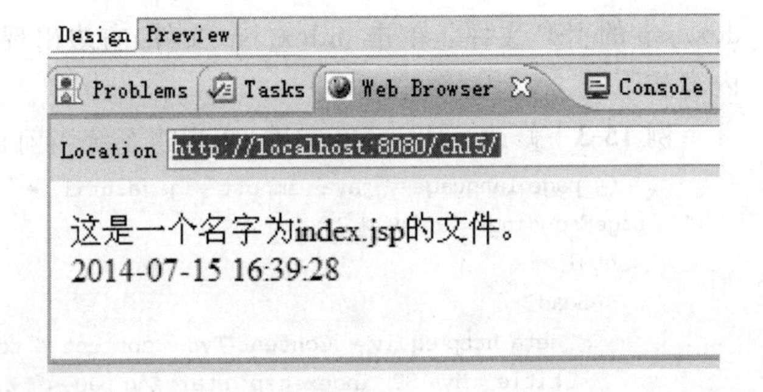

图 15-12　例 15-4 的运行结果

4. 启动 Tomcat，运行 Web 应用程序

在启动 Tomcat 服务器成功后，在 MyEclipse 的 Web Browser 窗口中输入命令：

```
http://localhost:8080/ch15/
```

将出现如图 15-12 所示的运行结果。也可以在 IE 浏览器中输入上面命令，得到的运行结果是一样的。

动手题：请将前面例 15-1 和例 15-2 生成的页面文件 h1.html 和 h2.html 添加到项目 ch15 的 WebRoot 文件下，然后把项目重新发布，重新启动 Tomcat，分别输入下面的运行命令：

```
http://localhost:8080/ch15/h1.html
```

和

```
http://localhost:8080/ch15/h2.html
```

观察浏览器的输出结果。

要注意的是：一个 Web 项目在发布后，或因为修改项目需要重新发布，通常需要重新启动 Tomcat 服务器后，才能观察到修改后的 Web 应用运行效果。

15.2.4 发布到 Tomcat 服务器上的 Web 应用的目录结构

在 Tomcat 6 服务器下，Web 应用程序通常部署在 Tomcat 6 的安装目录的子目录 webapps 中。webapps 目录下为每一个发布的 Web 应用创建一个上下文根目录(context root)，如上例中根目录名为 ch15. 在根目录下又创建了几个子目录。目录结构描述如表 15-1 所示。

表 15-1 Web 应用程序的标准目录描述

目 录	描 述
上下文根目录	该目录的名称由程序员创建。它是 Web 应用程序的根目录，如上例中的 ch15. 所有的 JSP、HTML 文档以及图像等支持文件都放在该目录或它的子目录下
WEB-INF	该目录包含 Web 应用程序部署描述符的文件 web.xml
WEB-INF\classes	该目录包含 Web 应用程序中使用的 servlet 类文件和其他支持的类文件(.class)。如果类放在包结构中，则完整的包目录结构从该目录开始
WEB-INF\lib	该目录包含 Java 存档文件(JAR)

15.3 JSP 指令

JSP 指令(directives)是那些发送给 JSP 容器的消息，它使程序员能够指定页面设置、包含其他资源中的内容和指定 JSP 中使用的定制标记库。指令由

```
<%@ 指令名 属性="值"%>
```

界定。JSP 指令是在 JSP 容器翻译转换时进行处理的，因此指令并不产生任何输出。表 14-2 描述了常用的三种类型的指令元素(Directives Elements)。

表 15-2 JSP 指令元素

指 令	描 述
page	定义页面的全局属性，供 JSP 容器处理
include	在 JSP 编译时插入一个包含文本或代码的文件
taglib	能够让用户以标记库的形式自定义新的标记，这些标记可用来封装功能，同时简化 JSP 的编码

15.3.1 page 指令

page 指令可以指定所使用的脚本语言、JSP 代表的 servlet 实现的接口、servlet 扩展的类

以及导入的软件包。

page 指令的使用格式：

<%@ page 属性 1="值 1" 属性 2="值 2"… %>

下面对 page 指令中的常用属性以及属性值进行讨论。

1. Language

声明脚本语言的种类，默认值为"java"。例如，

language="java"

2. import

为 JSP 页面需要导入的 Java 包的列表，该列表用逗号分隔。这些包可作用于程序段、表达式以及声明。语法格式为：

import="{package.class | package.* }, ..."

例如：

import"java.io.*, java.util.Date"

下面是 JSP 缺省导入的包，所以就不需要再导入。

java.lang.*, javax.servlet.*, javax.servlet.jsp.*

javax.servlet.http.*

3. method

它指定 Java 程序片段(Scriplet)所属的方法的名称，默认方法是 servlet 的 service 方法。该属性的有效值包括 service、doGet、doPost 等。例如：

<%@ page method="doPost" %>

4. contentType

指定响应结果的 MIME 类型，也就是指服务器发送给客户端时的内容编码，默认 MIME 类型是 text/html，默认字符编码为 ISO-8859-1。当浏览器中汉字显示乱码时，可利用此属性设置服务器发送给客户端时的内容编码为"utf-8"或"gb2312"或"gbk".

例如：

<%@ page contentType="text/html;charset=GB2312" %>

或

<%@ page contentType="text/html;GB2312" %>

5. pageEncoding

pageEncoding 是 JSP 文件本身的编码。例如：

<%@ page pageEncoding==GB2312" %>

6. session

指定 JSP 页是否使用 session。默认值为 true。格式为：

session="true | false"

7. buffer

指定隐含对象 out 所用的缓冲区的大小。默认值为 8 kb 的缓冲区，none 表示没有缓冲区。格式为：

buffer="none | 8kb | sizekb"

8. autoFlush

autoFlush="true | false"

设置隐含对象 out 所用的缓冲区在填满时是否自动地刷新(强制输出)。

9. errorPage

errorPage="relativeURL"

指定当发生异常时，客户请求被重新定向到哪个网页。

10. isErrorPage

```
isErrorPage ="true | false"
```

表示此 JSP Page 是否为处理异常错误的网页。

15.3.2 include 指令

include 指令用于在当前 JSP 页面中包含其他文件，被包含的文件可以是 JSP、HTML 或文本文件。包含的过程发生在将 JSP 翻译成 servlet 时，当前 JSP 和被包含的文件会融合到一起，形成一个 servlet，然后进行编译并运行。include 指令的语法格式：

```
<%@ include file="relativeURL" %>
```

其中，relativeURL 是和发出 include 指令的 JSP 页面相对的 URL。

例如，设当前页面文件和要包含的文件 java.htm 在同一路径下：

```
<%@ include file="./java.htm" %>
```

例 15-4 用 include 指令在 includeDirective.jsp 页面文件中包含另一个文件 java.htm 内容。

此例有三个文件：includeDirective.jsp、文本 java.htm 和图像 java.jpg。

includeDirective.jsp 的内容如下：

```
<%@ page language="java"  pageEncoding="utf-8" %>
<html>
<head>
<title>include directice</title>
<meta http-equiv="Content-Type" content="text/html; charset=utf-8">
</head>
<body >
<p>INCLUDE 指令插入另一文档</p>
<div align="center"> <%@ include file="./java.htm" %> </div>
</body>
</html>
```

java.htm 文件内容如下：

```
<img src="java.jpg" width="125" height="160" alt="java 图像" />
```

运行步骤如下。

(1) 将文件 includeDirective.jsp、java.htl 和 java.jpg 复制到项目 ch15 的 WebRoot 路径中。

(2) 将 ch5 项目重新发布。

(3) 重新启动 Tomcate 服务器完成后，在浏览器中输入：

```
http://localhost:8080/ch15/includeDirective.jsp
```

将显示如图 15-13 所示的请求结果。

使用INCLUDE指令插入一幅图像

图 15-13　例 15-2 程序的运行结果

15.3.3 taglib 指令:定义页面使用的标签库

taglib 指令用于指定 JSP 页面所使用的标签库及自定义标签的前缀。使用格式:

```
<%@ taglib uri="标签库 URI" prefix="标签前缀" %>
```

例如:要使用 JSP 的标准标签库(JSP Standard Tag Library,JSTL),语句为:

```
<%@ taglib uri="http://java.sun.com/jsp/jstl/core" prefix="c" %>
```

在使用标签库的标签时,要在标签名前面加上前缀。例如:

```
<c:out value="hello world"/>
```

此语句使用 JSTL 的标签 cout,其作用相当于 JSP 的表达式:<% = "hello world" %>。

注意:要在 Web 项目的 JSP 文件中正确使用标签库,必须把标签库的.jar 包加入到项目的库路径中。如 JSTL 的包为 jstl.jar。

15.4 JSP 脚本

JSP 常常提供页面动态生成的内容。程序员可利用脚本,在 JSP 中嵌入 Java 代码,以实现 JSP 动态生成的内容。

JSP 脚本元素有:声明(Declaration)、表达式(Expression)、脚本程序(Scriptlet)和注解。

15.4.1 声明

声明(declaration)用来在 JSP 页面中声明变量和定义方法。声明的语法格式:

```
<%! declaration;[declaration;]… %>
```

例如:

```
<%! int count = 0; %>
```

声明了一个名为 count 的变量并将其初始化为 0。

声明的变量仅在页面第一次载入时由容器初始化一次,初始化后在后面的请求中一直保持该值。声明是用来声明插入 servlet 类的方法和成员变量。

下面代码声明了一个变量和一个方法:

```
<%!
String color[] = {"red", "green", "blue"};
String getColor(int i){
return color[i];
}
%>
```

对于声明要注意以下一些规则:

(1) 声明必须以";"结尾;

(2) 可以直接使用在<% @ page %>中被包含进来的已经声明的变量和方法,而不需要对它们重新进行声明。

15.4.2 表达式

包含一个符合 Java 语法的表达式。表达式元素在运行后被自动转化为字符串,然后插入到这个表达式在 JSP 文件的位置显示。表达式是一个简化了的 out.println 语句。

表达式的语法格式为：

```
<% = expression %>
```

例如：

```
<% = count %>
```

当前时间：

```
<% =  (new java.util.Date()).toLocaleString() %>
```

当在 JSP 中使用表达式时请记住：不能用一个分号(;)来作为表达式的结束符。

15.4.3 ScriptLet

脚本的语法格式为：

```
<%  code fragment  %>
```

脚本 Scriptlet 是 Java 程序的一段代码，只要符合 Java 语法的语句都可以写在这里，它是在请求时期执行的，它可以使用 JSP 页面所定义的变量、方法、表达式或 JavaBeans。如果在 page 指令中没有指定 method 属性，则生成的代码默认为 service()的主体。

例如：

```
<% String gender = "female";
        if(gender.equals("female"))
{ %>
      She is a girl.
<% }else{ %>
      He is a boy.
<% } %>
```

这段代码只有两行"She is a girl."和"He is a boy."是 HTML 代码，其他代码都是 Java 脚本。

15.4.4 注解

JSP 注释用来对程序进行说明注释，不会显示在客户的浏览器中。JSP 注释的有三种形式：

```
<! --客户端注释-->
<! --<% = 客户端动态注释 %>-->
<%--服务器端注释--%>
```

客户端注释的例子：

```
<! -- This file displays the user login screen -->
```

这会在客户端的 HTML 源代码中产生和上面一样的数据

客户端动态注释的例子：

```
<! -- This page was loaded on <% = (new java.util.Date()).toLocaleString() %> -->
```

这会在客户端的 HTML 源代码中显示为：

```
<! -- This page was loaded on January 1, 2000 -->
```

服务器端注释写在 JSP 程序中，但不会发送给客户。注释标记的字符串在 JSP 编译时被忽略掉。服务器端注释的例子：

```
<%-- This comment will not be visible in the page source --%>
```

15.5 JSP 隐含对象

隐含对象(implicit object)是 JSP 预定义好的一组可以直接使用的对象，使用这些对象能方便地访问请求、响应或会话等信息。这些对象不用声明可以直接拿来使用，在从 JSP 转换为 servlet 时，它们被转换为对应的 servlet 类型。在 JSP 脚本段中，可以访问这些隐含对象来与 JSP 网页中的可执行 servlet 环境交互。

15.5.1 隐含对象介绍

JSP 的隐含对象有：request、response、out、session、pageContext、application、config、page 和 exception。每个隐含对象按照作用范围和持续时间不同分为 4 种不同的作用域：页内有效 page、请求有效 request、会话有效 session 和应用有效 application。

下面先讨论 4 种不同的作用域。

1. 页面作用域 page scope

页内有效是指对象创建后只能在当前 JSP 页面内被访问，所有页内有效的对象的引用都存储在页面上下文对象 pageContext（类型为 javax. servlet. jsp. PageContext）中。

2. 请求作用域 request scope

请求有效是指在处理同一个请求时，不同 JSP 页面创建的对象在这些页面内都可以访问的，这些请求有效的对象都存储在 JSP 页面的 request 对象中。

3. 会话作用域 session scope

从一个客户打开浏览器并连接到服务器开始，到客户关闭浏览器离开这个服务器结束，被称为一个会话。会话有效指在客户端和服务器端之间保持连接的这段时间内是有效的，这些信息存放在 session 对象中。

4. 应用程序作用域 application scope

应用有效是指 Web 服务器从开始运行一直到结束服务时有效，这些信息存放在 application（类型为 javax. servlet. ServletContext）对象中。

JSP 的 9 个常用隐含对象如表 15-3 所示。

表 15-3 JSP 隐含对象

隐含对象	作用域	描 述
application	应用程序作用域	允许 JSP 页面的 servlet 与包括在同一 Web 应用程序中的任何 Web 组件共享信息
session	会话作用域	用来传递客户的会话内容
request	请求作用域	通过 getParameter 方法可以获得相应的请求参数的值
config		类型 javax. servlet. ServletConfig，用于代码片段的配置选项。配置选项在 Web 应用程序描述符中指定
pageContext		该对象用于管理网页属性
out	页面作用域	类型 PrintWriter。使用此对象将内容输出到页面中
page		表示当前 JSP 实例的 this 引用
response		表示对客户的响应，类型为 HttpServletResponse
exception		使用 exception 对象处理异常和错误

在 JSP 脚本段中，可以通过访问这些隐含对象来与 JSP 网页的 servlet 可执行环境交互，而用不着深入了解太多的 servlet API 细节。

例如，使用 out 隐含对象，将某些内容输出到浏览器中：

```
<% out.println("Hello"); %>
```

例如，从请求对象 request 获取参数值：

```
<% String name = request.getParameter("name");
    out.println(name);
  %>
```

15.5.2 request 对象及其应用实例

request 对象的类型为 javax.servlet.ServletRequest，它包含了来自浏览器请求的相关信息，它封装了用户提交的信息，通过调用该对象相应的方法可以获取用户提交的信息。通过 request 对象的 String getParameter(String name)方法可以获得相应的请求参数的值。

例 15-5 用户在一个页面输入表单内容提交，在另一个页面显示用户提交的输入内容。

本例由 2 个文件 request1.jsp 和 display.jsp 组成。用户在浏览器中输入命令：

```
http://localhost:8080/ch15/request1.jsp
```

后，运行结果如图 15-14 所示。

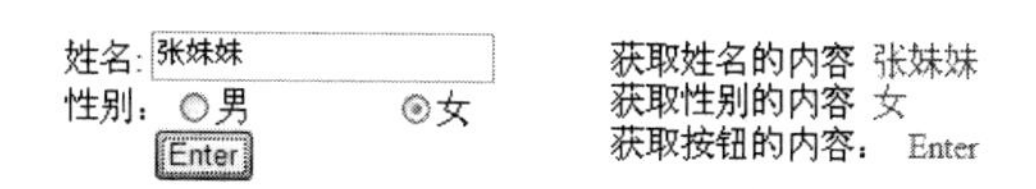

图 15-14 例 15-5 程序运行的输出结果

(1) request1.jsp 文件的页面表单中包含有三种类型的输入字段：text、radio 和 submit，表单提交的目标是 display.jsp，文件内容如下：

```
<%@ page  pageEncoding="utf-8" %>
<HTML>
<head>
    <meta http-equiv="Content-Type" content="text/html; charset=utf-8">
    <title></title>
  </head>
<BODY >
  <FORM action="display.jsp" method="post" >
    <div> 姓名：<INPUT type="text" name="userName"> </div>
  <div>性别：<INPUT type="radio" name="sex" value="男">男
      <INPUT type="radio" name="sex" value="女">女
    </div>
     <INPUT type="submit" value="Enter" name="submit">
  </FORM>
</BODY>
</HTML>
```

(2) display.jsp 文件的页面中，通过 request 对象的 getParameter()方法得到用户提交的数据。其中 request.setCharacterEncoding("UTF-8")是设置 request 对象的编码方式为 "utf-8"，以保证从 request 对象中读到的字段汉字内容能正常显示，而不是乱码。display.jsp 文件的内容如下：

```
<%@ page contentType="text/html;charset=utf-8" pageEncoding="utf-8" %>
<% request.setCharacterEncoding("UTF-8");
    String textContent = request.getParameter("userName");
    String button = request.getParameter("submit");
%>
```

```
<HTML><head>
    <meta http-equiv="Content-Type" content="text/html; charset=utf-8">
    <title></title>
  </head>
<BODY >
<div>获取姓名的内容<span style="font-size: 16px; color: #F00;">
    <%=textContent%></span> </div>
<div>获取性别的内容<span style="font-size: 16px; color: #F00;">
    <%=request.getParameter("sex")%></span></div>
<div>获取按钮的内容:<span style="font-size: 16px; color: #F00;"> <%=button%> </span></div>
</BODY>
</HTML>
```

15.5.3 session 对象及其应用实例

session 对象用来传递客户的会话内容,其类型为 javax. servlet. http. HttpSession。

session 对象的常用方法如下。

- public void setAttribute(String key,Object obj):将参数 Object 指定的对象 obj 添加到 session 对象中,并为添加的对象指定一个索引关键字 key。
- public Object getAttribute(String key):获取 session 对象中含有关键字 key 的对象。

例 15-6 将用户输入的用户名放置于 session 上下文中,并在其他网页显示该用户名。

本例是 request 和 session 的应用实例:当用户访问网站的时候,首先出现用户登录界面,登录成功后,用户数据被存放于 session 中,在用户访问其他网页时,这些数据就可以在需要的时候从 session 中取出。通常存放在 seesion 中的信息为同一用户访问同一网站的不同网页提供的共享信息。

本例由三个文件 getname. htm、savename. jsp 和 nextpage. jsp 组成。

(1) getname. htm:定义表单用于输入用户的名字,这个表单提交的目标是 savename. jsp。getname. htm 的代码如下:

```
<HTML>
  <head>
    <meta http-equiv="Content-Type" content="text/html; charset=utf-8">
    <title></title>
    </head>
<BODY>
<FORM METHOD="POST" ACTION="savename.jsp">
What's your name? <INPUT TYPE="TEXT" NAME="username" SIZE="20"/>
<input name="" type="submit" value="提交">
</FORM>
</BODY>/HTML>
```

(2) savename. jsp:此页面中读取 request 对象的用户名字的值,并写入 session 上下文中,再链接到另外一个网页 nextpage. jsp。saveName. jsp 代码如下:

```
<%@ page pageEncoding="utf-8" %>
<%
String name = request.getParameter( "username" );
session.setAttribute( "theName", name );
%>
```

```
<HTML>
<head>
    <meta http-equiv = "Content-Type" content = "text/html; charset = utf-8">
    <title>save name in session</title>
  </head>
<BODY>
<A HREF = "nextpage.jsp">link to nextpage.jsp</A>
</BODY>
</HTML>
```

(3) nextpage.jsp:完成从 session 中取出被保存的用户名字并显示它。其代码如下:

```
<%@ page  contentType = "text/html;utf-8" %>
    <HTML>
    <head>
    <meta http-equiv = "Content-Type" content = "text/html; charset = utf-8">
    <title></title>
  </head>
<BODY>
Hello, <% = session.getAttribute( "theName" ) %>
</BODY>
</HTML>
```

程序运行:在 IE 中输入命令:

```
http://localhost:8080:/ch15/getname.htm
```

将出现如图 15-15 所示的画面。

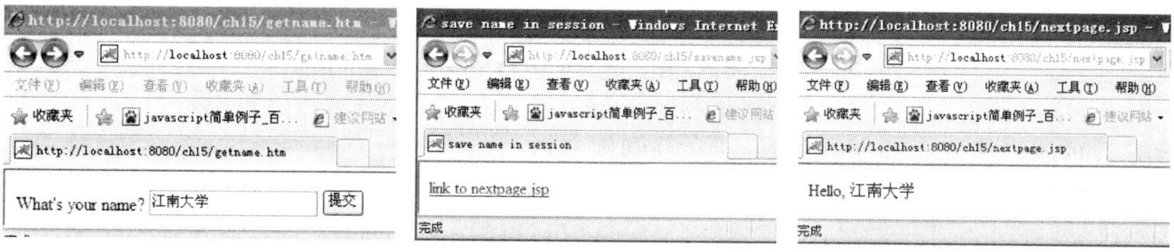

图 15-15 例 15-6 程序运行输出结果

15.6 EL 表达式

EL 表达式语言是 Express Language 的简称,是计算和输出 Java 对象的简单语言。

EL 表达式的语法结构为:

```
${expression}
```

这里的 expression 可以为一个变量或某个表达式。例如:

```
${3}<br/>
${4 + 6}<br/>
${"Hello!"}
```

将在页面输出:

```
3
10
Hello!
```

例如,表示输出变量 count 的值,EL 表示为:

```
${count}
```

相当于脚本表达式：

```
<% =count %>
```

又如，输出 user 对象的属性 name 的值，EL 表示为：

```
${user.name}
```

EL 可以实现对变量的简单访问，当没有指定变量的访问范围时，例如 ${count}，JSP 容器会依序从 Page scope、Request scope、Session scope、Application scope 去查找，如果找到则返回它的值；如果找不到则返回 null。

15.7 JSP 标准动作和应用实例

标准动作元素用于执行一些常用的 JSP 页面动作，例如：将页面转向、使用 JavaBean、设置 JavaBean 的属性等，JSP 容器在请求时处理动作。JSP 的标准动作元素概括在表 15-4 中。

表 15-4 JSP 标准动作

动作元素	描述
<jsp:include>	用于动态加载 HTML 页面或者 JSP 页面
<jsp:forward>	使 JSP 能够将请求的处理转发给另一个不同的资源。该动作终止当前 JSP 的执行
<jsp:plugin>	用于执行一个 applet 或 Bean，有可能的话还要下载一个 Java 插件用于执行它
<jsp:param>	用于传递参数
JavaBean 操作	
<jsp:useBean>	创建一个 JavaBean 实例并指定它的名字 ID 和作用域
<jsp:setProperty>	设定 JavaBean 实例的属性
<jsp:getProperty>	获取 Bean 实例的属性

本节先介绍<jsp:include>、<jsp:forward>、<jsp:param> 动作的属性和应用实例，然后介绍与 JavaBean 有关的动作<jsp:useBean> 、<jsp:setProperty>、<jsp:getProperty>的属性和应用实例。

15.7.1 <jsp:param> 动作

<jsp:param>动作用于传递参数，必须配合<jsp:include>、<jsp:forward>、<jsp:plugin>动作一起使用。

语法格式：

```
<jsp:param name = "name1" value = "value1"/>
```

可在一个页面中使用多个<jsp:param>来传递多个参数给动态文件。

15.7.2 <jsp:include> 动作

<jsp:include> 标准动作，用于在 JSP 页面动态包含其他页面。如果所包含的页面在请求期间发生变化，则下一次请求包含 <jsp:include>动作的 JSP 时，将包含变化的页面内容。而这一点与 include 指令是有区别的。“include 指令”(是一种静态包含)在转换时一次性地将页面内容复制到 JSP 中，如果所包含的页面内容发生变化，则使用“include 指令”的 JSP 将不

能反映出新的内容，除非重新编译该 JSP。

不带参数的<jsp:include> 动作的语法格式为：

```
<jsp:include page = "网页路径"  flush = "true"/>
```

其中，page 属性为网页的相对路径 URL(位于相同 Web 应用程序中)，flush 属性必须为“true”。

带参数的<jsp:include> 动作的语法格式：

```
<jsp:include page = "网页路径" flush = "true"  >
<jsp:param name = "name1" value = "value1"/>
<jsp:param name = "name2" value = "value2"/>
<jsp:include/>
```

这是包含两个参数传递的<jsp:include> 动作语句。在 jsp 页面中，可以利用下面的语句取得传递返回的参数：

```
request.getParameter("name1");
request.getParameter("name2");
```

例 15-7 <jsp:include> 动作的应用例子。

本例由两个 JSP 文件 includeAction.jsp 和 time.jsp 组成。在 includeAction.jsp 页面中用 include 动作包含 time.jsp。includeAction.jsp 文件内容如下：

```
<html>
    <head> <title>Using jsp:include action</title>     </head>
    <body>
        这是使用 jsp:include 动作包含的时钟
         <span style = "vertical-align: top">
            <%-- include  time2.jsp in this JSP --%>
            <jsp:include page = "time1.jsp"  flush = "true" />
         </span>
    </body>
</html>
```

time.jsp 文件的内容如下：

```
<div ><% = ( (new java.util.Date()) ).toLocaleString() %>  </div>
```

程序运行步骤如下。

在浏览器中输入：

```
http://localhost:8080/ch15/includeAction.jsp
```

将显示如图 15-16 所示的请求 includeAction.jsp 的结果。

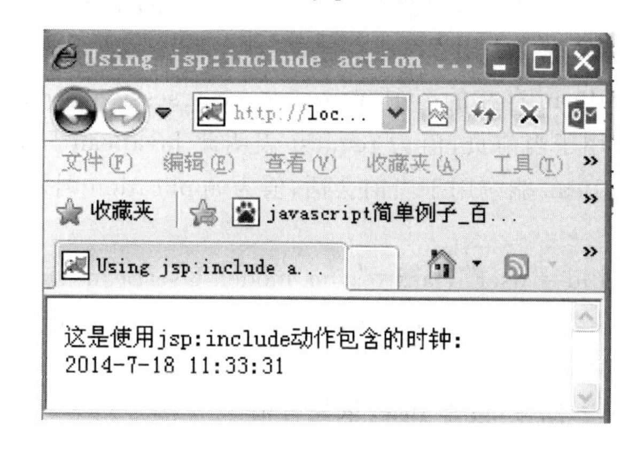

图 15-16 例 15-7 程序运行输出结果

15.7.3 <jsp:forward> 动作

<jsp:forward>动作，使 JSP 能够将请求的处理转发给另一个不同的 URL 资源。原来的 JSP 一旦转发请求，它就立即终止对该请求的处理，这时浏览器显示的页面导向另一个 HTML 页面或者 JSP 页面。该动作终止当前 JSP 的执行。

<jsp:forward>动作的语法格式：

```
<jsp:forward page = "网页路径"/>
```

网页路径是相对 URL(位于相同 Web 应用程序中)，请求将通过该 URL 发送给该资源。

当然，<jsp:forward>动作中也可以加入<jsp:param>参数，其设置和获得参数的方法与<jsp:include>类似。

例如：

```
<jsp:forward page = "/servlet/login">
<jsp:param name = "username" value = "jsmith" />
</jsp:forward>
```

15.7.4 <jsp:useBean>、<jsp:setProperty>、<jsp:getProperty>动作

<jsp:useBean>、<jsp:setProperty>、<jsp:getProperty>是与 JavaBean(实体)有关的动作。在 JSP 中调用 JavaBeans 时，需要创建一个 JavaBean 实例并指定它的名字和作用域。可通过 Bean 实例的名字设定或读取 Bean 的属性。

1. <jsp:useBean>动作：创建或定位一个 Bean 对象

<jsp:useBean>动作能够操作一个 Bean 对象。该动作或者创建一个 Bean 对象，或者定位一个现有 Bean 对象供 JSP 使用。

<jsp:useBean>动作的语法格式：

```
<jsp:useBean id = "beanInstanceName"
  scope = "page | request | session | application"
  class = "package.className"
/>
```

其中：

(1) id：指定该 JavaBean 实例的变量名，通过 id 可以访问这个实例。

(2) class：指定 JavaBean 的类名。如果需要创建一个新的实例，容器会使用 class 指定的类并调用无参构造方法来完成实例化。

(3) scope：指定 JavaBean 存在的范围：page、request、session、application，缺省值 page，表明此 JavaBean 只能应用于当前页；值为 request 表明此 JavaBean 只能应用于当前的请求；值为 session 表明此 JavaBean 能应用于当前会话；值为 application 则表明此 JavaBean 能应用于整个应用程序内。

例如：<jsp:useBean id="user" class="com.model.UserBean" scope="request "/>

在当前页面创建一个类型为 UserBean 的一个对象，名为 user，存在范围为 request；如果 user 已经存在就定位该对象。

2. <jsp:setProperty>动作：设定 Bean 的属性值

一个 bean 对象被创建后，可在 JSP 中设定 Bean 的属性值。<jsp:setProperty>动作用于设定 Bean 的属性值。

<jsp:setProperty>动作的语法格式为：

```
<jsp:setProperty
  name = "beanInstanceName"
    { property = "propertyName" value = "{string | <% = expression %>}" |
      property = " * " | property = "propertyName" [ param = "parameterName" ]
}
```

其中：

(1) name 指定 JavaBean 对象名，与 useBean 标准动作中的 id 相对应；

(2) property 指定 JavaBean 中需要赋值的属性名；

(3) value 指定要为属性设置的值，其值可为字符串 string 或脚本表达式

(4) param 指定请求中的参数名（该参数可以来自表单、URL 传递参数等），并将该参数的值赋给 property 所指定的属性。

设置 Bean 的属性值。可分为 3 种情形。

情形 1：property=" * "

储存在 JSP 用户输入的所有请求参数的值，用于匹配 Bean 中同名的属性。在 Bean 中的属性的名字必须和 request 对象中的请求参数名一致。

情形 2：property="propertyName" [param="parameterName"]

使用 request 中的一个参数值来指定 Bean 中的一个属性值。

情形 3：property="propertyName" value="{string | <%= expression %>}"

使用指定的值来设定 Bean 属性。这个值可以是字符串，也可以是表达式。

当然，一个 bean 对象被创建后，也可在 JSP 中通过脚本调用 Bean 的 set 方法来设置属性的值，一般调用语句格式为：<% Bean对象名. set属性(值) %>。

3. <jsp:getProperty>动作：读取 Bean 属性

一个 bean 对象被创建后，可在 JSP 中读取 Bean 的属性值，用于显示在页面中。

<jsp:getProperty>动作将获得 Bean 的属性值，并可以将其使用或显示在 JSP 页面中。在使用<jsp:getProperty>之前，必须用<jsp:useBean>创建它。

<jsp:getProperty>动作的语法格式：

```
<jsp:getProperty name = "beanInstanceName" property = "propertyName" />
```

上述<jsp:getProperty>动作的属性含义描述如下：

(1) name="beanInstanceName"

用于指定 bean 的名字。在<jsp:getProperty>中的 name 的值应当和<jsp:useBean>中 id 的值相同。

(2) property="propertyName"

用于指定 Bean 的属性名。

例如：

```
<jsp:useBean id = "calendar" scope = "page" class = "employee.Calendar" />
<h2>
Calendar of <jsp:getProperty name = "calendar" property = "username" />
</h2>
```

一个 bean 对象被创建后，也可在 JSP 中通过脚本调用 Bean 的 get 方法来获取属性的值，一般调用格式为：Bean 对象名. get 属性名()。

例 15-8 在 JSP 中使用 JavaBean。

此例将用户登录的信息(用户名和口令)定义成 Bean。首先要编写实体类 LoginData. java 和 JSP 文件 useBean. jsp。在 useBean. jsp 中,创建类 LoginData 的 Bean 对象,起名为 login,并设置和获取 login 对象的属性值,显示在 HTML 文档中。

(1) 定义 Bean 类

一个实体类的声明,一般包括:声明一组域变量,声明每个域变量的设置值和访问值的方法,构造方法是无参的。LoginData. java 的代码如下:

```
package beans;
public class LoginData
{  //定义 Bean 属性
   private String Name = "";
   private String Pwd = "";
    // 设定 Bean 属性的方法
   public void setLoginName(String name);
   { this.Name = name; }
   public void setPassword(String pwd)
   { this.Pwd = pwd; }
   //取得 Bean 属性的方法
   public String getLoginName()
   { return this.Name; }
   public String getPassword()
     { return this.Pwd; }
}
```

(2) 编写 JSP 页面,创建和使用 Beans 实例

在 useBean. jsp 中,创建类 LoginData 的 Bean 对象起名为 login,并设置和获取 login 对象的属性值,显示在 HTML 文档中。

文件 useBean. jsp 的源代码如下:

```
<%@ page contentType="text/html; charset=GB2312" %>
<%@page import="beans.LoginData" %>
<HTML>
  <HEAD>
    <TITLE>使用 Beans</TITLE>
  </HEAD>
  <BODY>
   <H2> <CENTER> <FONT SIZE = 5 COLOR = blue>使用 Beans</FONT>  </CENTER> </H2>
   <jsp:useBean id="login" scope="session"
    class="beans.LoginData"/> <jsp:setProperty name="login"
      property="loginName" value="系统管理员"/>
   <%
    login.setPassword("123456"); //调用 Bean 对象的方法, 设定属性
   %>
   <Font color = red>LoginName</Font>属性值为
   <Font color = blue>
     <jsp:getProperty name="login" property="loginName"/>
   </Font><BR>
   <Font color = red>Password </Font>属性值为
   <Font color = blue>
     <%--以调用 Bean 对象方法的方式取得属性--%>
     <% = login.getPassword() %></Font>
  </BODY>
```

```
</HTML>
```

程序开发步骤如下：

(1) 在 Eclipse 环境下，创建一个 Web project，名字为 ch15；

(2) 创建包名为 beans，在包 beans 中创建类 LoginData.java，并输入内容；

(3) 在项目 ch15 的 WebRoot 下，创建 JSP 文件名为 useBean.jsp，并输入内容；

(4) 将项目 ch15 发布到 Tomcat6 服务器中去；

(5) 重新启动 Tomcat 服务器完成后，在浏览器中输入：

```
http://localhost:8080/ch15/useBean.jsp
```

将显示如图 15-11 的请求结果。

图 15-17　例 15-8 运行结果

15.8　基于 JavaBean 的客户信息管理系统的开发

例 15-9　客户信息登记和信息浏览。

客户的基本信息有姓名、地址、Email。客户通过 Web 浏览器，可填写客户的信息，并进行提交。提交后客户的信息将存放在 study 数据库的 guests 表中，并通过 JSP 提供 Web 浏览。为了完成这些功能，我们必须先在 MySQL 数据库服务器上创建客户信息表 guests 的结构；编写访问数据库的 Web 应用程序并部署在 Web 服务器上。

15.8.1　在 MySQL 的 Study 数据库上创建客户信息表 guests

打开 MySQL 命令行窗口(MySQL Command Line Client)，通过运行下面的脚本命令：

```
use study;
create table guests (
  firstName varchar (20)   NOT NULL ,
  lastName varchar (20)   NOT NULL ,
  email varchar (50)      NOT NULL PRIMARY KEY
);
```

在 study 数据库中创建了表 guests，此表由三个字段组成，且以 Email 字段作为主键(PRIMARY KEY)，即要求 guests 表中的每条记录 Email 的值都不得相同。

15.8.2　创建实体类

创建 2 个实体类：GuestBean.java 放在包 beans，连接数据库的类 DBConnection.java 放

入包 db。

(1) 实体类 GuestBean.java:对应着数据库的基表 guests。

(2) 实体类 GuestDataBean.java:封装了对数据库的所有操作,包括连接、添加记录、浏览记录和关闭连接。

GuestBean.java 的实现代码如下:

```
package beans;
public class GuestBean {
   private String name, addr, email;
public String getName() {
    return name;
}
public void setName(String name) {
    this.name = name;
}
public String getAddr() {
    return addr;
}
public void setAddr(String addr) {
    this.addr = addr;
}public String getEmail() {
    return email;
}
public void setEmail(String email) {
    this.email = email;
}
}
```

GuestDataBean 类的设计思路如下。

(1) 构造方法 GuestDataBean():用于创建数据库连接对象并赋给引用 connection,创建 SQL 语句的对象并赋给引用 statement。这两个引用变量被类中其他三个方法 getGuestList、addGuest 和 finalize 使用。

(2) 方法 getGuestList():对表 guests 执行 SQL 查询语句,返回 ArrayList 列表对象。

(3) 方法 addGuest():完成插入一个顾客记录到表 guests 中。

(4) 方法 finalize():是 GuestDataBean 对象撤销之前(对象占有的空间被释放之前),系统调用的方法。finalize 方法完成的功能是关闭数据库连接。此方法中可能抛出的异常在该方法体中进行了捕获处理。

GuestDataBean.java 类的源代码如下:

```
package beans;
import java.sql. * ;
import java.util. * ;
import db.DBConnection;
public class GuestDataBean extends DBConnection{
   private Connection connection;
   private Statement statement;
   // construct GuestDataBean object
   public GuestDataBean() throws Exception    {
       connection = getConnection();
       statement = connection.createStatement();
```

```
    }
    // return an ArrayList of GuestBeans
    public List <GuestBean >getGuestList() throws SQLException
    {  ResultSet results = null;
       List<GuestBean>  guestList = new ArrayList<GuestBean >();
       results = statement.executeQuery( "SELECT name, addr, email FROM guests" );
       while ( results.next() ) {
          GuestBean guest = new GuestBean();
          guest.setName( results.getString( 1 ) );
          guest.setAddr( results.getString( 2 ) );
          guest.setEmail( results.getString( 3 ) );
          guestList.add( guest );
        }
       return guestList;
    }
    // insert a guest in guestbook database
    public int addGuest( GuestBean guest ) throws SQLException
    { int result = 0;
    statement  = connection.createStatement();
    result = statement.executeUpdate( "INSERT INTO guests ( name, addr, email ) VALUES ( '"
        + guest.getName() + "', '" + guest.getAddr() + "', '" + guest.getEmail() + "' )" );
        return result;
    }
    // close statements and terminate database connection
    protected void finalize()
    {  // attempt to close database connection
       try {
          statement.close();
          connection.close();
       }
       // process SQLException on close operation
       catch ( SQLException sqlException ) {
          sqlException.printStackTrace();
       }
    }
}
```

创建连接数据库的类DBConnection.java放入包db中。其代码参照第14章中的14.5.2的内容。

15.8.3 创建JSP页面文件

JSP页面文件由guestBookLogin.jsp、guestBookView.jsp和guestBookErrorPage.jsp三个文件组成。

1. guestBookLogin.jsp 的设计

其功能是显示一个表单,供用户输入客户信息:姓名、住址和Email。在用户提交该表单时,会再次请求guestBookLogin.jsp,以确保所有输入字段用户全部输入。如果三个字段值全部输入,则guestBookLogin.jsp将请求发送给guestBookView.jsp进行处理。

guestBookLogin.jsp的实现代码如下:

```
<%@ page pageEncoding = "gb2312"  contentType = "text/html;gb2312" %>
```

```
<%@ page errorPage = "guestBookErrorPage.jsp"  %>
<%-- beans used in this JSP --%>
<jsp:useBean id = "guest" scope = "page"
  class = "beans.GuestBean" />
<jsp:useBean id = "guestData" scope = "request"
  class = "beans.GuestDataBean" />
<html>
<head>
  <title>Guest Book Login</title>
  <meta http-equiv = "Content-Type" content = "text/html; charset = gb2312">
  <style type = "text/css">
     body {
        font-family: tahoma, helvetica, arial, sans-serif;
     }
     table, tr, td {
        font-size: .9em;
        border: 3px groove;
        padding: 5px;
        background-color: #dddddd;
     }
  </style>
</head>
<body>  
  <jsp:setProperty name = "guest" property = "*" />
  <% String name = request.getParameter("name");
     String addr = request.getParameter("addr");
     String email = request.getParameter("email");
    if ( name == null || addr == null || email == null ) {
  %>
    <form method = "post" action = "guestBookLogin.jsp">
          <p>输入客户信息</p>
          <table>
        <tr><td>姓名</td>
              <td> <input type = "text" name = "name" />
              </td>
           </tr>
           <tr>
              <td>地址</td>
              <td><input type = "text" name = "addr" /> </td>
           </tr>
           <tr>
              <td>Email</td>
              <td>
                 <input type = "text" name = "email" /> </td>
           </tr>
           <tr>
              <td colspan = "2"> <input type = "submit" value = "Submit" />
              </td>
           </tr>
          </table>
       </form>
  <% }  // end if
```

```
        else {
          name = new String( name.getBytes("iso8859-1"));
          guest.setName(name); //调用 Bean 对象的方法，设定属性 firstName
          addr = new String( addr.getBytes("iso8859-1"));
          email = new String( email.getBytes("iso8859-1"));
          guest.setAddr(addr); //调用 Bean 对象的方法，设定属性 lastName
          guest.setEmail(email); //调用 Bean 对象的方法，设定属性 mail
          guestData.addGuest( guest );
    %>
    <%-- forward to display guest book contents --%>
     <jsp:forward page = "guestBookView.jsp" />
    <%
    }  // end else
    %>
    </body>
</html>
```

代码解释如下。

(1) 文件开始头处,用 page 指令来定义 JSP 中页面的可用信息。语句:

```
<%@ page errorPage = "guestBookErrorPage.jsp" %>
```

设置该页面所有未捕获的异常将发送给 guestBookErrorPage.jsp 页面进行处理。这与在 GuestDataBean 类中产生的异常未进行处理而向上传递抛出,达到了呼应处理。接下来的两个<jsp:useBean>动作语句:

```
<jsp:useBean id = "guest" scope = "page"
   class = "beans.GuestBean" />
<jsp:useBean id = "guestData" scope = "request"
   class = "beans.GuestDataBean" />
```

设置了加载或定位两个 Bean 对象,并命名为 guest 和 guestData,这两个 bean 对象,其中一个对象 guest 供本页中的 JSP 元素使用,另一个对象 guestData 供用户的整个请求使用。

(2) 语句:

```
<jsp:setProperty name = "guest" property = "*" />
```

是<jsp:setProperty>动作,设置 Bean 对象 guest 三个属性的值:用请求对象 request 中 name、addr 和 email 三个参数值,去设置 guest 对象的三个同名属性的值。这里对象 guest 中的三个属性的名字和 request 对象中的三个请求参数名字一致。

(3) 将请求对象中的汉字信息写入 MySQL 数据库前需要进行代码转换,否则会出现数据库中汉字内容乱码。

```
name = new String( name.getBytes("iso8859-1"));
guest.setName(name);
...
guestData.addGuest( guest );
```

完成插入一个顾客记录到表 guests 中。完成后将请求发送给 guestBookView.jsp 进行处理。

2. JSP 文件 guestBookView.jsp 的设计

其功能是读取 guests 表中的所有记录,并在 Web 上浏览。

guestBookView.jsp 的代码如下:

```
<%@ page pageEncoding = "gb2312"  contentType = "text/html;gb2312" %>
<%@page errorPage = "guestBookErrorPage.jsp" %>
<%@ page import = "java.util.*" %>
<%@ page import = "beans.*" %>
```

```
<%-- GuestDataBean to obtain guest list --%>
<jsp:useBean id = "guestData" scope = "request"
  class = "beans.GuestDataBean" />
<html>
  <head>
     <title>Guest List</title>
       <meta http-equiv = "Content-Type" content = "text/html; charset = gb2312">
     <style type = "text/css">
        body {
           font-family: tahoma, helvetica, arial, sans-serif;
        }
        table, tr, td, th {
           text-align: center;
           font-size: .9em;
           border: 3px groove;
           padding: 5px;
           background-color: #dddddd;
        }
     </style>
  </head>
  <body>
     <p style = "font-size: 2em;"> 浏览客户信息</p>
     <table>
        <thead>
           <tr>
              <th style = "width: 100px;">姓名</th>
              <th style = "width: 100px;">地址</th>
              <th style = "width: 200px;">Email</th>
           </tr>
        </thead>
        <tbody>
        <%
           List guestList = guestData.getGuestList();
           Iterator guestListIterator = guestList.iterator();
           GuestBean guest;
           while ( guestListIterator.hasNext() ) {
              guest = ( GuestBean ) guestListIterator.next();
        %>  <tr>
                <td><%= guest.getName() %></td>
                <td><%= guest.getAddr() %></td>
                <td>
                   <a href = "mailto:<%= guest.getEmail() %>">
                      <%= guest.getEmail() %></a>
                </td>
              </tr>
        <%
        }
        %> <%-- end scriptlet --%>
        </tbody>
     </table>
  </body>
</html>
```

代码解释如下。

(1) List guestList = guestData.getGuestList();

调用 guestData 对象的 getGuestList()方法，返回 guests 表中全部记录，并转换成 list 对象

(2) Iterator guestListIterator = guestList.iterator();

```
GuestBean guest;
 while ( guestListIterator.hasNext() ) {
          guest = ( GuestBean ) guestListIterator.next();
  … }
```

定义了与列表 guestList 相关的迭代算子 guestListIterator，用于对列表 guestList 中的元素逐个进行处理。

3. guestBookErrorPage.jsp 的设计

该页面是为其他页面进行出错处理的，即其他页面所有未捕获的异常将发送给 guestBookErrorPage.jsp 页面进行处理。

语句：

```
<%@ page isErrorPage = "true" %>
```

设置该页面为错误处理页面。

guestBookErrorPage.jsp 的代码实现如下：

```
<%@ page isErrorPage = "true" %>
<%@ page import = "java.util.*" %>
<%@ page import = "java.sql.*" %>
<html>
   <head>
      <title>Error! </title>
      <style type = "text/css">
         .bigRed {
            font-size: 2em;
            color: red;
            font-weight: bold;
         }
      </style>
   </head>
   <body>
      <p class = "bigRed">
        occurred while interacting with the guestbook database.
      </p>
      <p class = "bigRed">
         The error message was:<br />
         <% = exception.getMessage() %>
      </p>
      <p class = "bigRed">Please try again later</p>
   </body>
</html>
```

该例子的项目开发步骤如下：

(1) 在 Eclipse 环境下，创建一个 Web project，名字为 ch15；

(2) 将 MySQL 的 JDBC 驱动包加入到项目的库路径中；

(3) 创建包名为 beans，在包 beans 中创建类 GuestBean.java 和 GuestDataBean.java 并

输入内容;创建包名为 db,在包 db 中创建类 DBConnection. java;

(4) 在项目 ch15 的 WebRoot 下,创建三个 jsp 文件名:guestBookLogin. jsp、guestBook-View. jsp 和 guestBookErrorPage. jsp,并输入内容;

(5) 将项目 ch15 发布到 tomcat 6 服务器中去;

(6) 重新启动 Tomcate 服务器完成后,在浏览器中输入:

```
http://localhost:8080/ch15/guestBookLogin. jsp
```

将显示如图 15-18 所示的请求结果。图 15-18 中的最右边一张子图对应的是 guestBookErrorPage. jsp 的显示界面:当用户在 guestBookLogin. jsp 的请求表单中输入信息并提交后,当前用户注册的信息会被写到表 guests 中。在写入时,发现在表 guests 里已经有一条记录的 email 字段的值和当前用户要注册的信息相同,因为在表 guests 上预先已定义了主键,所以写入不成功,系统会抛出 SQLException 的异常对象,该异常对象中封装了出错的信息。guest-BookErrorPage. jsp 页面被调用,将出错的信息显示在客户端上。

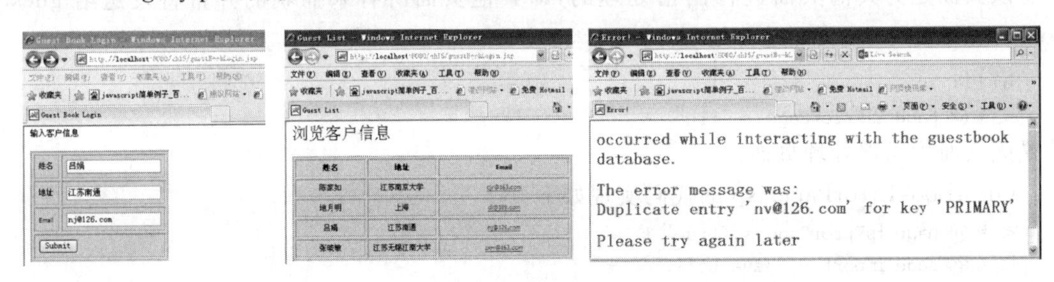

图 15-18 例 15-9 客户信息注册和浏览的输出结果

15.9 Struts 2. x 基础

15.9.1 Struts 2. x 的体系结构

Struts 2 是使用 WebWork 的设计原理,吸收了 Struts1. x 的部分优点,但不是使用 Struts1. x 的设计概念。Struts 2 内设了多个拦截器(Interceptors)和拦截器栈(由多个拦截器形成的链),将用户的 Web 请求进行拦截处理,如国际化、转换器、校验等,从而提供了更加丰富的功能。以 WebWork 为核心的 Struts 2 的体系结构如图 15-19 所示。Struts 2 框架的基本处理流程如下。

(1) 浏览器发送一个请求。

(2) 这个请求经过一系列的过滤器(filter)后,核心控制器 FilterDispatcher 根据请求以决定调用合适的 Action。

(3) WebWork 的拦截器自动对该请求应用通用的功能,如验证等。

(4) 调用 Action 的 execute 方法,该 execute 方法先获取用户请求参数,然后执行某种访问数据库的操作。实际上,Action 是一个业务控制器,它会调用业务逻辑组件来处理用户的请求。

(5) Action 的 execute 方法处理结果信息输出到浏览器中,支持多种形式的视图。

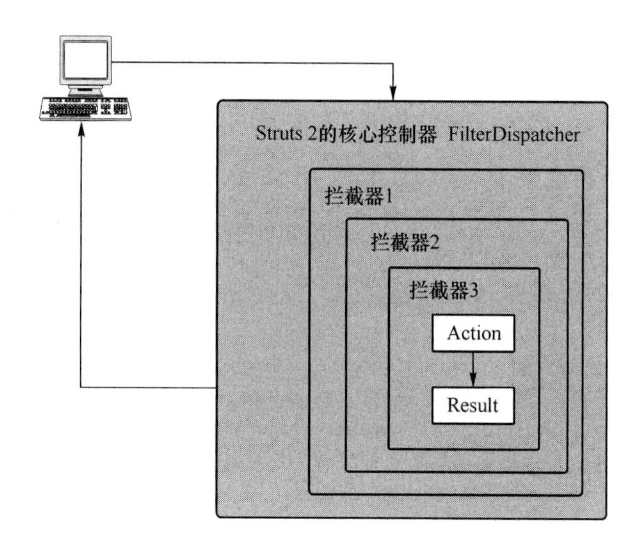

图 15-19　Struts2 的体系结构简图

15.9.2　Struts 2.x 的配置文件

Struts 2 配置文件有：web. xml、Struts. xml 和 Struts. properties。

1. web. xml 配置文件

web. xml 是针对核心控制器 FilterDispatcher 和一些自定义的过滤器的配置，它存放在 Web 应用的子目录 WEB-INF 下。web. xml 部分代码片段如下：

```
<? xml version = "1.0" encoding = "UTF-8"? >
<web-app version = "2.5" >
  <welcome-file-list>
    <welcome-file>index. jsp</welcome-file>
  </welcome-file-list>
<! --配置核心控制器 FilterDispatcher 的名字 struts2 -->
  <filter>
    <filter-name>struts2</filter-name>
    <filter-class>
        org. apache. struts2. dispatcher. ng. filter. StrutsPrepareAndExecuteFilter
    </filter-class>
  </filter>
  <filter-mapping>
    <filter-name>struts2</filter-name>
    <url-pattern>/ * </url-pattern>
  </filter-mapping>
  </web-app>
```

2. struts. properties 全局属性配置文件

struts. properties 是配置 Struts 2 全局属性的文件。该文件包含了一个系列的 key-value 对象，它定义了 Struts 2 的每个属性 key 和该 key 对应的属性值 value。struts. properties 文件通常放在 Web 应用的 WEB-INF/classes 路径下。struts. properties 的配置片段如下：

```
truts. devMode = false
struts. action. extension = do,action
struts. i18n. reload = true
```

```
struts.ui.theme = simple
struts.locale = zh_CN
struts.serve.static.browserCache = false
```

3. struts.xml 核心配置文件

struts.xml 文件是用户视图 View(页面文件)和业务控制模块 Action 之间的联系桥梁。开发者可通过修改该配置文件来适应业务需求,很好地适应了 MVC 的程序设计架构。

struts.xml 文件的配置片段如下:

```
<?xml version="1.0" encoding="UTF-8" ?>
<struts>
<package name="top" extends="struts-default" namespace="/">
    <!-- 定义处理请求 URL 为 login.action 的 Action -->
      <action name="login" class="com.oa.action.LoginAction" method="login">
          <!-- 定义处理结果字符串和资源之间的映射关系 -->
          <result name="back_index">/frameset.jsp</result>
          <result name="login">/index.jsp</result>
      </action>
   </package>
</struts>
```

下面对配置文件中的元素进行说明。

(1) package 元素

用于定义包。其中属性 name:用来指定包的名字,Struts 2 中的每个 package 元素的包名字不能同名。extends:用来指定该包继承其他包。namespace:可选属性,用来指定该包的命名空间。

(2) action 元素

定义了处理请求 URL 与 Action 的对应关系。其中属性 name 定义了 URL 名字;属性 class 是 action 的类名;method 是属性要调用的类实例的方法名,如果缺省就表明方法名为 execute()。

(3) result 元素

定义调用 action 的方法后返回的处理结果字符串和要转向的资源 URL 之间的映射关系。

上例中 action 元素定义了 URL 请求为 longin.action 时,将调用类 com.oa.action.LoginAction 实例的方法 login(),该方法的执行会返回一个字符串:当返回的字符串值为"back_index"时,将跳转到页面/frameset.jsp;当返回的字符串值为"login"时,将跳转到页面/index.jsp。

15.9.3 Action 动作类的定义

Action 是 Struts 2 编程的核心部分,反映了对 Web 应用程序的功能需求。Action 是继承父类 ActionSupport 的子类。下面给出一个声明 Action 动作类的主要代码示例:

```
package com.oa.action;
import com.slide.oa.managers.impl.UserManagerImpl;
import com.slide.oa.model.User;
import java.util.Map;
import com.opensymphony.xwork2.ActionContext;
import com.opensymphony.xwork2.ActionSupport;
```

```
public class LoginAction extends ActionSupport {
   private static final long serialVersionUID = -7859034849358151831L;
   private  String   username;
   private  String   password;
   //登录提交的动作
   public String login()throws Exception {
       ActionContext actionContext  =  ActionContext.getContext();
       Map session  =  actionContext.getSession();
       try {
         /* 根据用户提交表单数据 username 和 password,检查数据库中是否有此用户,若存在返回数
            据库中的完整用户信息,生成 User 对象  */
         if (存在情形) {
            session.put("login", user); //将用户信息放入 session 中
            return "back_index";
         }
         else  //不存在情形
              return "login";
   }
  //注销登录的动作
  public String logout()throws Exception   {
      ActionContext actionContext  =  ActionContext.getContext();
      Map session  =  actionContext.getSession();
      session.clear();  //把该用户的 session 信息清除
      return "login";
  }
  public String getUsername() {
      return username;
  }
  public void setUsername(String username) {
      this.username  =  username;
  }
  public String getPassword() {
      return password;
  }
  public void setPassword(String password) {
      this.password  =  password;
  }
}
```

前台用户提交的表单页面 index.jsp,主要代码为:

```
<%@ page language="java" contentType="text/html;charset=utf-8"%>
<!—页面中要使用 Struts2 的标签库-->
<%@ taglib uri="/struts-tags" prefix="s"%>…
<form action="login.action" method="post" >
用户名:<INPUT type="text" name="username" value="${login.username}">
密码:<INPUT type="password"  name="password"  >
<input type="submit" name="saveButton" value="登录">
</form……
```

定义 Action 的类 LoginAction 的解释说明如下。

(1) 声明了实例变量 username、password,以及这两个实例变量的 get 和 set 方法,并且两个实例变量与 URL 请求页面表单中的输入字段同名。这样就可以在 LoginAction 的方法中直接访问前台页面的用户输入数据内容。

(2) 声明了 2 个 Action 控制组件的方法 login()和 logout()。这两个方法在 Structs. xml 文件中作为 2 个不同的 action 进行配置。

15.9.4 MyEclipse 8.5 环境下如何开发 Structs 2.x Web 应用

1. 新建一个 Web 项目

在 MyEclipse 主窗口的菜单中，通过单击“File—>New—>Web Project”，在弹出窗口中填写项目的名称，例如 persens。

2. 给 Web 项目添加 Struts 2.x 支持

将鼠标定位到 Package Explorer 视窗的项目名称 persens 上，右击，在弹出的菜单上依次选择“MyEclipse—>Add Struts Capabilities…”，在弹窗中选择 Struts 的版本为 Struts 2.1，如图 15-20 所示。单击“Next”按钮进入下一个窗口，选择要加入的支持类库，库的选择视项目需求而定，可以在后续开发过程中陆续加入，这里按默认选择“Struts 2 Core Libraries”。单击“finish”。完成这些操作后你会发现系统在当前项目的 src 目录下添加了一个配置文件 struts. xml，这个文件就是 Struts 2 的核心配置文件。

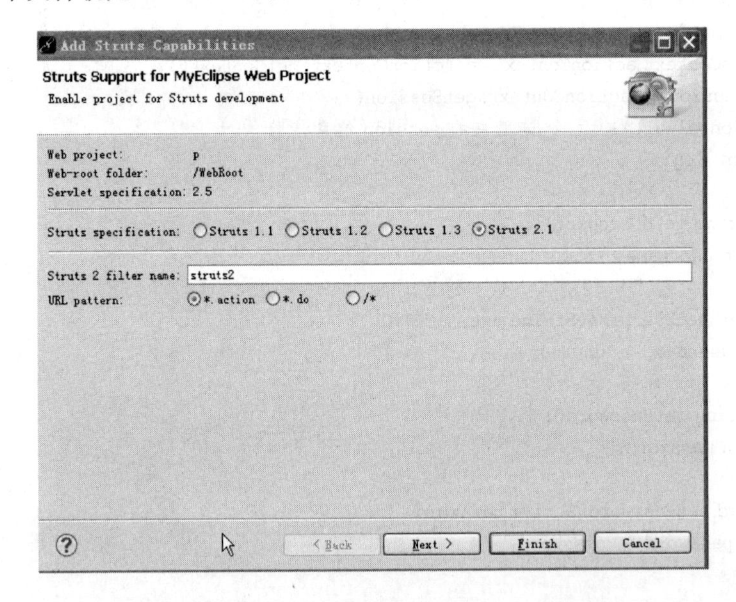

图 15-20 给 Web 项目添加 Struts 2 支持的界面

3. 编写后台 Java 类文件

按应用需求，创建相应的包放入 src 中；编写相应的 Java 类的源文件，放入 src 的相关包中。这些 Java 类文件按功能调用层次模型，一般分为实体类、存取服务类、Action 类等，最好将它们放入不同的包中。

4. 编写前台 JSP 页面文件

按应用需求，编写 JSP 文件放入项目的 WebRoot 或 WebRoot 的子文件夹中。这里要注意的是：如果在页面中要使用 Struts 2 的标签或 JSP 的标准标签 JSTL，必须在页面代码开始部分引入它们的标签库，即添加代码：

```
<%@ taglib uri="/struts-tags" prefix="s" %>
<%@ taglib prefix="c"  uri="http://java.sun.com/jsp/jstl/core" %>
```

上面的第一行引入 Struts 2 的标签库，第二行引入 JSTL 标签库。在页面其后的标签语句中，当标签的前缀是“c:”表示使用 JSTL 标签，当标签的前缀是“s:”表示使用 Struts 2 标签。

页面的实例代码结构如下：

```
% @ page language = "java" contentType = "text/html; charset = utf-8"
    pageEncoding = "utf-8" % >
< % @ taglib uri = "/struts-tags" prefix = "s" % >
< % @ taglib prefix = "c"  uri = "http://java.sun.com/jsp/jstl/core" % >
<html>
<head>
<meta http-equiv = "Content-Type" content = "text/html; charset = utf-8">
<link href = "style/oa.css" rel = "stylesheet" type = "text/css">
<script language = "javascript" src = "script/public.js"></script>
<title>....</title>
</head>
<body> ...... </body>
</html>
```

5. 完成配置文件的编写

修改三个配置文件 Web.xml、Struts.xml 和 Struts.properties。

6. 将 Web 项目发布到服务器，启动 Web 服务器并运行 Web 应用

请参见 15.2.3 节介绍的步骤，完成 Web 应用的运行和测试。

15.10 基于 Struts 2.x 的用户账号管理系统开发

本节实现的用户账号管理系统，是一个基于 JSP＋Struts 2.x＋MySQL 的 Web 应用实例。要求存储用户注册的基本信息，包括：用户账号（username）、口令（password）、账号创建时间（createTime）、账号有效时间等。具备的基本功能有：系统管理员能完成用户账号添加、修改、删除、列表查询；为了保证账号和密码的安全性，设置用户账号密码的有效时间为 30 天，有效期将到时，提示用户修改；前台登录正确进入系统的用户能修改自己口令等功能。

（请读者注意：由于书的篇幅所限，本节中给出的程序代码有些是不完整的，每个程序完整的代码请参见课件中的例子。）

15.10.1 创建 MySQL 的数据库 persons

打开 MySQL 命令行窗口（MySQL Command Line Client），运行下面的脚本命令：

```
create database persons;
use persons;
DROP TABLE IF EXISTS `t_user`;
CREATE TABLE `t_user` (
  `id` int(11) NOT NULL auto_increment,
  `username` varchar(255) NOT NULL,
  `password` varchar(255) NOT NULL,
  `createTime` datetime default NULL,
  `expireTime` datetime default NULL,
  PRIMARY KEY  (`id`),
  UNIQUE KEY `username` (`username`)
);
INSERT INTO `t_user` VALUES (1,'admin','admin','2014-07-20 09:35:38','2014-08-20 09:35:38');
```

将在 persons 数据库中创建表 t_user，此表由五个字段组成，其中 id 字段的值由系统自动生成,且以 id 字段作为主键(PRIMARY KEY)。

15.10.2 实体层的设计

1. 定义实体类 User 和 Message

定义实体类 User,对应着数据库基表 t_user,放在包 com. slide. oa. model 中。User. java 的实现代码如下:

```
package com.slide.oa.model;
import java.util.Date;
public class User {
    private int id;
    private String username = "";
    private String password = "";
    private Date createTime;
    private Date expireTime;
    private String expireTimes = ""; //字符串型的日期,用于前台页面的日期显示
    public int getId() {   return id;   }
    public void setId(int id) {
        this.id = id;   }
    public String getUsername() {   return username;   }
    public void setUsername(String username) {
        this.username = username;
    }
    public String getPassword() {
        return password;
    }
    public void setPassword(String password) {
        this.password = password;
    }
    public Date getCreateTime() {
        return createTime;
    }
    public void setCreateTime(Date createTime) {
        this.createTime = createTime;
    }
    public Date getExpireTime() {
        return expireTime;
    }
    public void setExpireTime(Date expireTime) {
        this.expireTime = expireTime;
    }
    public String getExpireTimes() {
        return expireTimes;
    }
    public void setExpireTimes(String expireTimes) {
        this.expireTimes = expireTimes;
    }
}
```

2. 消息提示类 Message

消息提示类 Message 是用于定义消息提示子窗口中的显示内容。Message. java 的代码

如下：

```
package com.oa.obj;
public class Message {
    private String messageContent;    //消息提示内容
    private String skipPath; //消息提示后,用户按“返回”后,定义要返回的页面地址
    public Message() {
        this.messageContent = messageContent;
        this.skipPath = skipPath;
    }
    public String getMessageContent() {
        return messageContent;
    }
    public void setMessageContent(String messageContent) {
        this.messageContent = messageContent;
    }
    public String getSkipPath() {
        return skipPath;
    }
    public void setSkipPath(String skipPath) {
        this.skipPath = skipPath;
    }
}
```

15.10.3 数据库存取服务层的设计

定义连接数据库的类 DBConnection.java 放在包 com.oa.conn 中。

DBConnection.java 的代码实现如下：

```
package com.oa.conn;
import java.io.IOException;
import java.sql.*;
import com.slide.oa.managers.impl.SystemException;
public class DBConnection {
    protected Connection conn = null;
    protected Statement stmt = null;
    protected ResultSet rs = null;
    private static String username = "root";
    private static String password = "root" ;
    private static String driverClassName = "com.mysql.jdbc.Driver";
    private static String url =
"jdbc:mysql://localhost:3306/persons? useUnicode = true&amp;characterEncoding = UTF-8";
    public  Connection getConnection() throws Exception {
      try{
        Class.forName(driverClassName).newInstance();
        conn = DriverManager.getConnection
         (url + "&user = " + username + "&password = " + password);

        }
        catch(Exception e){
            e.printStackTrace();
            throw new SystemException(" MysQl connection exeception." );
        }
```

```java
            return conn;
        }
    //SQL 通用查询功能
    public void execQuery(String sql)throws Exception {
            try {
                if (conn == null)
                  conn = getConnection();
                  stmt = conn.createStatement();
                  rs = stmt.executeQuery(sql);
            } catch (Exception se) {
                se.printStackTrace();
            }finally {
            }
        }
        public void closeDBConnection() throws SQLException {
            if (conn!= null) {
                try {
                    conn.close();
                } catch (SQLException e) {
                    e.printStackTrace();
                    throw new SystemException("Failed to close connection");
                }
            }
        }
    }
```

UserManagerImpl.java:封装了对数据库表 t_user 的所有操作:包括账号的添加、修改口令、删除;查询某一账号的信息和账号信息列表。

UserManagerImpl 类的设计说明如下。

(1) 构造方法 UserManagerImpl():用于创建数据库的连接。

(2) 方法 User login(String username, String password,Map session):登录验证用户在数据库中是否存在,若存在,则将该用户信息写入 ssession 中。

(3) 方法 boolean save(User user) throws Exception:添加账号记录。

(4) 方法 boolean delete(int id):删除某一账号。

(5) 方法 boolean changePassword(int userId, String newPassword,Map session):用户修改口令后保存 ,并将 session 中的用户登录信息修改。

(6) 方法 boolean changePassword(int userId, String newPassword):管理员初始化口令后保存到数据库中。

(7) List<User> getUserList( ):从数据库查询得到所有账号列表。

UserManagerImpl.java 的实现代码如下:

```java
package com.slide.oa.managers.impl;
import  com.oa.conn.DBConnection;
import java.sql.SQLException;
import java.text.SimpleDateFormat;
import java.util.ArrayList;
import java.util.Calendar;
import java.util.Date;
import java.util.List;
import java.util.Map;
```

```
import com.slide.oa.managers.impl.SystemException;
import com.slide.oa.model.User;
public class UserManagerImpl extends DBConnection {
    public UserManagerImpl() throws Exception
    {
        try {
          getConnection();
        }
        catch (SQLException e) {
         e.printStackTrace();
                throw new SystemException ("Database connection exeception ." );
        }
  }
  //登录验证用户是否存在,若存在该用户信息写入 ssession 中
  public User login(String username, String password,Map session)throws Exception {
   User user = null;
   String sql =  "select * from t_user where username  =" + "'" + username + "'";
   try {
         stmt  = conn.createStatement();
         rs  =  stmt.executeQuery(sql);
         if (rs.next()) {
                    user = new User();
                    user.setId(rs.getInt("id"));
                    user.setUsername(rs.getString("username"));
                    user.setPassword(rs.getString("password"));
                    user.setCreateTime(rs.getDate("createTime"));
                    user.setExpireTime(rs.getDate("expireTime"));
         }
  }catch (Exception e) {
          e.printStackTrace();
          throw new SystemException ("Exeception for select t_user tables.");
  }finally {
        if (stmt!= null)
            stmt.close();
      if (rs!= null)
            rs.close();
      closeDBConnection();
  }

  if(user == null){
        throw new SystemException("没有这个用户!");
  }
  if(user.getExpireTime() != null){
        Calendar now  =  Calendar.getInstance();  //现在时间
        Calendar expireTime  =  Calendar.getInstance();  //失效时间
        expireTime.setTime(user.getExpireTime());
        if(now.after(expireTime)){  //如果现在在失效时间之后
            session.put("login", user);
            throw new SystemException
          ("用户密码已过期! <a href = 'user/update_password.jsp'>修改密码</a>");
      }
  }
```

```
        return user;
    }
    //添加账号记录
    public boolean  save(User user) throws Exception {
        boolean flag = false;
         Date now = new Date();
         SimpleDateFormat ss  =  new SimpleDateFormat("yyyy-MM-dd HH:mm:ss");
         String today  =  ss.format(now);
          int m = now.getMonth();
         now.setMonth(m + 1);
         String nextday  =  ss.format(now);
         String sql  =  "insert into T_user (username,password,createTime,expireTime) values ('"
                    + user.getUsername() + "','"
                    + user.getPassword() + "','"
                    + today + "','"
                    + nextday + "')";
           System.out.println(sql);
        try {
             stmt  =  conn.createStatement();
             int result  =  stmt.executeUpdate(sql);
             if (result != 0) {
               flag =  true;
             }
        }catch (Exception e) {
             e.printStackTrace();
             flag = false;
             throw new SystemException (e.toString());
        }finally {
               closeDBConnection();
        }
        return flag;
    }
      //删除某一账号
        public boolean  delete(int id) throws Exception {
            boolean flag = false;
            String sql ;
            sql  =  "delete from  T_user  where id = " + id;
            try {
                stmt  =  conn.createStatement();
                int result = stmt.executeUpdate(sq);
                if (result != 0) {flag =  true;
                 }
            }catch (Exception e) {
                e.printStackTrace();
                flag = false;
                throw new SystemException (e.toString());
            }finally {
              closeDBConnection();
            }
            return flag;
     }
    //修改口令后保存,并将session中的用户登录信息修改
```

```
public boolean changePassword(int userId, String newPassword,Map session)
  throws Exception {
      boolean flag = false;
      User user = null;
      String sql =  "select * from t_user where id =" + userId;
      try {
        stmt = conn.createStatement();
        rs = stmt.executeQuery(sql);  //rs 变量从父类继承,返回查询结果
        if (rs.next()) {
                  user = new User();
                  user.setId(rs.getInt("id"));
                  user.setUsername(rs.getString("username"));
                  user.setPassword(rs.getString("password"));
                  if(user.getPassword().equals(newPassword))
                        throw new SystemException("不能和旧密码相同");
                  user.setPassword(newPassword);
                  Date now = new Date();
                  SimpleDateFormat ss = new
                     SimpleDateFormat("yyyy-MM-dd HH:mm:ss");
                  String today = ss.format(now);
                  user.setCreateTime(ss.parse(today));
                  int m = now.getMonth();
                  now.setMonth(m + 1);
                  String nextday = ss.format(now);
                  user.setExpireTime(ss.parse(nextday));
                  sql = "UPDATE t_user set password = '" + newPassword +
                  "' ,createTime = '" + today + "',expireTime = '" + nextday +
                  "' where id = " + userId;
                  stmt = conn.createStatement();
                  int result = stmt.executeUpdate(sql);
                  if (result!= 0) {
                        session.remove("login");
                        session.put("login", user);  flag = true;
                  }
        }
  }catch (Exception e) {
      e.printStackTrace();
      throw new SystemException (e.toString());
  }finally {
          if (rs!= null)
              rs.close();
          if (stmt!= null)
              stmt.close();
          closeDBConnection();
    }
    return flag;
}
//初始化口令后保存
public boolean changePassword(int userId, String newPassword)throws Exception {
      boolean flag = false;

      Date now = new Date();
```

```
    SimpleDateFormat ss = new SimpleDateFormat("yyyy-MM-dd HH:mm:ss");
    String today = ss.format(now);
    int m = now.getMonth();
    now.setMonth(m + 1);
    String nextday = ss.format(now);
    String sql = "UPDATE t_user set password = '" + newPassword +
    "',createTime = '" + today + "',expireTime = '" + nextday +
    "' where id = " + userId;
    try {
                stmt = conn.createStatement();
                int result = stmt.executeUpdate(sql);
                if (result!= 0) {   flag = true;   }
    }catch (Exception e) {
        e.printStackTrace();
        throw new SystemException (e.toString());
    }finally {
        if (rs!= null)     rs.close();
        if (stmt!= null)    stmt.close();
        closeDBConnection();
    }
    return flag;
}
//所有账号列表
public List<User>  getUserList( ) throws Exception {
    List<User> list = new ArrayList<User>();
    String sql;
    sql = "select * from t_user;";
    User user = null;
    try {
        execQuery(sql);  //rs 变量从父类继承,返回查询结果
        while (rs.next()) {
                user = new User();
                user.setId(rs.getInt("id"));
                user.setUsername(rs.getString("username"));
                user.setPassword(rs.getString("password"));
                user.setCreateTime(rs.getDate("createTime"));
                user.setExpireTime(rs.getDate("expireTime"));
                if(rs.getDate("expireTime")!= null)
                  user.setExpireTimes(new SimpleDateFormat
                   ("yyyy-MM-dd").format(rs.getDate("expireTime")));
                list.add(user);
        }
    }catch (Exception e) {
        e.printStackTrace();
        throw new SystemException ("Exeception for select t_Organization tables.");
    }finally {
        closeDBConnection();
    }
    return list;
  }
}
```

15.10.4 模型控制层 Action 的设计

模型控制层由三个文件组成：LoginAction.java、IndexAction.java 和 UserAction.java。

LoginAction.java 定义用户登录和注销的动作 Action，IndexAction.java 定义进入主界面框架集(FrameSet)的 3 个 frame 的动作 Action，UserAction.java 定义系统管理和个人信息维护的动作 Action.。

1. LoginAction.java 文件的代码

```
package com.oa.action;
import com.slide.oa.managers.impl.UserManagerImpl;
import com.slide.oa.model.User;
import java.util.Map;
import com.opensymphony.xwork2.ActionContext;
import com.opensymphony.xwork2.ActionSupport;
public class LoginAction extends ActionSupport {
  private static final long serialVersionUID = -7859034849358151831L;
  private  String  username;
  private  String  password;
  private  String  error = "";
  //登录提交动作
  public String login()throws Exception {
      ActionContext actionContext = ActionContext.getContext();
      Map session = actionContext.getSession();
      try {
        UserManagerImpl userManagerImpl = new UserManagerImpl();
        //String pass = EncryptUtil.encrypt((this.password.trim()));//md5 加密
        String pass = this.password.trim();//md5 加密
        User user = userManagerImpl.login(username.trim(),pass,session);
        if (user!= null) {
            session.put("login", user);
            return "back_index";
        }
        else
            return "login";
      }
      catch (Exception ex) {
            error = ex.getMessage();
            return "login";
      }
  }
  // 注销动作
  public String logout()throws Exception    {
      ActionContext actionContext = ActionContext.getContext();
      Map session = actionContext.getSession();
      session.clear();
      return "login";
  }
  public String getUsername() {
      return username;
  }
```

```
    public void setUsername(String username) {
        this.username = username;
    }
    public String getPassword() {
        return password;
    }
    public void setPassword(String password) {
        this.password = password;
    }
    public String getError() {
        return error;
    }
    public void setError(String error) {
        this.error = error;
    }
}
```

2. IndexAction.java 文件的代码

```
package com.oa.action;
import java.text.SimpleDateFormat;
import java.util.Date;
import java.util.Map;
import com.opensymphony.xwork2.ActionContext;
import com.opensymphony.xwork2.ActionSupport;
import com.slide.oa.model.User;
public class IndexAction extends ActionSupport {
    private static final long serialVersionUID = -7859034849358151830L;
    private  String  error = "";
    private  String loginInfo = "";
    //打开 outlook 界面动作
    public String outlook() throws Exception {
        try {
          return "success";
        }
        catch (Exception ex) {
            error = ex.getMessage();
            return "fail";
        }
    }
  //打开 main 界面动作
  public String  main() throws Exception {
      ActionContext actionContext = ActionContext.getContext();
      Map session = actionContext.getSession();
      User user = (User)session.get("login");
      return "success";
  }
  //打开 header 页面动作
  public String  header() throws Exception {
      ActionContext actionContext = ActionContext.getContext();
      Map session = actionContext.getSession();
      User user = (User)session.get("login");
      Date dd = new Date();
      SimpleDateFormat ss = new SimpleDateFormat("yyyy-MM-dd");
```

```
        String today = ss.format(dd);
        StringBuilder strInfo = new StringBuilder("当前用户:");
        strInfo.append(user.getUsername());
        strInfo.append("    今天是:");
        strInfo.append(today);
        loginInfo = strInfo.toString();
        return "success";
    }
    public String getError() {
        return error;
    }
    public void setError(String error) {
        this.error = error;
    }
    public String getLoginInfo() {
        return loginInfo;
    }
    public void setLoginInfo(String loginInfo) {
        this.loginInfo = loginInfo;
    }
}
```

3. UserAction.java 的代码

```
package com.oa.action;
import java.util.List;
import java.util.Map;
import com.oa.obj.Message;
import com.opensymphony.xwork2.ActionContext;
import com.opensymphony.xwork2.ActionSupport;
import com.slide.oa.managers.impl.UserManagerImpl;
import com.slide.oa.model.User;
public class UserAction extends ActionSupport {
  private static final long serialVersionUID = -7859034849358151833L;
  private Integer id;
  private String password;
  private String username;
  private Message mes = new Message();
  private User user;
  private List<User> persons;
  private String name;
  //用户登录时系统发现口令过期,转向修改口令界面提交后动作
    public String change_password() throws Exception {
      ActionContext actionContext = ActionContext.getContext();
      Map session = actionContext.getSession();
      mes = new Message();
      try {
        UserManagerImpl manager = new UserManagerImpl();
        if(manager.changePassword(id, password.trim(),session))
        {   mes.setMessageContent("密码修改成功!");
            mes.setSkipPath("./index.jsp");
        }
        else {
            mes.setMessageContent("密码修改失败!");
```

```
            mes.setSkipPath("./user/update_password.jsp");
        }
      }
      catch(Exception ex)
      {  mes.setMessageContent(ex.getMessage());
         mes.setSkipPath("./user/update_password.jsp");
      }
      return "success";
    }
    //主界面的修改口令输入界面动作
    public String change_password1Input() throws Exception {
        return  "success";
    }
    ///修改口令界面提交后动作
    public String change_password1() throws Exception {
      ActionContext actionContext = ActionContext.getContext();
      Map session = actionContext.getSession();
      //mes = new Message();
      try {
        UserManagerImpl manager = new UserManagerImpl();
        if(manager.changePassword(id, password.trim(),session))
        {  mes.setMessageContent("密码修改成功!");
           mes.setSkipPath("./main.action");
        }
        else {
            mes.setMessageContent("密码修改失败!");
            mes.setSkipPath("./change_password1Input.action");
        }
      }
      catch(Exception ex)
      {  mes.setMessageContent(ex.getMessage());
         mes.setSkipPath("./change_password1Input.action");
      }
      return "success";
    }
    //管理员设置初始化口令动作
    public String init_password() throws Exception {
      ActionContext actionContext = ActionContext.getContext();
      Map session = actionContext.getSession();
      password = "a123456";
      try {
        UserManagerImpl manager = new UserManagerImpl();
        String password = "root";
        if(manager.changePassword(id, password.trim()))
        {  mes.setMessageContent(
           "<font color = 'black'>设置初始密码成功!</font>");
           mes.setSkipPath("./listUser.action");
        }
        else {
            mes.setMessageContent("设置初始密码失败!");
        mes.setSkipPath("./listUser.action");
        }
```

```
        }
        catch(Exception ex)
        {  mes.setMessageContent(ex.getMessage());
           mes.setSkipPath("./listUser.action");
        }
        return "success";
    }
      //管理员进入添加账号界面动作
      public String userInput() throws Exception {
        return "success";
      }
      //管理员添加账号表单提交后动作
      public String add_user() throws Exception {
        boolean flag = false;
        user = new User();
        user.setUsername(username);
        user.setPassword(password.trim());
        try {
              flag = new UserManagerImpl().save(user);
        }
        catch(Exception ex)
        {
          flag = false;
        }
        if (flag) {
                   mes.setMessageContent("<font color = 'black'>恭喜你,保存成功啦! </font>");
           mes.setSkipPath("./listUser.action");
          }
          else {
                mes.setMessageContent("很抱歉,保存保存失败! 请重新修改");
           mes.setSkipPath("./userInput.action");
        }
       return "success";
      }
//管理员删除账号动作
public String delUser() throws Exception {
  boolean flag = new UserManagerImpl().delete(id);
  if (flag) {
      mes.setMessageContent("<font color = 'black'>恭喜你,删除成功啦! </font>");
      mes.setSkipPath("./listUser.action");
  }
  else {
      mes.setMessageContent("很抱歉,删除失败! 请重新操作");
      mes.setSkipPath("./listUser.action");
  }
  return "success";
    }
    //管理员浏览所有用户信息列表动作
    public String listUser() throws Exception {
      persons = new UserManagerImpl().getUserList();
      return "success";
    }
```

```
    public String getPassword() {
        return password;
    }
    public void setPassword(String password) {
        this.password = password;
    }
    public Message getMes() {
        return mes;
    }
    public void setMes(Message mes) {
        this.mes = mes;
    }
    public String getName() {
        return name;
    }
    public void setName(String name) {
        this.name = name;
    }
    public Integer getId() {
        return id;
    }
    public void setId(Integer id) {
        this.id = id;
    }
    public String getUsername() {
        return username;
    }
    public void setUsername(String username) {
        this.username = username;
    }
    public User getUser() {
        return user;
    }
    public void setUser(User user) {
        this.user = user;
    }
    public List<User> getPersons() {
        return persons;
    }
    public void setPersons(List<User> persons) {
        this.persons = persons;
    }
}
```

15.10.5 汉字编码过滤器的设计

为了解决 Struts2 中的中文乱码问题，设计了字符编码过滤器，以统一处理系统中的数据编码。设计的过滤器类 EncodingFilter 实现 Filter 接口，将提交数据的编码统一转换为 UTF-8，该代码位于包：com. web. filter 下。EncodingFilter. java 的代码实现如下：

```
package org.css.filter;
import java.io.IOException;
```

```
import javax.servlet.Filter;
import javax.servlet.FilterChain;
import javax.servlet.FilterConfig;
import javax.servlet.ServletException;
import javax.servlet.ServletRequest;
import javax.servlet.ServletResponse;
public class EncodingFilter implements Filter {
    protected String encoding = null;
    protected FilterConfig filterConfig = null;
    protected boolean ignore = true;
    public void init(FilterConfig filterConfig) throws ServletException {
        try {
            freemarker.log.Logger
                    .selectLoggerLibrary(freemarker.log.Logger.LIBRARY_NONE);
        } catch (ClassNotFoundException e) {
            e.printStackTrace();
        }
        this.filterConfig = filterConfig;
        this.encoding = filterConfig.getInitParameter("encoding");
        String value = filterConfig.getInitParameter("ignore");
        if (value == null)
            this.ignore = true;
        else if (value.equalsIgnoreCase("true"))
            this.ignore = true;
        else
            this.ignore = false;
    }
    public void doFilter(ServletRequest request, ServletResponse response,
            FilterChain chain) throws IOException, ServletException {
        if (ignore || (request.getCharacterEncoding() == null)) {
            String encoding = selectEncoding(request);
            if (encoding != null)
                request.setCharacterEncoding(encoding);
        }
        chain.doFilter(request, response);
    }
    public void destroy() {
        this.encoding = null;
        this.filterConfig = null;
    }
    protected String selectEncoding(ServletRequest request) {
        return (this.encoding);
    }
}
```

15.10.6 设置配置文件

1. 配置文件 web.xml

web.xml 文件的内容：

```
<?xml version="1.0" encoding="UTF-8"?>
<web-app version="2.5"
```

```
    xmlns = "http://java.sun.com/xml/ns/javaee"
    xmlns:xsi = "http://www.w3.org/2001/XMLSchema-instance"
    xsi:schemaLocation = "http://java.sun.com/xml/ns/javaee
    http://java.sun.com/xml/ns/javaee/web-app_2_5.xsd">
  <welcome-file-list>
    <welcome-file>index.jsp</welcome-file>
  </welcome-file-list>
  <!-- 解决所有请求路径下的编码问题 -->
    <filter>
        <filter-name>EncodingFilter</filter-name>
        <filter-class>org.css.filter.EncodingFilter</filter-class>
        <init-param>
            <param-name>encoding</param-name>
            <param-value>UTF-8</param-value>
        </init-param>
        <init-param>
            <param-name>forceEncoding</param-name>
            <param-value>true</param-value>
        </init-param>
    </filter>
    <listener>
        <listener-class>
            org.apache.commons.fileupload.servlet.FileCleanerCleanup
        </listener-class>
    </listener>
    <filter-mapping>
        <filter-name>EncodingFilter</filter-name>
        <url-pattern>/ * </url-pattern>
    </filter-mapping>
  <filter>
    <filter-name>struts2</filter-name>
    <filter-class>
        org.apache.struts2.dispatcher.ng.filter.StrutsPrepareAndExecuteFilter
    </filter-class>
  </filter>
  <filter-mapping>
    <filter-name>struts2</filter-name>
    <url-pattern>/ * </url-pattern>
  </filter-mapping>
  </web-app>
```

2. 配置文件 struts.properties

struts.properties 文件的内容如下：

```
struts.devMode = false
struts.action.extension = do,action
struts.i18n.reload = true
struts.ui.theme = simple
struts.locale = zh_CN
struts.serve.static.browserCache = false
struts.url.includeParams = none
struts.multipart.maxSize = 52428800
struts.multipart.saveDir = /tmp
```

3. 配置文件 struts. xml

配置文件 struts. xml 文件的内容如下：

```
<? xml version = "1.0" encoding = "UTF-8" ? >
<! DOCTYPE struts PUBLIC "-//Apache Software Foundation//DTD Struts Configuration 2.1//EN" "http://struts.apache.org/dtds/struts-2.1.dtd">
<struts>
    <package name = "web" extends = "struts-default" namespace = "/">
     <global-results>
            <! -- 配置全局结果,用于检查进入界面的身份校验结果返回 -->
            <result name = "unhandledException">/login.action</result>
            <! --  skip.jsp 增加跳转的延迟过渡,后转向/login.jsp-->
            <result name = "login">/skip.jsp</result>
    </global-results>
      <! -- 登录动作的 result 的关联页面 -->
       <action name = "login" class = "com.oa.action.LoginAction" method = "login">
            <result name = "back_index">/frameset.jsp</result>
            <result name = "login">/index.jsp</result>
       </action>
       <! -- 注销         -->
       <action name = "logout" class = "com.oa.action.LoginAction" method = "logout">
            <result name = "login">/skip.jsp</result>
       </action>
       <! -- 主界面 FrmeSet 中左 Frame        -->
       <action name = "outlook" class = "com.oa.action.IndexAction" method = "outlook">
            <result name = "success">/outlook.jsp</result>
       </action>
       <action name = "header" class = "com.oa.action.IndexAction" method = "header">
            <result name = "success">/header.jsp</result>
       </action>
       <action name = "main" class = "com.oa.action.IndexAction" method = "main">
            <result name = "success">/main.jsp</result>
       </action>
     <! -- 用户账号管理-->
       <action name = "listUser" class = "com.oa.action.UserAction" method = "listUser">
           <result name = "success" >/user/index.jsp</result>
       </action>
       <action name = "userInput" class = "com.oa.action.UserAction" method = "userInput">
           <result name = "success" >/user/user_input.jsp</result>
       </action>
       <action name = "add_user" class = "com.oa.action.UserAction" method = "add_user">
           <result name = "success" >/message.jsp</result>
       </action>
       <action name = "delUser" class = "com.oa.action.UserAction" method = "delUser">
           <result name = "success" >/message.jsp</result>
       </action>
       <action name = "init_password" class = "com.oa.action.UserAction" method = "init_password">
           <result name = "success" >/message.jsp</result>
       </action>
       <action name = "change_password1Input" class = "com.oa.action.UserAction" method = "change_password1Input">
```

```
            <result name="success" >/user/update_password1.jsp</result>
        </action>
        <action name="change_password1" class="com.oa.action.UserAction" method="change_
password1">
            <result name="success" >/message.jsp</result>
        </action>

        <action name="listUser" class="com.oa.action.UserAction" method="listUser">
            <result name="success" >/user/index.jsp</result>
        </action>
    </package>
</struts>
```

15.10.7 用户界面层的设计

在前台用户界面层的设计上,JSP 页面中应用 CSS 和 JavaScript 技术,应用 JSP 的标准标签库 jstl 和 struts2.x 标签库。整个网站的页面设计要求表现风格一致、功能明确。

1. 用户登录首页

用户登录首页的页面文件 index.jsp 的运行效果如图 15-21 所示。在此页面的表单中用户输入内容,单击“进入系统”图像按钮后,调用 index.action 动作进行用户验证,验证通过后进入系统主界面。如果验证不通过或用户口令有效期失效,将给出提示信息并处理。

图 15-21 用户登录首页显示效果

index.jsp 的代码如下:

```
<%@ page language="java" contentType="text/html;charset=utf-8" %>
<%@ taglib uri="/struts-tags" prefix="s" %>
<html>
<head>
<meta http-equiv="Content-Type" content="text/html; charset=utf-8">
<LINK href="style/login.css" type=text/css rel=stylesheet>
<title>用户账号管理系统</title>
</head>
<BODY class="PageBody" leftMargin=0 topMargin=0 MARGINHEIGHT="0"
MARGINWIDTH="0">
<CENTER>
```

```
<form action = "login.action" method = "post" >
<TABLE height = "100 % " cellSpacing = 0 cellPadding = 0 border = 0>
  <TBODY>
  <TR> <TD></TD></TR>
  <TR> <TD>
      <TABLE cellSpacing = 0 cellPadding = 0 border = 0>
        <TBODY>
              <TR class = UpTr height = 65>
          <TD width = 20></TD>
          <TD colSpan = 2><IMG src = "images/login/logo.gif"
            border = 0></TD></TR>
        <TR height = 3>
          <TD background = images/login/logo_under_line.gif
          colSpan = 3></TD></TR>
        <TR class = DownTr>
          <TD></TD>
          <TD>
            <TABLE height = 204 cellSpacing = 0 cellPadding = 0 border = 0>
              <TBODY>
              <TR height = 50>
                <TD colSpan = 3></TD></TR>
              <TR height = 18>
                <TD width = 290
                background = images/login/userLogin.gif
                colSpan = 3></TD></TR>
              <TR>
                <TD class = LoginLine width = 2></TD>
                <TD width = 286>
                 <TABLE height = "100 % " cellSpacing = 0 cellPadding = 0 width = "100 % "
                  border = 0>
                    <TBODY>
                    <TR height = 10>
                      <TD colSpan = 6></TD></TR>
                    <TR>
                      <TD class = ItemTitleFont align = right width = 80
                        height = 25>用户名:</TD>
                       <TD width = 100><INPUT class = inputFrm name = "username" id = "user-
name" value = " ${login.username}"></TD>
                      <TD align = middle rowSpan = 2><input type = "image" src = "images/log-
in/userLogin_button.gif"
                        border = 0   ></TD></TR>
                    <TR>
                      <TD class = ItemTitleFont align = right height = 25>密码:</TD>
                      <TD><INPUT class = "inputFrm" type = "password"
                    name = "password"   id = "password" ></TD></TR>
                    <TR>
                      <TD class = "ItemTitleFont" style = "color:red" align = "center" colspan = "
3" > ${error}</TD>
                      </TR></TBODY></TABLE></TD>
                <TD class = LoginLine width = 2></TD></TR>
              <TR height = 11>
                <TD width = 290
```

```
            background = images/login/userLogin_down.gif
            colSpan = 3></TD></TR></TBODY></TABLE></TD>
        <TD width = 228><IMG src = "images/login/logo_bg.gif"
          border = 0></TD></TR>
          </TBODY></TABLE></TD>
    </TBODY></TABLE>
    </form>    </CENTER></BODY> </html>
```

2. 系统主页界面 frameset.jsp

系统主页界面设计时，采用框架集布局：把浏览器窗口划分为左(outlook.jsp)、上(header.jsp)、中(main.jsp)三部分区域。网页的最左区域显示系统的导航菜单；上方区域显示当前登录的用户名和当前日期，以及注销按钮；中间区域为主要工作区，显示当前用户单击导航菜单后要访问的内容。系统运行主界面的效果如图 15-22 所示。

图 15-22 系统主页界面显示效果

frameset.jsp 的关键性部分代码如下：

```
<frameset rows = "80, * " framespacing = "0" frameborder = "0" border = "0">
  <frame src = "header.action" SCROLLING = "No" noresize = "noresize">
    <frameset cols = "150, * " framespacing = "0" framespadding = "0" frameborder = "0" border = "0">
        <frame SRC = "outlook.action" NAME = "Links" SCROLLING = "No" noresize = "noresize">
    <frameset rows = "15, * " framespacing = "0" frameborder = "0" border = "0" >
        <frame SRC = "space.jsp"  SCROLLING = "No"  NAME = "msg" >
        <frame SRC = "main.action" NAME = "main" >
    </frameset>
  </frameset>
</frameset>
```

网页的最左区域的导航菜单，对应的网页文件是 outlook.jsp，在此文件中利用 JavaScrpts 显示系统的多级导航菜单。生成多级导航菜单的部分 JavaScript 代码如下：

```
var o = new createOutlookBar
  ('Bar',0,0,screenSize.width,screenSize.height,'#6680CA','red');//'#000099')
var p0 = new createPanel('p0','系统管理');
o.addPanel(p0);
p0.addButton('netm.gif','账号列表','parent.main.location = "listUser.action"');
p0.addButton('netm.gif','添加账号','parent.main.location = "userInput.action"');
var p1 = new createPanel('p1','个人信息维护');
o.addPanel(p1);
p1.addButton('netm.gif','修改密码'
  ,'parent.main.location = "change_password1Input.action"');
p1.addButton('netm.gif','退出登录','parent.main.location = "logout.action"');
```

3. 账号列表页面 /user/index.jsp

用户单击导航栏的子菜单“系统管理—>账号列表”，将调用 listUser.action 动作，访问后台数据库，返回全部用户信息的列表 list，然后转向页面文件/user/index.jsp 显示结果。页面运行的效果如图 15-23 所示。在表格内容的每一行显示了每一个用户的账号信息，还提供两个按钮“删除账号”和“初始化口令”。用户单击 “删除账号”按钮，将弹出账号确认提示框，单击“确定”后真正删除当前行账号。用户单击 “初始化口令”按钮，将当前行的账号的口令设置为初始值“a1234567”。

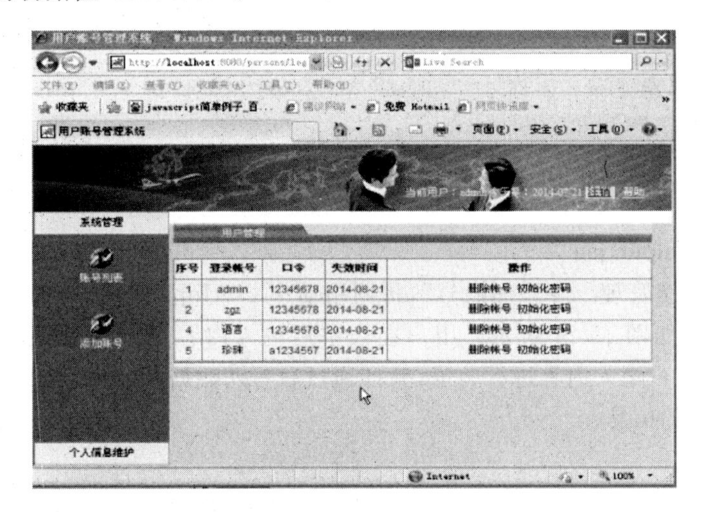

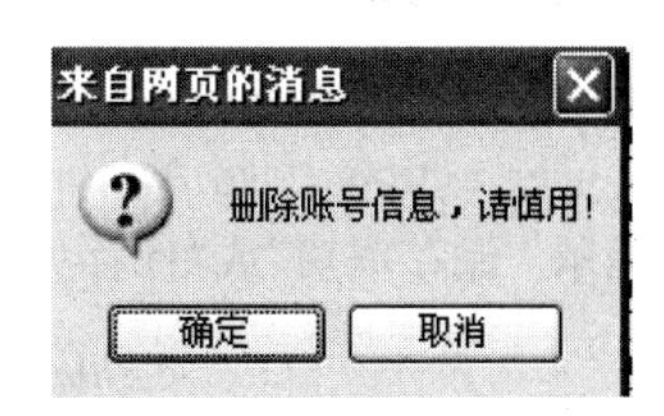

图 15-23　账号列表页面显示效果

index.jsp 的部分关键性代码如下：

```
<%@ page language="java" contentType="text/html; charset=utf-8"
    pageEncoding="utf-8"%>
<%@ taglib uri="/struts-tags" prefix="s"%>
<%@ taglib prefix="c"  uri="http://java.sun.com/jsp/jstl/core" %>
......
<table width="100%" border="0" cellPadding="0" cellSpacing="1" bgcolor="#6386d6">
<!-- 列表标题栏 -->
<tr bgcolor="#EFF3F7" class="TableBody1">
<td width="5%" height="37" align="center"><b>序号</b></td>
<td width="10%" height="37" align="center"><B>登录账号</B></td>
<td width="10%" height="37" align="center"><b>口令</b></td>
<td width="10%" height="37" align="center"><b>失效时间</b></td>
<td width="45%" height="37" align="center"><b>操作</b></td>
</tr>
<!-- 列表数据栏内容 -->
<c:if test="${! empty persons}">
<c:forEach items="${persons }" var="person">
<tr bgcolor="#EFF3F7" class="TableBody1" onMouseOver="this.bgColor = '#DEE7FF';" onMouse-
Out="this.bgColor='#EFF3F7';">
<td align="center" vAlign="center">${person.id }</td>
<td align="center" vAlign="center">${person.username }</td>
<td align="center" vAlign="center">${person.password }</td>
<td align="center" vAlign="center">${person.expireTimes}</td>
<td align="center" vAlign="center">
```

```
<c:if test = "${person.username!= null&&person.username!=''}">
  <a href = "javascript:del('./delUser.action? id = ${person.id}',
    '删除账号信息,请慎用!');" target = "main"> 删除账号</a>
  </c:if>
<a href = "./init_password.action? id= ${person.id}" >初始化密码</a>
</td> </tr>
</c:forEach>
</c:if>
<! -- 在列表数据为空的时候,要显示的提示信息 -->
<c:if test = "${empty persons}">
   <tr>
<td colspan = "8" align = "center" bgcolor = "#EFF3F7" class = "TableBody1" onMouseOver = "this.bg-
Color = '#DEE7FF';" onMouseOut = "this.bgColor = '#EFF3F7';">
没有找到相应的记录</td> </tr>
</c:if>
</table>
```

4. 添加账号页面 /user/user_input.jsp

用户单击导航栏的子菜单"系统管理—>添加账号",将调用动作 userInput.action 后,转向 user_input.jsp 页面,显示效果如图 15-24 所示。用户在表单后输入内容,提交的目标是 add_user.action,由此 Action 对用户输入内容进行验证,验证通过后写入数据库中,并返回前台提示 "保存成功"或"保存失败"的信息。

图 15-24　添加账号页面显示效果和结果返回状态

```
<%@ page language = "java" contentType = "text/html; charset = utf-8"
    pageEncoding = "utf-8" %>
<%@ taglib uri = "/struts-tags" prefix = "s" %>
<html><head>
<meta http-equiv = "Content-Type" content = "text/html; charset = utf-8">
<link href = "style/oa.css" rel = "stylesheet" type = "text/css">
<script language = "javascript" src = "script/public.js"></script>
<title>分配用户账号</title></head>
<body><center>
<form action = "add_user.action" method = "post">
<input type = "hidden" name = "person" value = "${person}">
<TABLE class = "tableEdit" border = "0" cellspacing = "1" cellpadding = "0" style = "width:580px;">
<TBODY>
<TR>
  <td align = "center" class = "tdEditTitle">分配用户账号</TD>   </TR>
<TR>  <td>
     <table class = "tableEdit" style = "width:580px;" cellspacing = "0" border = "0" cellpadding = "0">
       <tr>
       <td class = "tdEditLabel" >用户账号</td>
       <td class = "tdEditContent"><input type = "text" name = "username">  </td>
     <td class = "tdEditLabel" >初始密码</td>
     <td class = "tdEditContent"><input type = "text" name = "password" value = "a1234567"></td>
  </tr>
    </table>
```

```
<! -- 主输入域结束 -->
</td></TR>/TBODY></TABLE>
<TABLE>
        <TR align = "center">
          <TD colspan = "3" bgcolor = "#EFF3F7">
            <input type = "submit" name = "saveButton"
                class = "MyButton" value = "保存用户账号">
            <input type = "button" class = "MyButton"
                value = "返回"
            onClick = "window.parent.main.location = './listUser.action';">
            </TD></TR></TABLE>
</form> </CENTER></BODY> </html>
```

5. 修改密码页面 update_password1.jsp

用户单击导航栏的子菜单"个人信息维护—>修改密码"，将调用动作 change_password1Input..action 后，转向 update_password1.jsp 页面，显示效果如图 15-25 所示。用户在表单后输入内容，提交的目标是 change_password1，执行此 Action 动作将对用户输入内容进行验证，验证通过后写入数据库中，并返回前台提示"密码修改成功"或"密码修改失败"的信息。

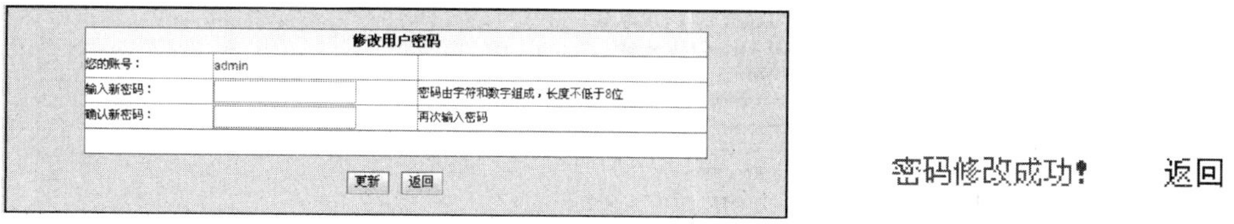

图 15-25　修改密码页面显示效果和结果状态

6. 退出

用户单击导航栏的子菜单"个人信息维护—>退出登录"，将调用 logout.action，把用户登录时写入 session 的信息清除干净后，转向跳转页面 skip.jsp，此页面出现延迟 1 秒后，直接跳转到系统登录的首页。

7. 消息提示信息子窗口的设计 message.jsp

用户账号管理系统中，用户需要将所有操作的操作结果状态返回给客户端，这里采用了统一的 message.jsp 页面来返回结果的状态，只是页面显示的结果状态内容是可变的，返回的页面链接地址也是可变的。为此，系统在 Action 层定义了类型为 Message 的 mes 对象，不同的 Action 动作执行后，赋给 mes 的两个属性变量 messageContent 和 skippath 不同的内容，然后 Action 转向 message.jsp 页面时，message.jsp 页面显示的内容是动态变化的，但显示的风格是一致的。message.jsp 文件的代码如下：

```
%@ page language = "java" contentType = "text/html;charset = utf-8" %>
<%@ taglib uri = "/struts-tags" prefix = "s" %>
<html>
<head><meta http-equiv = "Content-Type" content = "text/html; charset = utf-8" />
<title>结果提示</title></head>
<body><p></p>
<center>
<font color = "red" style = "font-size: 12px;" target = "main"> ${mes.messageContent}</font>
<label>    
```

<a href='${mes.skipPath}? first=${first}' style="text-decoration: none; font-size: 12px;">返回</a></label></center></body></html>

在 MyEclipse 环境下该项目的开发、发布和运行过程：

在第 15 章的例子源代码中，ch15 文件夹下的子文件夹名 persons 是 MyEclipse 8.5 环境下本例的一个完整项目。

1. 打开 MyEclipse 的主界面，在"Package View"视图上右击，在弹出的菜单中单击"Import…->Existing Projects into workspace"，选择 persons 存放的路径，将 persons 项目导入 MyEclipse 中。完成导入后项目 persons 的 Package Explore 视图如图 15-26 所示。

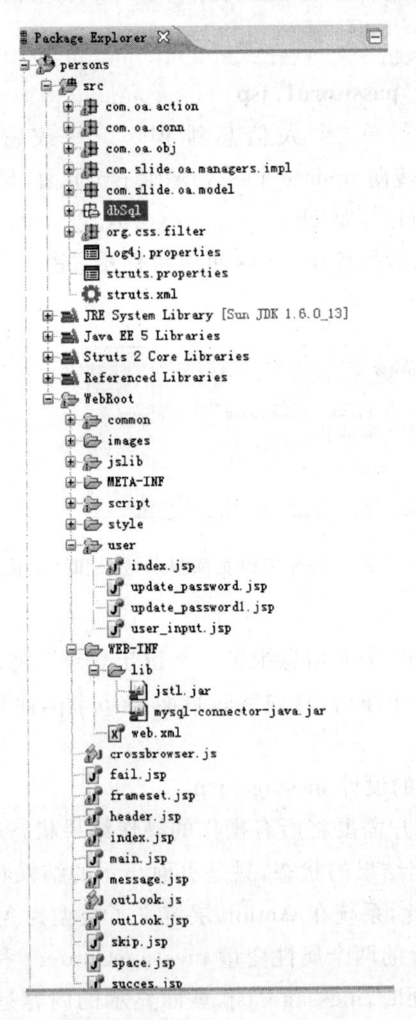

图 15-26 persons 项目的 Package Explore

2. 发布 persons 项目到 Tomcat6 服务器上。

3. 启动 Tomcat 服务器运行。

4. 在浏览器窗口输入命令：

http://localhost:8080/persons/

15.11 小　结

JSP 是 Servlet 技术的一种扩展。JSP 是一种实现静态 HTML 和动态 HTML 混合编码的技术。页面静态部分的 HTML 内容可以使用的 Web 制作工具编写，而页面动态部分的代码由 JSP 元素组成。JSP 元素包括指令、脚本、动作和标记库。这些 JSP 元素将根据不同的请求，产生网页的动态部分。

JSP 结合 JDBC 技术存取数据库服务器上的数据。

JavaBeans 在 JSP 程序中用来封装事务逻辑，可以很好地实现业务逻辑和前台程序的分离，使得系统具有更好的健壮性和灵活性。

Struts 2. x 技术的编程基础包括：Struts 2. x 的架构、Action 的定义、页面文件中标签库的使用、配置文件的书写。Struts 2. x 结合 JSP 技术能很好地开发一个具备 MVC 模式的 Web 应用。

习　题

15.1　填空题

1）JSP 有四个关键的元素：________、________、________和________。

2）与 JSP 编程有关的类和接口位于________和________包中。

3）隐含对象有四种作用域：________、________、________和________。

4）________动作，通过将属性 property 指定为“*”，能将请求参数与 Bean 中同名的属性进行匹配。

5）执行 JSP 的服务器组件通常称为________。

6）JSP 脚本元素有________、________、________和________。

15.2　试述一个 JSP 的执行过程？

15.3　编写一个 JSP 页面，使用脚本元素显示 8 次“Hello world!”的信息。

15.4　编写一个包含一张图片的 HTML 文件。

15.5　编写一个包含系统日期和时间的 JSP 文件。

15.6　编写一个包含一张表格(1 行(2 列)的 JSP 文件，表格的第一列显示提示信息，第 2 列是用 include 指令包含的另一个 HTML 文件(此文件是包含一张图片的 HTML 文件)。

15.7　编写一个包含一张表格(1 行(2 列)的 JSP 文件，其中第 1 列是显示提示信息，第 2 列是用<jsp include>动作包含的另一个 JSP 文件(此文件是包含系统日期和时间的 JSP 文件)。

15.8　结合第 14 章创建数据库 Study 数据库，编写 Web 应用程序，完成下列功能：

1）显示所有学生的基本情况。包括学号、姓名、性别、年龄、所在系，按学号排序。以二维表格形式显示。

2）显示所有课程的基本情况。包括课程号、课程名、学分。以二维表格形式显示。

3）查询某个学生的某门课的成绩。要求用户通过表单输入学号、课程号，表单提交后，显示数据库的查询结果。

参考文献

[1] Martin Kalin. 面向对象程序设计——Java 语言描述. 孙艳春，等，译. 北京：机械工业出版社，2002.

[2] 飞思科技产品研发中心. Java2 应用开发指南. 北京：电子工业出版社，2002.

[3] 孙一林. JAVA 语言程序设计. 北京：清华大学出版社，2001.

[4] 印旻. JAVA 语言与面向对象程序设计. 北京：清华大学出版社，2000.

[5] 张猋. Jbuilder 5 实用教程. 北京：清华大学出版社，2002.

[6] P. J. Deitel. Java 程序设计教程. 施平安，施惠琼，柳赐佳，译. 5 版. 北京：机械工业出版社，2002.

[7] Cay S. Horstmann Gary Cornell. Java 核心技术卷 1. 王建华，董志敏，杨保明，等，译. 5 版. 北京：机械工业出版社，2003.

[8] Cay S. Horstmann Gary Cornell. Java 核心技术卷 2. 王建华，董志敏，杨保明，等，译. 5 版. 北京：机械工业出版社，2003.

[9] Bruce Eckel. Java 编程思想. 京京工作室，译. 2 版. 机械工业出版社，1999.

[10] 周颢，网络编程语言 JSP 实例教程. 北京：电子工业出版社，2002.

[11] 张卫民，等. Java 语言与 WWW. 北京：人民邮电出版社，1997.

[12] 陈烨，赵文武. JBuilder 7 编程实作指南. 北京：希望电子出版社，2003.

[13] Deitel H. M.. 高级 JAVA 2 大学教程. 钱方，等，译. 北京：电子工业出版社，2003.

[14] E. Reed Doke，John W. Satzinger，SuSan Rebstock Williams. Java 面向对象应用程序开发(影印本). 北京：清华大学出版社，2003.